MULTIPLE USE OF FORESTS AND OTHER NATURAL RESOURCES

FORESTRY SCIENCES

Volume 61

The titles published in this series are listed at the end of this volume.

Multipe Use of Forests and Other Natural Resources

Aspects of Theory and Application

Edited by

FINN HELLES,
PER HOLTEN-ANDERSEN
and
LARS WICHMANN

*The Royal Veterinary and Agricultural University,
Department of Economics and Natural Resources,
Copenhagen, Denmark*

KLUWER ACADEMIC PUBLISHERS
DORDRECHT / BOSTON / LONDON

A C.I.P. Catalogue record for this book is available from the Library of Congress.

ISBN 0-7923-5859-7

Published by Kluwer Academic Publishers,
P.O. Box 17, 3300 AA Dordrecht, The Netherlands.

Sold and distributed in North, Central and South America
by Kluwer Academic Publishers,
101 Philip Drive, Norwell, MA 02061, U.S.A.

In all other countries, sold and distributed
by Kluwer Academic Publishers,
P.O. Box 322, 3300 AH Dordrecht, The Netherlands.

Printed on acid-free paper

Printed in the Netherlands.

List of Contents

Foreword vii

Preface
Helles, F., Holten-Andersen, P. and Wichmann, L. ix

Chapter I Survey

Berck, P.
Why are the uses multiple? 3

Qi, Y. and Gilles, J. K.
Modelling ecosystem processes and patterns for multiple-use management 14

Boon, T. E. and Helles, F.
Descriptive indicators of sustainable forest management 23

Price, C.
Discounting in the context of multiple objectives 34

Price, C.
Stated and revealed preference analysis: Cost–benefit analysis, democracy and multiple-objective decision-making 46

Chapter II Multi-Criteria Decision-Making: Theory and Methods

Bogetoft, P.
Choosing between models of choice: On the selection of MCDM planning modes 69

Korhonen, P.
Multiple objective linear programming in supporting forest management 85

Kangas, J.
The analytic hierachy process (AHP): Standard version, forestry application and advances 96

Chapter III Stakeholder Involvement

Agrell, P.
Multiple objective optimisation in agro-ecological land use planning. 109

Brukas, V., Hjortsø, N., Tarp, P. and Boon, T. E.
Public participation in forest management planning supported by multi-criteria decision-making techniques 122

Chapter IV Diversity Objectives and Option Value

Lin, C-R. and Buongiorno, J.
Managing forests for tree and landscape diversity
Part I: Predictions with Markov-chain models 137

Lin, C-R. and Buongiorno, J.
Managing forests for tree and landscape diversity
Part II: Optimisation with Markov decision process model 158

Abildtrup, J. and Strange, N.
Groundwater protection and forest management - an option value analysis 175

Chapter V Spatial Objectives

Kurtilla, M.
How to integrate spatial objectives into private forest planning in Finland 187

Jensen, J. J.
A spatial optimisation model for wetlands restoration planning integrating
GIS and linear programming 194

Strange, N. and Meilby, H.
A cellular automaton approach to optimising the choice of tree species. 205

Chapter VI Cost Efficiency

Asmild, M., Hougaard, J. L. and Kronborg, D.
A method for comparison of efficiency scores: A case study of Danish
dairy farms 220

Tarp, P., Helles, F. and Zajac, S.
Biological insecticides contra insect growth regulators in Polish forestry
- a financial cost-benefit analysis 228

List of Authors 255

Foreword

In 1996 a major six-year research programme, 'Economic Optimisation of Multiple-Use Forestry and Other Natural Resources' was implemented at Department of Economics and Natural Resources, The Royal Veterinary and Agricultural University (KVL), Copenhagen. The research is funded by KVL; The Danish Agricultural and Veterinary Research Council; The Danish Research Academy; The Danish Forest and Landscape Institute; The Danish Forest and Nature Agency; and The Danish Environmental Protection Agency.

The overall objective of the research programme is to enhance the economic theory of sustainable multiple-use forestry and landscape management planning. Emphasis is on decision-making / management planning from an *economic* point of view, the basic criterion being *rationality* as implemented by application of Operations Research methods with regard to sustainable and multiple use of forests and other natural resources in the landscape.

The research programme benefits from collaboration agreements with University of California at Berkeley, Department of Agricultural and Resource Economics, and Oregon State University, Department of Forest Resources.

As part of the research programme, a second international conference and workshop was held 6 - 12 August, 1998 at KVL, with the title: '2nd Berkeley-KVL Conference on Natural Resource Management - Design and Implementation of Multiple-Use Management'. This event was financed by The Danish Research Academy. Some of the papers presented were selected for peer-reviewing and subsequent publishing. The outcome is the present book in which no paper has been previously published.

The editors express their sincere appreciations to the many persons who have contributed to the realisation of this book: participants in the research programme, and in the conference and workshop; the authors; the reviewers. Finally, The Danish Research Academy is acknowledged and thanked for a very flexible and competent cooperation regarding the financing of the international conference and workshop which formed the background of this book.

Copenhagen, April 1999.

Finn Helles, Per Holten-Andersen, Lars Wichmann

Preface

Finn Helles, Per Holten-Andersen and Lars Wichmann

The papers in this book deal with decision-making / management planning of the use of forests and other natural resources in the landscape, and they all relate to multiple use and sustainability. Generally, an economic approach is used to decision-making / management planning, as implemented by Operations Research methods, but the multiple-use /sustainability problem is also discussed in broader contexts, such as ecosystem management, discounting , preference analysis, and policy formulation. Some papers put emphasis on theory, others on application.

Decision-making / management planning of the use of forests and other natural resources in the landscape is a fascinating research field. Irrespective of how much research is carried out on theory and implementation, the knowledge will presently not keep pace with demands. This unceasing challenge is not only due to increasing demands from resource owners and society, but also to the scientific accomplishments themselves. The resources must in general be managed for multiple uses and such management must also be sustainable. Multiple use and sustainability are dynamic - and somewhat elusive - concepts, and the decision-making / management systems must consequently be flexible. Operations Research methods have proved to be effectual tools for decision-making and management planning, and they are continuously developed. There is a need for putting more emphasis on the ecological structures and processes underlying the production of outputs, for realising that multiple use is a normative concept, and for developing public participation in the decision-making process.

This book does not pretend to show the entire state-of-the-art, but nonetheless it is claimed to illustrate many recent accomplishments.

Chapter I: Survey deals with setting and theory. *Peter Berck*'s review of how recent events in the American West have imposed different objectives upon forest managers, is disquieting reading. A new planning environment for multiple use has arisen, dominated by biologists and leaving little room for traditional planning based on optimisation. The paper by *Ye Qi* and *J. Keith Gilless* is important in outlining the challenges which the new paradigm of ecosystem management presents to natural resource modelling. The timber growth and yield models traditionally used in forestry are inadequate for addressing the concerns of ecosystem management. Greater reliance will have to be placed on process-based ecosystem models, and these models must be integrated with socioeconomic models and decision analysis models in GIS-based spatial frameworks, i.e. spatial decision-support systems. *Tove Enggrob Boon and Finn Helles* argue that indicators of sustainable forest management must of necessity be normative because there is no 'one best way' to ensure sustainable forestry. Rather, the concept of sustainable forestry implies forest policy formulation, and therefore the operationalisation provides a common framework for discussing the distribution of scarce resources within forestry and in relation to society. *Colin Price* insists that the inclusion of many objectives other than profit maximisation should raise questions about using a financially based rate to discount all costs and benefits. There is hardly a more effective method than discounting for compromising sustainable development. In a second paper, *Colin Price* argues that as compared with multi-objective techniques, cost-benefit analysis values a restricted set of options and gives minimal space for interaction with decision makers. Its utility maximising assumption is refuted by actual behaviour in public and private arenas. Yet, utility maximisation for those who bear the outcomes of decisions is a worthy objective, and alternative decision techniques arguably over-emphasise the unimportant values and the perspectives of those who construct decisions.

Chapter II: Multi-Criteria Decision-Making - Theory and Methods starts with a paper by *Peter Bogetoft* who stresses the need for synthesising the Multiple Criteria Decision Making planning modes. A simple taxonomy for such modes is proposed, depending on the overall organisation of the investigation, communication and choice activities involved. The multiple costs and benefits of these activities are identified, and the relative importance of the costs and benefits is related to the planning mode and to the contextual characteristics. This provides the basis for a contextual theory of the choice between alternative MCDM modes. The theory is founded on procedural rationality. The paper's emphasis is on methodology, but it is suggested that work along the lines shown could eventually lead to a broader view on actual decision-making, planning and organisational design. *Pekka Korhonen* gives a brief introduction to Multiple Objective Linear Programming and discusses its possibilities in the area of forest management. An overview is provided of basic concepts, formulations, and principles of solving MOLP problems, and as the intervention of the decision-maker is always required, the behavioural aspects are also considered. By means of a case study picked from forest literature it is demonstrated how a MOLP approach may be used to support forest planning. *Jyrki Kangas* presents the standard version of the Analytical Hierarchy Process which has been found especially suitable for environmental planning with multiple criteria. Some experiences of the use of the AHP in Finnish forest management planning are discussed, advantages as well as drawbacks, and research efforts carried out in Finland to develop forestry decision support methods utilising some of the ideas and techniques applied in the AHP are presented. The HERO heuristic optimisation method, specially developed for tactical planning mainly at area or forest holding level, is expounded.

Chapter III: Stakeholder Involvement. *Per J. Agrell* expounds an interactive decision support system for agro-ecological integrated land-use planning and management in developing countries, an optimisation system which operates with multiple objectives in a flexible and user friendly fashion. An application of the model to a region in Kenya demonstrates its benefits and limitations. The model fulfils the fundamental requirement of being explicitly able to include analysis of several objectives, without incorporating the irrelevant preferences of the analyst. Features that allow back-tracking and regret are also supported in this application. *Vilis Brukas, Nico Hjortsø, Peter Tarp and Tove E. Boon* suggest a methodology for integrating economic efficiency and user group preferences in multiple-criteria forest management decision-making. The methodology includes: i) a heuristic decision-making approach; ii) the use of an optimisation model, comprising political, economic, and biological aspects; and iii) the use of a Geographical Information System to facilitate communication between user groups and administration. The methodology is implemented on a short-term management problem in Lithuania.

Chapter IV: Diversity Objectives and Option Value. *Ching-Rong Lin* and *Joseph Buongiorno*'s paper falls in two parts. In Part I, a Markov-chain model is applied to the Wisconsin maple-birch forests, incorporating stochastic biological growth and prices, to predict the consequences of natural and human disturbances on the diversity of tree size and species, diversity of landscape, and income. The frequency distribution of stand states throughout the maple-birch forest type measures the landscape diversity at different stages of forest development. It is found, e.g. that continuing the current management regime would lead to a steady-state maple-birch forest with twice the landscape diversity of an unmanaged forest, but producing a low income. Part II extends the analysis with a Markov decision process model to seek optimum management policies. Among the results are that an unconstrained economic harvesting policy gave more than five times the returns of the current management regime, with similar long-run landscape diversity, and diversity of tree size and species less than ten per cent lower. *Jens Abildtrup and Niels Strange* provide an outline analysis of the economics of converting the 'natural' tree species beech into Christmas tree production. In Denmark, such production is the more profitable according to conventional NPV calculations. However, this production demands the application of fertiliser and pesticides which may lead to irreversible contamination of drinking water. An option value approach, taking the uncertain future values of non-contaminated groundwater and Christmas tree production into account, will lead to a more conservative conversion strategy.

Chapter V: Spatial Objectives. *Mikko Kurttila* emphasises that spatial analysis is needed in forest planning that is addressing questions of biodiversity management and landscape ecology. The needs concern the relations between landscape structure and viability of species populations. Multi-objective planning must be applied where decision alternatives are assessed subjective to all objectives and also to overall priority. The study proposes a practical framework to include spatial objectives into Finnish private forest planning, using indices that describe the structure of the landscape. *Jan Juul Jensen* presents a spatial optimisation model for planning the restoration of riparian wetlands. It is a Binary Integer Programming model including criteria regarding costs, nitrate retention and landscape ecology. The model overcomes the problem that Geographic Information Systems are well suited to analyse and visualise spatial relationships between the many aspects of restoration sites selected on the basis of multiple criteria, but are lacking options for solving spatial optimisation problems. The model approach is based on a loose coupling between GIS and LP, the idea being to take advantage of the specific strengths of each of the two tools - spatial analysis and optimisation, respectively. *Niels Strange and Henrik Meilby* present a cellular automaton approach to optimising the choice of tree species in afforestation. In a case study area, four land use alternatives are considered, viz. pasture, beech, Norway spruce, and oak. The task of finding the optimal solution constitutes a tremendous combinatorial problem which no heuristic is able to solve. However, the authors show that appropriately formulated cellular automata may converge towards near-optimum solutions with a comparatively low number of iterations, and they conclude that a major advantage of self-organising algorithms is that the overall maximisation problem is solved as a decentralised optimisation problem where the use of the individual forest components is optimised in parallel. In that respect, self-organising algorithms contrast with the centralistic top-down approach known from most traditional optimisation methods.

Chapter VI: Cost Efficiency. *Mette Asmild, Jens Leth Hougaard and Dorte Kronborg* state that the analysis of capacity cost efficiency is an important tool for managerial decision-making in general. They apply Data Envelopment Analysis to the capacity cost efficiency problem of Danish milk producers, a technique which may well be appropriate to problems of more immediate relevance to forests and other natural resources. Assuming that the efficiency scores are right censored log-normally distributed, the influence of potential external factors affecting the efficiency may be investigated. *Peter Tarp, Finn Helles and Stanislaw Zajac* claim that strategic decision-making within integrated forest pest management is often mistakenly based on short-term treatment cost evaluation. For analysing pest management in the Polish state forestry, a modelling framework is built with a long-term perspective focussed on the environmental impact of two groups of control agents: biological insecticides and insect growth regulators. The authors link the economic evaluation of the damage done by the defoliating pest to the effect of changes of the eco-physiological processes dependent on the activity of the decomposing soil fauna biomass. Unlike the biological insecticide, the insect growth regulator has considerable negative effect on the soil fauna activity and consequently on the growth and yield of the stand. A sensitivity analysis of the incremental cost is performed by use of a long-term LP model, leading to a financial cost-benefit analysis. It is found that yield reduction may well make a biological insecticide more profitable than an insect growth regulator, in spite of the treatment cost of the former being considerably higher.

Chapter I

Survey

Why are the Uses Multiple?

Peter Berck

Abstract

This paper is largely about the objectives of multiple-use management and how those objectives have and will change over time. The first section of the paper provides a review of recent events in the American West with an emphasis on how different objectives were imposed upon private and public forest managers. The second section discusses political economic models that could lead to these changes in objectives. The third and last section discusses changing objectives in terms of a stochastic model and the use of planning tools.

Keywords: Multiple use, sustained yield, endangered species

1. Introduction

Forests are managed for multiple uses because multiple uses are unavoidable. Management is the manipulation of the forest to produce different mixes among the uses. This manipulation is undertaken to achieve a set of objectives. This paper is largely about the objectives of multiple-use management and how those objectives have and will change over time. The first section of the paper provides a review of recent events in the American West with an emphasis on how different objectives were imposed upon private and public forest managers. The second section discusses political economic models that could lead to these changes in objectives. The third and last section discusses changing objectives in terms of a stochastic model and the use of planning tools.

Management decides the relative importance of forest objectives. For example, the amount of water that comes from forests is not very sensitive to management. The quality of the water, however, is quite sensitive and depends upon whether or not silt is dumped into streams and whether clear cuts have effective buffers. Recreation is another form of forest use. In the United States, where the expanses of land are quite large, hikers physically cannot be excluded, so there is always some recreational use. Of course, they could be encouraged with facilities or discouraged with fences. Wildlife lives in the forest anyway. Clear cuts are a management regime that favours game while lack of clear cuts favours old-growth habitat for animals, such as spotted owls. There is no way to avoid multiple use in forests. There are, however, ways to make some uses be favoured over others. Which prescriptions are used depend upon the goals for which the forest is managed.

2. Postwar Goals and Policies

The question in forest planning is largely which use shall be master or how to trade off one use against another. In the American West, it is politics that drives multiple-use management. These politics have had three distinct regimes in the postwar period. From about the war until the early 1980s, timber was king. There was a very large housing boom in the United States, and the forests were thought of mostly in terms of their ability to provide wood. In the mid-1980s, there was considerable concern expressed through political and legal channels for maintaining old-growth forests. In the early 1990s, there was a new regime in which the Spotted Owl was king. No longer were forests planned for timber with other uses to be handled later. Now, the forests were to be planned to maintain old-growth habitat. With this change in objectives, there were also some management changes. The tools that are used have adapted to the politics, and the managers and planners have adapted to the politics. Viewing this over the long run, one often wonders whether or not the planner matters here.

The early postwar period was characterised by a political agreement that timber was king and that the purpose of the forests was to be, in the words of Gifford Pinchot, the first forester in the United States: tree farms. During the early period, informal tools and discretion, i.e., professional judgement of foresters, were the most important determinants of how the forests were used.

In this heyday of old-style planning and up to approximately 1990, the planning job could be characterised in a very simple way. There was a planner, and the planner was a professional forester and often a forest supervisor. The planner had intimate knowledge of the resource - knowledge that was not held by outsiders or easily written down. The planner's job was to figure out what the owner, in this case, the United States, wished to be done with the forest. It was not necessary to undergo much of a planning or preference elicitation exercise to determine the preferred use. Indeed, there were other uses, such as recreation, but timber seemed to be the overriding consideration. The owner also supplied the capital, and the planner needed to know how much money the owner would supply to actually cut the forest down as national forests were run as money-losing operations. The planning job was classic. The planner determined the preferences and budget and then found the best plan among the feasible plans. This type of exercise was easily amenable to mathematical programming formulations. But, in the early days, there was no need to do so. Professional judgement would have sufficed.

The catch in this idealistic planning algorithm was that an owner was needed. A close substitute would be a wide consensus on the appropriate goals and a political willingness to let the planner determine the goals within that consensus. In other words, the American people had faith in the U.S. Forest Service (USFS). Before approximately 1970, management of forests was not so contentious and these conditions were certainly true.

The Multiple Use Sustained Yield (MUSY) Act of 1960 was a high point of the influence of the USFS. The Act codified what the USFS was doing anyway and named a set of multiple uses: recreation, range, timber watershed, wildlife, and fish. The Act stated that no specific use could predominate and that a high level of annual output should be maintained without impairment of the productivity of the land. This was very broad guidance, and it gave the Agency a considerable amount of freedom. The USFS could operate, more or less, as it wished under MUSY of 1960. This was so simple because there were substantial areas that were untouched by cutting. The last frontier had not been reached.

Though not in the Act, but very much in USFS literature, were the Agency objectives. One was community stability - thought to be jobs. Conveniently, this is coincident with no profits. Supply of fiber (i.e., wood) was another important objective. In recreation, big game and fish (both of which are hunted) were predominant. Scenic drives were also mentioned quite often. And hiking had not yet come into its own. These objectives were harmonious: More wood was more jobs and open forest, and more open forest was more game. There was very little conflict.

3. Wilderness

The beginning of the end for this period was the Wilderness Act of 1964. Indeed, the USFS had designated wilderness on its own. It saw wilderness as a worthwhile use of its lands but mainly lands that were on high and inhospitable terrain. The USFS did not appreciate being told that these areas were off limits to timber and other activities by law. In addition, they were not pleased when they were told to study additional lands for inclusion in the wilderness system. These large single-purpose reserves went against the multiple-use grain. Perhaps, more important, the planner would not be able to decide which lands to reserve. Congress was now taking an active hand in land-use management.

The transformation of the management of the national forests came about by three Acts and were not considered radical: National Environmental Policy Act (NEPA), Resources Planning Act (RPA), and Endangered Species Act (ESA).

The NEPA of 1970 simply states that, before a major federal action can be taken, the Agency must get public comment on issues to be considered; make a plan [Environmental Impact Statement (EIS)] and several alternate plans, including a no-action alternative; get public comment on the plans; and choose a preferred alternative. This was not considered radical legislation but merely a way to require the Agency to carefully think about its actions. The Agency was, under this law, going to retain its discretion to choose its preferred alternative.

The RPA of 1974 was, again, in this vein although the idea was to produce plans specifically for national forests. It was a top-down act starting at the national level where goals would be set. These targets would then be sent to regions and, ultimately, to forests for fulfilment. The forests were to produce plans to meet their parts of the target. The Act was a way for the USFS to get long-term agreement with Congress because these were five-year plans. Presumably, goals would be made clear and Congress would appropriate the money to meet these goals. For the forest industry, it also gave a clear mandate to produce wood - something that the Wilderness Act surely had not. The RPA did not work because people did not agree on the goals. Environmentalists wanted more wildland than the USFS was planning to provide. The conflict was forced to a head by a lawsuit over clear-cutting in the Monongahela National Forest. The result was adverse to the USFS. That decision resurrected language in the Organic Act, which is 100 years old, that made it necessary to consider each tree before cutting. In effect, it made the preparation of forest sales throughout the United States far too expensive to be carried out. There was a very clear need for new legislation that would repeal the offending provision in the Organic Act.

The National Forest Management Act of 1976 (NFMA) was a political compromise. In the markup of that Act was the Sierra Club and the timber industry, and they both gave their assent. There were two key elements in the Act aside from the repeal of the offending clause. The first requirement was that the forest be managed for nondeclining flow. This was meant by the environmentalists to conserve old-growth although even a cursory analysis would show that it would only delay the day when the last bit of old timber was cut out. The second requirement was that trees were not to be cut until a majority of the stand had reached the culmination of mean annual increment (CMAI). This was meant to put teeth into the idea of sustained yield. Stands would be allowed to reach a reasonable age before they were harvested. The Act, however, is ecological meaningless. The trees at CMAI are still too small to support an old-growth ecosystem or even a very diverse forest.

The ESA of 1973, again, was thought to be innocuous. It stated that the Fish and Wildlife Service would have to produce a list of animals that were either in danger of being extinguished or threatened with being extinguished. Any animal on that list could not be taken even on private land. The word 'take' was later construed to include the removal of habitat. The Act specifically required a list of the habitats to be protected in order to keep the animal in existence. This Act, more than any other, put the whip hand in the environmentalists' hands. It has also spawned a very complicated legal problem regarding taking. When is it that the regulation of private property becomes the confiscation of private property? Takings, under the U.S. Constitution, obligate the government to pay for the property. Indeed, this is the major issue in the management of private forest land in the West.

Together, these Acts provided for participation and litigation. The RPA was radical in that it provided for interdisciplinary teams to make the forest plans. At the beginning, it was unclear whether or not these teams or the forest supervisors would have the final say. Later, the regulations made it clear that the forest supervisors would have the final say. There was a period for public comment, and the public comment needed to be recorded and taken seriously. In addition, all actions that were taken needed to be fully disclosed to the public. The ESA gave the public the right to sue to protect the endangered animals. These Acts gave the public the ability to see the inner workings of the Agency, including its mistakes, and science that could be debated if not overturned and to sue, based upon the need to protect the animals, or perceive mistakes in the procedures used by the USFS.

Under NFMA and RPA, it was formal planning for multiple use, and it was carried out by linear programming. It had been known for some time that forests could be programmed in this manner. Daniel Navon wrote an early code for this purpose although it was the model of Johnson and Scheurman (1977) that is widely used. The basic idea was to maximise the present value of timber subject to CMAI, nondeclining flow, and other constraints. Other constraints included a need not to destroy the habitat of endangered species - the Spotted Owl, of course, being the most famous of the species.

Type of Site, j
Many 'birthdays'
$h^j(t,s)$
t is calendar time
s is birthday of stand
h is acres harvested
v(t) is cut at t
$D^j(t-s)$ is volume per acre
$v=\sum_j \sum_s D^j(t-s)\, h^j(t, s)$
Max present value
of P times V
s.t. biology
$v(t+1) \geq v(t)$
$t-s > CMAI$ or $h = 0$

Figure 1 Simple Forest Planning

Initial Acres = Acres Cut over all time

$$A^j(s) = \sum_t h^j(t,s)$$

W is what is left standing

$$w^j(s,t) = A^j(s) - \sum_t h^j(t,s) = \sum_s h^j(t,s) - \sum_a h^j(a,t)$$

Cut acres regrow and are recut

$$\sum_s h^j(t,s) = \sum_a h^j(a,t)$$

Figure 2 The Biology Constraint

The simple forest-planning algorithm is the well-known Johnson and Scheurman algorithm. Figures 1 and 2 include the mathematics. Here, there are many different sites labelled 'j' and many 'birthdays' for stands labelled s. The t is calendar time, so $h^j(t, s)$ is the amount that is harvested from site j at time t for stands that were first planted in year s. Similarly, $D^j(t - s)$ is a volume table, giving the volume per acre of function h. Letting v be cut, it is simply the sum over sites and birthdays of volume per acre times acres harvested. In the simple model, the present value of P times V can be maximised. The constraints are the biology, nondeclining flow, and CMAI. The biology is just conservation of acres. Initial acres are the same as acres cut over all time. Therefore, any acres that are cut are replanted; they regrow and then must be cut as well. It is convenient to keep track of what is left standing because what is left standing is habitat for animals, particularly, old-growth habitat, which is denoted by W. Berck and Bible (1984) provide the mechanics of expanding the Johnson and Scheurman model to the case where W is accounted for. To give more meaning to the model, one must decide what types of sites one has.

Traditionally, of course, it would be different species and site classes. Less traditionally, it would be critical locations. Land near streams or in visual buffers could simply be assigned another site class and prohibited from certain treatments. These would be constraints, such as 'do not cut type j' or 'keep n per cent of the forest at some age or above.' These are all constraints on forest stock or W. Treatments would be added as well, e.g., commercial thinning and precommercial. It is the thinning treatments that have become controversial in the case of the Northern Sierras.

The traditional problem with this mode of planning was that the plans on the computer did not correspond very well to the plans on the ground. This was described as finding the cut. The plans were not spatial. They simply stated that a certain amount of acres of a particular forest type needed to be cut in this period. The foresters still had to find a specific parcel to cut. It was difficult to see what the cumulative effect of their decision would be because mapping technology was poor. Taking one of these plans and mapping out what it would mean over 100 years would be a tedious exercise. Taking 50, 60, or 100 plans that were made in the process of reaching a preferred alternative was an impossible exercise. Hrubes (1981) wrote a Berkeley dissertation on the problem of finding the cut. What he found was that the cutable land base was much smaller than the planned land base. Some of the reasons, in those years, included streams, Indian burial grounds, and needed habitat. The difference between the truly cutable land base and the computer plans was only discovered when one went out to find the cut.

The second class of problems with planning, which remains to this day, was called the allowable cut effect. Consider a forest with both old-growth, which is valuable, and high-altitude, unroaded, low-density stands, which are not valuable. A profit-maximising, nondeclining flow plan would include cutting the valuable stands in the early period and cutting the unvalued stands toward the end of the planning horizon. As long as equal amounts were cut both at the beginning and end and in between, the nondeclining even-flow constraints would have been met. This process would be fine except, after the old-growth was cut, the forest would be replanned. When the forest was replanned, a present-value maximising plan would not specify cutting the remote unprofitable trees. Thus, the second set of planning done after the valuable old-growth was cut would call for a lower yield than the first set of planning. Under nondeclining even flow, the cut would, in fact, decline. The USFS was perfectly satisfied with the process. A great deal of wood would be available if plans were held with commitment. Environmentalists detested the process. They preferred to see the old-growth remain, and the allowable cut effect caused the old-growth to be cut down sooner. This fundamental problem of lack of commitment to plans persists to this day in American forest planning.

Another problem was that forest plans took a great deal of time. At the beginning, as an observer, I believe that they took a great deal of time because the planning-requirements process was difficult. In hindsight, it does not appear so innocent. The old plans, before NFMA, were used during the waiting period. The old plans called for a great deal more cutting of timber than did the NFMA plans. Once it was clear that new plans would call for less timber, industry and their political allies, particularly the Reagan Republicans, did not want the plans to be final. Environmentalists obliged them by obstructing plans with lawsuits.

4. Spotted Owls

The Spotted Owl is an owl that lives in older trees although that is somewhat debated in California where they have been found living in trees of much younger age. In 1986, well before the Spotted Owl was listed as a threatened species, the USFS drafted an EIS on the Spotted Owl that would lead to a 5 per cent reduction in timber cut. The final EIS in April, 1988, called for slightly more cutting. This level of protection was not enough to comply with the law to protect the Spotted Owl. In March, 1989, Judge Dwyer issued an order that restrained the USFS from offering 139 planned sales, which was, essentially, its entire programme in the West. In June, 1989, the Spotted Owl was proposed as a threatened species. In June, 1990, it was listed. These actions were forced by lawsuits. The 1990 listing was defective in that it not include a listing of the critical habitat of the Spotted Owl. The ESA states that a list of the habitat to be protected in order for the animal to survive must be published in the Federal Register. The Fish and Wildlife Service had declined to do that.

Congress became quite upset with the situation as there was no timber being cut and that meant, in the states of Washington and Oregon, there was considerable dislocation. Congress was no stranger to direct control of the forest. They had been using the notion of allowable sale quota as far back as Carter, a Democrat and supposed friend of the environment, who asked the USFS to massively increase the amount of timber cut to provide for a housing boom. Again, in approximately 1984, Congress intervened, this time, to save the timber industry. In the late 1970s, bids for federal timber, sold on five-year contracts, had large inflationary expectations built in. Restrictive monetary policy dramatically lowered the inflation rate, and the companies found that they had contracted to pay more than the timber was worth. In 1984, Congress released them from their obligations without full penalties. It was no surprise in 1989 that the Hatfield-Adams bill directly prescribed the amount of timber to be sold for 1989-90. It prescribed it at 9.6 billion board foot. It streamlined the appeals process; made the EIS not subject to immediate judicial review; prohibited the use of temporary restraining orders or preliminary injunctions on fiscal, 1990, timber sales; set very strict deadlines for judicial review; and allowed the use of special masters to further speed up the court system. In short, Congress said that the USFS must cut more timber than the ESA and the USFS's plans would permit.

The Act also called for a study of the now pesky Spotted Owl. The future chief of the USFS, Jack Ward Thomas, was one of the leading members of Interagency Scientific Committee that produced this study in 1990. This study called for reducing harvests in the owl area by 30 per cent to 40 per cent. This was a far cry from the Hatfield-Adams congressional response that harvests were to be kept up at any cost. In order to comply with the orders of the federal courts, the Fish and Wildlife Service, under the Reagan Administration, which was extremely anti-environmental, was forced to comply with the ESA and list critical habitat for the Spotted Owl. It used the report by Thomas and, basically, enshrined those recommendations in law, listing as habitat 11.6 million acres in which 3 million were private. At this point, the Spotted Owl had moved from being a pest or a minor issue in multiple-use forestry to being *the* use for these forests. There was, indeed, a small counterattack by the Reagan Administration in which several Bureau of Land Management sales, not nearly as many as they had hoped, were exempted from the ESA by a group called the God Squad. The God Squad is a common parlance for the group that is permitted to allow a species to be extirpated. This being the federal government, one study followed another. The next study was known as FEMAT. Again, Thomas was one of the leading blights. The study gave many possible options for managing the forests in the West and decided in favour of option nine in response to the President's summit in April, 1993. In option nine, timber cut would go to 2 billion board foot, then 1.7 billion board foot, and then decline to nearly 1 billion board foot in the long run. The average was 1.2 billion board foot over 10 years. That is approximately a 90 per cent reduction from all-time highs and, certainly, about a 90 per cent reduction from the Hatfield-Adams congressional mandate. The FEMAT went further. It called for an ecosystem management plan for holistic, adaptive planning of the national forests. This was to be something new and different. Adaptive management meant that there was to be experimentation and that local communities and agencies would have some control of the experimentation. Of course, FEMAT option nine still protected the Spotted Owls.

President Clinton's Forest Plan is, basically, option nine and calls for less timber and more attention to the ecosystem. The President went even further in his restructuring of forest priorities, he replaced the planner. Jack Ward Thomas, the biologist, and now Mike Dombeck, a biologist, became head of the USFS. The timber beast was no more. The planning process is also being replaced as this talk is being written. It appears very likely that the linear programme that we know as Forplan will be replaced. It may be true that there will be no role for programming. It is very certain that the use of Geographical Information Systems (GIS) will be required.

Congress, being Republican and not environmentally inclined, objected strenuously to the President's Plan and passed what was known as the Salvage Ride. This bill was in effect for two years and was a response both to destructive fires and to the very strongly declining cut. Under the Salvage Ride, the USFS was to double the cutting of dead and dying trees in national forests over the next 18 months. It had always been policy to sell green trees with dead and dying trees so as to cause the dead and dying trees to pay their way out of the woods. Thus, the salvage rider actually called for the cutting of green trees as well. The Agency would be unhindered by the ESA and other laws protecting wildlife, and timber sales would be exempt from court challenge during this time. Clearly, Congress was in the driver's seat again and, by using the Appropriation Bill, had directly countered the President's Forest Plan and the current plans of the USFS.

The short summary of the planning objectives on public land is a trend toward preservation, and against timber, on which is superimposed a great vacillation of purpose. The story of private lands continues to unfold.

In Northern California, there is another animal that is worth mentioning - the Marbled Murrelet. They were listed as threatened in 1992. They nest in older redwood trees. In addition to the murrelets, in California, various species of trout and salmon are listed as endangered. What has caused the most trouble is that these endangered species live on private land as well as public land.

The Headwaters Forest and the area around it are very controversial. It involves the nesting areas of the murrelet and the density of old-growth redwood. The densest part is the Headwaters Forest, which is currently owned by the Pacific Lumber Company and is being sold, as I speak, to the United States and the state of California. The Headwaters Deal is complicated. One part of it is simple: The United States and the state of California buy the forest for $380 million. It becomes complicated when, in exchange, both entities agree to a habitat conservation plan for the remainder of Pacific Lumber's holdings. In effect, by selling the Headwaters Forest to the State at less-than-market value, Pacific Lumber has been allowed to file a favourable habitat plan for its remaining land and to be free of hindrance by the ESA in its other woods. Of course, environmentalists complain bitterly about this and the Plan may fall apart. Should the purchase not be consummated and the habitat conservation plans not be approved, Pacific Lumber will sue the United States and the State and claim that they have been allowed no economic use of their property. They may well win such a lawsuit, with the effect that the ESA would no longer be enforceable on private land. The goals on private lands are out of the control of the private landowners and are very much a matter of negotiation and litigation.

Another piece of the Clinton Plan looks like a stakeholder process. The idea is that the politicians would get interested parties into the room, and these parties would bargain with each other. Of course, they would be bargaining in the shadow of the law. Then, the parties would reach an agreement that the politicians would be able to implement without fear of political reprisal.

An example of the stakeholder process is the Quincy Library Group and the resultant proposed legislation. Local environmentalists and the local timber dominant firm, Sierra Pacific Industries, are trying to get Congress to accept their view of how the area around Quincy should be managed. The local view had a lot to do with jobs, timber, and fire. It cannot be emphasised how great an effect large forest fires have had on the people who live outside of the cities of California. The National Conservation organisations were not well represented, partially because they did not believe that anything that could happen in a library in Quincy would have anything to do with what happened in Washington where they live. The big issue, here, is forest condition. The locals want thinning. They want to reduce fire risk, and they want the volume of the lumber produced by the thinning, particularly, jobs. The thinning regimes would be an experiment in adaptive management if the bill were to pass. Again, there is a conflict of goals. Local 'stakeholders' have a very different view than national organisations, and it is unclear who will prevail and for how long.

With that background, I would like to try to make some economic sense of this record of multiple use and find some of the implications for planning.

5. Stock, Flow, and Jobs

There is a new emphasis on stock. The Agency and Administration have gone over to the view that it is protecting wildlife; reducing susceptibility to fire; and creating healthy, natural, or diverse forests, possibly getting them back to pre-European conditions, that are the real goals. These goals are quite at loggerheads with the Republican and congressional points of view, which emphasise the flow of services, particularly, jobs and timber. Of course, jobs make much better politics than timber. Sample and Le Master (1992) looked at four studies of how many jobs would be lost. This was before the full effect of the owl was actually known. Their study found that thousands of jobs in the pacific northwest would be lost, and those job losses were about two parts from the Interagency Scientific Committee recommendations on the owl and, perhaps, one part from either reduced harvests from federal forest plans, technological change, or any of these other possibilities. There are many problems with this point of view. A basic problem is that it is always stated in terms of jobs, and it should be stated in terms of job years. When old-growth is cut down, it can be done only once. Therefore, jobs per board foot make some sense, but they need to be spread out over many years. There are other technical problems. Jobs are, of course, a demand for a factor of production, and factor demands are unlikely to be linear in output. They are also unlikely to be constant over time because of technical progress. Another area of contention is indirect jobs. The standard procedure is to see in an input/output table how many additional jobs there would be if there were one additional basic timber job. If this calculation is done, two or three jobs is the outcome. Done more carefully, according to Stewart (1993), a lower number is derived because taken into account are the number of jobs that have been caused by, for example, transfer payments for retirees that have no part in the productive process. My coauthors, Chris Costello, Louise Fortmann, and Sandy Hoffmann, and I have examined the same problem using regression techniques. What we found is extremely different. We found that an additional job in the timber industry costs jobs in the other sector. That is, in Northern California, if one has one more job in the timber industry, after two years, one will only have 80 per cent of one job left and that the nontimber sector will have shrunk. I do not believe, based upon this work, that timber drives the economy, at least in Northern California counties, or that jobs are anything more than a stalking horse for industry profits.

6. Political Economy

An economist making sense out of politics can be done in four different ways: a game, median voter theorem, money and votes, or a political business cycle.

As a game, there would be two players - an environmentalist and a timber beast. The condition of the forest, say, the rotation age, and how much of the forest is used for timber as opposed to environment would be the issues of concern. The player interested in timber, the timber beast, would like to choose his rotation age to maximise the value of his return given the interest rate and a chance that some of his wood would be taken away from him and used for environmental purposes. As the chosen rotation age goes down, the chance that the government expropriates the property for environmental uses rises. The environmentalist lobbies for reservations of timber, which are expropriations. Lobbying is costly. The timber beast prefers a low rotation age while the environmentalist prefers a high rotation age. If this problem is worked out, and it has been worked out by Chris Costello, it shows that the more aggressive the environmentalist the more the timber beast will be likely to choose a low rotation age. Therefore, the condition of the woods will deteriorate as the probability of them being taken for environmental purposes increases.

In 1976, another then student, Haim Shalit, in an unpublished paper, looked at this problem as the median voter theorem. There are a number of technical assumptions that one would need to make, but the answer is that the USFS will do what the median voter prefers. A generalisation is Peltzman's

(1976) model in which both money and votes matter. Here, the timber beast wants the government to cut timber. Cutting, however, is unpopular with the voters. The timber beast can remedy this by donating money for political campaigns. Voters respond to political campaigns. As long as the timber beast gives the government (i.e., politicians) enough money so that they can be reelected, they will cut the timber against the will of the environmentalist.

Rogoff (1990) is the author of choice regarding the political business cycle. Presidents last only four years until reelection and congressmen, two. That gives them very short time horizons and induces a political business cycle. The idea is to take the downturn at the beginning of the term so that the recovery will be well underway by reelection. For instance, the Senator from Washington, Hatfield, needs rural votes. If he can shift timber harvest to the present, then there will be good times in the State and he will, most likely, be reelected. The fact that there will be a severe restructuring later in his term (six years), or even in the term of someone else, is really of no consequence to him.

7. Consequences: A Stochastic Model

A different view of the problems of the planning process and why multiple use is contentious would be in a simple stochastic model. I have in mind Granger's (1988) model of a stochastic objective function. In Granger's model, there are goals, in our case, timber and owls, which will be called GT and GO. And there are achievements, T and O. I have discounted these at a rate of interest and summed over time. Figures 3 and 4 has the formal model. The b is the price of owls relative to timber. The constraint set is that of the biology, and this would work perfectly if the constraints were linear, which they would be if the Forplan were used. Recent politics would be, in a sense, a change in the goal for owls. The goal was low and now it is high or the goal for owls and timber can both change. Recent history represents not just changes in the goals but an oscillation in the goals. The ESA, Congress, and the President all win for awhile.

Goals for timber, owls

GT, GO

$$\min \sum_t (T\text{-}GT)^2 + b(O - GO)^2 (1+r)^{-t}$$

r is interest rate

b is price of owls relative to timber

s.t. biology, other constraints as before

Clinton's selection of Option 9:

GT = 1.2 billion, long run; GO high.

Recent History: a shift in the goals

Figure 3 Simple Stochastic View

There are, of course, real stochastic elements, such as fire, which stochastically assign acres to new 'birthdays' without the benefit of harvests. A piece by Johnson et al. (1996) does approximately this. Trees do not always reach the one-year-older age classes, e.g., Bud Worms may destroy them. And the relation between owls and habitat has a random element. Indeed, there is no certainty whatsoever after 30 years of what the relation would be between the prescriptions made and the number of owls that would be there. The idea that these are, indeed, stochastic equations is quite realistic.

$$V = \min .5\ \{(T_1 - T^G)^2 + E\ b(O_2 - O^G_2)^2\ (1+r)^{-1}\}$$
$$\text{s.t. } O_2 = (O_1 - T_1)\ d + e$$
$$O^G_2 = O^G_1 + n$$
$$T_1* = \frac{1+r}{1+r+\delta^2 b}\left[\frac{\delta b}{1+r}\left(O_1\delta - O_1^G\right) + T_1^G\right]$$
$$V = \tfrac{1}{2}\left\{\left(T_1^* - T^G\right)^2 + \frac{b}{1+r}\left[\delta\left(O_1 - T_1^*\right) - O_1^G\right]^2 + \frac{b}{1+r} Var\left(\varepsilon + v\right)\right\}$$

Figure 4 Two period Stochastic

The horror of the stochastic model is that, since the goals change with the political winds, so will the plans. The plans will be, in this model, a linear function of the goals. Therefore, whatever type of random fluctuations the targets have, the goals will have the same fluctuations.

The implications for planning are simple. The planning exercise will be done and redone. There will be no commitment to carry out the plans. They will shift, for awhile, with the political winds. Old-growth dependent species are on a one-way trip. Once the habitat goes, it does not matter how high the goal is set in later plans. It cannot be brought back. In this type of model, there will be endless plans to save this species, and they will lose out one piece at a time. The planner will look like a fool.

The new planning environment for multiple use also reflects the primacy of the biologist. Since the day when the Interagency Scientific Committee and Thomas became the important elements in Western forest planning, the biologist has, quite literally, ruled the roost. Biologists will do their planning with the tools that are amenable to them. In this case, they would be mapping tools because biologists wish to see how different areas where their animals live relate to other areas. They will plan the condition with these mapping tools to get the forest condition that they want although they will run endless simulation models to show how the forest condition will relate to the animal populations. These plans will, ultimately, be made 'by eye.' What is leftover for optimisation is nearly nothing, and the traditional planner will find that he has a very minor role.

In short, a GIS that is run by biologist planners, with an interest in planning forest condition or stock, will drive whatever is left of traditional programming-type planning and not vice versa.

References

BERCK, P. and BIBLE, T. 1984. Solving and Interpreting Large-Scale Harvest Scheduling Problems by Duality and Decomposition. Forest Science 30:173-182.

GRANGER, C. 1988. Causality, Cointegration, and Control. Journal of Economic Dynamics and Control 12:551-559.

HRUBES, R.J. 1981. The USDA Forest Service in the Environmental Era: Institutional and Programmatic Change.

JOHNSON, K.N. and SCHEURMAN, H.L. 1977. Techniques for Prescribing Optimal Timber Harvest. Forest Science Monograph No. 18.

JOHNSON, K.N., SAPSIS, D., BAHRO, B. AND GABRIEL, J.T. 1996. Methodology for Simulating Forest Growth, Fire Effects, Timber Harvest, and Watershed Disturbance under Different Management Regimes. Chapter 5. In: Sierra Nevada Ecosystem Project Final Report to Congress: Status of the Sierra Nevada. Written and edited by the SNEP Science Team and Special Consultants. Davis: Centers for Water and Wildland Resources, University of California.

PELTZMAN, S. 1976. Toward a More General Theory of Regulation. Journal of Law and Economics 19:211-240.

ROGOFF, K. 1990. Equilibrium Politician Budget Cycles. American-Economic-Review 80(1):21-36.

SAMPLE, A.V. and LE MASTER, D.C. 1992. Economic Effects of Northern Spotted Owl Protection. Journal of Forestry 90(8):31-35.

STEWART, W.C. 1993. Predicting Employment Impacts of Changing Forest Management in California. University of California, Berkeley.

Modelling Ecosystem Processes and Patterns for Multiple-Use Management

Ye Qi and J. Keith Gilless

Abstract

The progression of natural resource management is reviewed from a focus on the output of products considered independently to function-oriented ecosystem management. The challenges of this paradigm shift for ecosystem modelling are identified, including needs to simulate interactions among multiple components, effects of climatic change and stochastic disturbances, long-term ecosystem development, and human dimensions. The shortcomings of the timber growth and yield models traditionally used in forestry are identified and evaluated. The potential value of process-based ecosystem models as tools for simulating ecosystem processes and patterns is discussed in light of recent developments in landscape ecology, GIS and spatial modelling. The need to integrate ecosystem models, socio-economic models, and decision analysis models in integrated spatial decision-support systems should determine the future direction of natural resource modelling.

Keywords: Ecosystem management, spatial decision-support system, GIS, adaptive management, forest management

1. The Rise of Ecosystem Management

Natural resource management in the United States underwent a paradigm shift in the 1990s. The traditional focus on individual product outputs gave way to concern for the functioning, diversity and sustainability of whole ecosystems (Kessler et al. 1992, Christensen et al. 1996). This change was in response to intensified and diversified demands on natural resources. The Multiple-Use and Sustained Yield Act of 1960 (MUSY) focused on the outputs of individual resources from the national forests and rangelands, including game species, recreation, timber, water, and wilderness. The management direction in the MUSY proved to be inadequate to address societal demands. Many important functions of forests and other ecosystems were not sufficiently considered in management plans and practices. Recognition of the inadequacy of traditional multiple-use management led to a series of efforts to improve the ways that renewable natural resources are managed. In 1989, US Forest Service (1989) proposed its New Perspectives for Managing the National Forest System as 'a different way of thinking about managing the national forests and national grasslands, emphasising ecological principles, to sustain their many values and uses' (Robertson 1991). In a landmark report, Forestry Research: A Mandate for Change, a National Research Council Committee (NRC 1990) concluded that forest management systems needed to be developed that both produced commodities and maintained and improved environmental values. These efforts finally led to the rise of ecosystem management. In 1992, Forest Service defined ecosystem management as 'the use of an ecological approach to achieve multiple-use management of the national forests and grasslands by blending the needs of people and environmental values in such a way that the national forests and grasslands represent diverse, healthy, productive, and sustainable ecosystems' (Robertson 1992).

Debate on what ecosystem management is and should be has helped to popularise the concept among the public, government, and academia. A committee on the scientific basis of ecosystem management, set up by the Ecological Society of America, compiled a long list of definitions by various authors, and provided a version of its own as 'management driven by explicit goals, executed by policies, protocols, and practices, and made adaptable by monitoring and research based on our best understanding of the ecological interactions and processes necessary to sustain ecosystem composition, structure, and function' (Christensen et al. 1996). By 1994, 16 Federal Agencies had nominally adopted an ecosystem management approach to natural resource management. Yaffee et al. (1996) was able to documented 105 ecosystem management projects in the United States.

As an approach to management of natural resources, ecosystem management recognises the diversity of ecosystem functions and services, and the importance of underlying processes, structures and patterns for sustaining these functions and services. It represents a fundamental shift from output-oriented management to function-oriented management. True multiple-use management can only be realised under such a paradigm. An ecosystem management approach also presents challenges to the way that we model ecosystems. Critical examination of the existing models and modelling approaches is a starting point to meeting these challenges.

2. The Challenge of Ecosystem Management

2.1 Interactions among multiple components of ecosystems

Multiple use of an ecosystem depends on the functioning of multiple ecosystem components as parts of a system. Traditional approaches tended to deal with these components separately. Timber management relied on growth and yield models to predict timber volumes in different time periods. Wildlife conservation focused on the population dynamics of wildlife species. Forest hydrology focused on water yield and quality. But these various ecosystem components are related to each other in complex ways. In management of the forested watersheds in Sierra Nevada, for example, land managers face a series of interrelated problems and decisions with respect to fire hazards for forests and structures, water quality and yield, conservation of the Northern Spotted Owl and other threatened and endangered species, and timber harvesting. Single purpose models of timber growth and yield, wildlife population dynamics, or water yield and quality are incapable of addressing these problems. It is and will be necessary to integrate these models of individual ecosystem components to balance diverse societal concerns.

2.2 Energy transformation, nutrient cycling and water balance

Energy transformation, nutrient cycling and water balance are fundamental processes that constitute the basis of all ecosystem functions. Nevertheless, these processes are often ignored in forest management models. This is understandable since managers are generally interested in the final results of energy transformation, carbon fixation and allocation, not the processes themselves. But the intimate relationships between timber harvest and ecosystem processes have long been recognised (Bormann and Likens 1994). Timber harvesting has important ecological implications. It affects water yield and quality by decreasing water holding capacity and increasing water runoff. It disturbs the nutrient cycling by unbalancing nutrient absorption and decomposition. Carbon sequestration is affected by removal of vegetative cover. Harvesting alters wildlife habitats and fire hazards. In turn, changes in ecosystem processes feed back to affect timber production. Maintaining these processes is the key to sustaining the productivity of ecosystems. Therefore, it is essential that the dynamics of energy, nutrients and water be explicitly included in ecosystem management models.

2.3 Ecosystem-climate interactions

Climate plays a key role in ecosystem functioning, yet it is often excluded from forestry models. The climate system is dynamic and assuming a steady state is potentially misleading.

Climatic variability can be detected at many scales ranging from seasonal, inter-annual, decadal, to centennial. Climatic variability affects the growth and development of all living organisms and their interactions, and is a major source of discrepancy in the predictions of forest growth and yield. Yeh (1997) found that +/-20% of the variability in timber volume growth is caused by climatic variability. In extreme years the figure may be as high as +/-80%. A direct approach to this problem is to incorporate climate variables into growth and yield models. With the growing concerns over regional and global climate change, ecosystem-climate interactions will likely become a more important issue for research and practice. On a large scale, the changing patterns of climate in many regions of the world may have profound impact on ecosystem processes and functions (Qi and Gong 1996). In turn, the of ecosystem responses and changes in land cover and land uses may be important feedback elements for local, regional, or even global climate systems (Dale et al. 1991, Henderson-Sellers et al. 1993).

2.4 Ecosystem development

Ecosystems always undergo dynamic changes. Apart from diurnal and seasonal changes, ecological succession, or ecosystem development, must also be considered in models designed to predict long-term ecosystem dynamics. Predictive models in forestry are often based on statistical relationships between age and other properties (e.g. height and diameter) of trees or stands. Ecosystem studies in recent decades have revealed the complexity and variety of ecological processes and structures affecting ecological succession. For example, a typical sequence of a conifer forest in the Pacific Northwest includes at least four phases: establishment, self-thinning, transition and shifting mosaics (Spies 1997). The entire process may well take hundreds to thousands of years. During the course of ecosystem development, the biological composition, structure, biogeochemical cycle, soil and water properties may change significantly. Some of the functions of ecosystems can only be achieved in the later successional stages. Our understanding of ecological succession has been continuously improved since the time of Clements (1916). The climax concept is now less emphasised as we gain more knowledge on the varying nature of mature ecosystems. Ecologists now conceptualise succession as a non-equilibrium spatial process of disturbance, birth, death, dispersal, and growth under changing environmental conditions (Peet and Christensen 1980; Glenn-Lewin and van der Maarel 1992). Theoretical concept of succession has therefore moved away from simple generalisations towards more complex constructs. Site-specific predictions about vegetation dynamics cannot be made from broad general theory; rather, they require a more complex scientific framework and specific information about disturbance, environment, and species biology (Pickett et al. 1994; Spies 1997). The complexity of ecosystem development cannot be captured by the stand dynamics models that assume a steady state. New knowledge and models are needed in this area. Understanding the ecosystem succession is not only important to long term plans for timber production, but also important to preservation of biodiversity, since some species rely on certain developmental stages or spatial configurations of ecosystems.

2.5 Disturbance

Disturbances play an important role in shaping ecosystem structure and function. Disturbances to ecosystems include wind damage, fire, pests and diseases, droughts and flood (Pickett and White 1985). The occurrence and impact of such disturbances is usually stochastic. For example, fire is a major disturbance to the forest ecosystems in Sierra Nevada. Fire occurrence and impacts in the region have been documented and analysed for decades. But few studies have been conducted that incorporated fire in simulation of long-term dynamics of the forest ecosystems in the region (SNEP 1996). Human interventions often further complicate the effects of disturbances.

Decades of fire suppression in the region has resulted in heavy accumulation of fuels, and enhanced the probability of severe fires.

2.6 Human dimensions

Modellers often have to choose between realism and simplicity of their models when dealing with human factors. Some resource modellers hesitate to include human variables in their models for fear of losing objectivity. Yet human activities are essential components of any managed ecosystem and cannot be avoided. It is important to know how forest stands respond to silvicultural prescriptions such as fertilisation, thinning and pruning. Social and economic forces influence the structure and function of natural and cultural landscapes. Furthermore, the social, economic, and ecological systems are all related and affect one another much as the biophysical components of ecosystem do.

An important lesson to be learnt from past resource management efforts is that our knowledge about the ecosystem that we manage is provisional, and our management plans are based on provisional knowledge. Yet management plans are often made as if they were based upon knowledge that is complete or sufficient. Realisation of this problem allows us to design management plans so that they can be adapted as more knowledge is obtained from the practice of management. Such an adaptive approach to management requires that human activities be explicitly recognised in ecosystem process models (Lee 1993).

2.7. Spatial scale, pattern and process

Recognising the importance of modelling processes in ecosystem management may require us to incorporate spatial patterns at multiple scales in ecosystem management models. Spatial heterogeneity is ubiquitous to all ecosystems. The microenvironments, and the biota therein, vary from one location to another. Processes in one location are often distinct from elsewhere. A pattern in space emerges due to these differences in composition, structure and processes of different locations. This emerged spatial pattern constitutes a land mosaic at greater spatial scale at which new functions emerge. These emerged patterns and functions, in turn, can affect the structures and processes at the smaller scales. Consider a forest as habitat for wildlife. In an even-aged, single species plantation, few habitat types are expected because of its homogeneity, even if the forest is large. However, if the forest consists of a number of stands that differ in age, species composition, density, or structure, it should be suitable for many more wildlife species. Interactions between process and pattern can be complex. Scale is a critical issue in dealing with process-pattern relationship. Since the ecosystems and landscapes are organised in hierarchical structures in which processes and patterns are linked and viewed at different scales, a modelling approach has been proposed based on the concept of hierarchical patch dynamics to address process, pattern and scale simultaneously in an integrated ecosystem model (Wu and Levin 1994).

Spatial scales are often treated implicitly rather than explicitly in forest management models. A forest stand is often the modelling unit for growth and yield models. A management unit in a linear programming model may include a number of forest stands and stand types. The size of the stands or harvest openings, as well as the scale of a management unit, are important parameters for ecosystem management.

Spatial configuration is related to scale. Harvest adjacency restrictions illustrate the importance of spatial configuration. Barrett et al. (1998) showed that graph theory could be used to address both spatial scale (in terms of harvest openings) and configuration. Hof and Bevers (1998) provide additional examples of optimisation under consideration of spatial configuration.

3. Development of Ecosystem Models

Ecosystem models have evolved through several stages. The traditional growth and yield models are still probably the best means for predicting timber outputs. To address other aspects of forest ecosystem dynamics, however, they must be supplemented with other types of ecosystem models reflecting ecosystem studies in ecology. Together these approaches can serve as a basis for simulating and predicting long-term ecosystem dynamics and can be integrated into a spatial framework which can serve as the foundation for a spatial decision-support system.

3.1 Timber growth and yield models

Growth and yield models are among the earliest forest ecosystem models, and are still most commonly used to predict the timber volume and stand growth in forestry (Davis et al. 1987). Growth and yield are usually calculated as functions of stem diameter, tree height, and crown size. Since this group of models were designed for support of timber production, non-timber aspects of forests were originally not of concern (Munro 1974). Ecological and biological processes are often excluded. Although competition among individual trees may be considered, interactions among ecosystem components are generally not modeled. The effects of climate and other environmental variables are rarely explicitly recognised. Environmental variables are often implicitly represented using an integrated index of site quality, which is usually assumed to be constant for a given site. The relationships in these models are usually established based on long-term observation of sample sites, and thus high degree of accuracy can be attained within the range that the sample plots can represent. However steady state models can perform only so well in predicting of growth in a changing environment. These features seriously limit the models to address the challenges of ecosystem management.

3.2 Gap models

As a special group of individual tree based models, gap models have added some new features to forest dynamics models, particularly with respect to inclusion of environmental variables. JABOWA (Botkin et al. 1972), considered to be the first gap model (Shugart 1998), added light and temperature to the diameter growth equations. This important addition made it possible to simulate the response of tree growth to the change of environmental variables. Since light conditions and microclimate for each individual tree change over time as a result of tree growth, the model allowed for consideration of feedback from tree growth to the environment. In theory, such a model could be used to simulate not only growth and yield in short term, but also the ecosystem succession over long time when site condition and microclimate are changing.

The early gap models stimulated the development of a new approach to ecosystem modelling recognising the environmental effects on biological processes are simulated. JABOWA included such ecosystem processes as birth, death, and growth of trees. Subsequent work has demonstrated the potential of this approach. Pastor and Post (1986) used a gap model to examine the effects of climate, soil moisture, nitrogen availability and light conditions on tree growth. Uncertainty can be reflected in such models through the use of stochastic functions. As the approach is used to more cases, the functions representing ecological processes have become more sophisticated.

The primary unit of a gap model is the individual tree. To run gap models for a forest stand, detailed information is needed about both the stand and the individual trees in the stand. This makes it difficult for gap models to be applied to large areas. The computational requirement in keeping track of changes in a large stand also contributes to what may be called the 'dimension problem'.

Despite the advance they represent, gap models were still developed largely for simulation of tree and stand growth, and non-tree components of the ecosystems were not the focus of simulations (but see exceptions in Shugart 1998). The gap model approach is reductionist in its focus on the individuals in a system, with the properties of the system being derived from aggregation of the properties of individuals. Aggregation error and error propagation are therefore of concern. Furthermore, gap models do not deal explicitly with spatial patterns, although a pattern emerges as a result of the spatial distribution of individuals.

3.3 Ecosystem process models

Ecosystem process models, or process-based-ecosystem models, differ from gap models and classical timber growth and yield models in their emphasis on the system as a whole, in which the ecosystem interacts with its environment. The ecosystem's components and their interactions with one another and the environment are modeled as flows of material, energy and information.

The ecosystem approach eases the dimension problems confronting gap models by reducing the number of components to be simulated, but retaining the fundamental focus on processes. It is this simplification that allows detailed treatment of processes such as energy flow, nutrient cycling, and water balance. Ecosystem/climate interaction and ecological succession are also treated as easily as with a gap model approach.

Spatial patterns rise from the site-specific information on composition, structure, and processes of ecosystems. The systems approach facilitates the hierarchical classification of the ecosystem according to its structural and spatial organisation. In addition, the flexibility in defining an ecosystem's boundary allows modelling at multiple scales.

Ecosystem process models were first developed in late 1960s, and major advance was made beginning mid-1980s. Ecosystem management concerns at the watershed or regional scales has been the driving force for the rapid development of such models (McNulty et al. 1994). In recent years, global change modelling efforts have helped to bring ecosystem process models closer to biophysical and biogeochmical models (Tian et al. 1998). Much research is needed, however, in order for them to be extensively applied to ecosystem management.

4. Spatial Decision-support System: Needs and Opportunities for Synthesis

Modelling ecosystem processes and patterns is critical to natural resource management. But ecosystem models alone are not sufficient to answer all the important questions. The systems we face involve ecosystems influenced, and sometimes controlled, by economic and social forces. Ecosystem patterns, processes and long-term dynamics often reflect the interactions of all these influences. It is therefore necessary to integrate the ecological models with socio-economic models in order to address the human dimensions in ecosystem management, and in particular with decision analysis models.

This integration will often entail combining simulation and optimisation models. While results from simulation models are often used in optimisation models, using the information from optimisation models as a reference to simulation models is also of value (Hof and Bevers 1998).

The importance of spatial patterns and scales and the need for site-specific information for management call for spatially explicit modelling. Geographic Information Systems (GIS) can provide the geographically referenced data and serve as tools for spatial modelling. GIS for ecosystem management will include data on ecosystem dynamics, socio-economic dynamics, as well as the baseline data such as topography, climate, transportation etc. GIS, ecological models, socio-

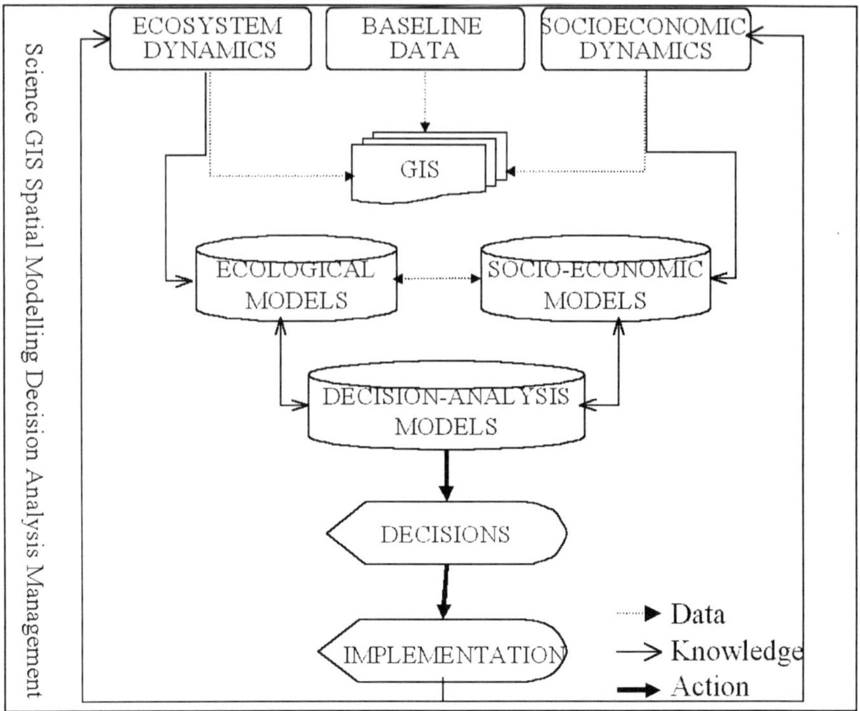

Figure 1 Diagram of a spatial decision-support system

economic models, decision analysis models together will form the core of future spatial decision support systems (SDSS) (Figure 1).

The SDSS concept (Figure 1) is consistent with the concept of adaptive management. With a SDSS, management decisions are made based on the data and models. Implementation of decisions provides feedback increasing knowledge of the ecosystems being managed and of the socio-economic systems they are part of. New knowledge is incorporated in the subsequent decision-making and management actions.

5. Concluding Remarks

The rise of ecosystem management presents new challenges to ecosystem modelling. The timber growth and yield models traditionally used in forestry are inadequate to address the concerns of ecosystem management. Greater reliance will have to be placed on process-based ecosystem models of the sort developed primarily for simulating ecosystem processes and patterns. The rapid recent development of landscape ecology, GIS and spatial modelling greatly enhanced the opportunities for synthesis. To address the human dimensions in ecosystem management, it will be necessary to integrate ecosystem models, socio-economic models, and decision analysis model in GIS-based spatial frameworks, i.e. spatial decision-support systems.

References

BARRETT, T. M., GILLESS, J. K. AND DAVIS, L. S. 1998. Economic and fragmentation effects of clearcut restrictions. Forest Science 44:569-577.

BORMANN, F. H. AND G. E. LIKENS. 1994. Pattern and process in a forested ecosystem: Disturbance, development and the steady state based on the Hubbard Brook ecosystem study.

BOTKIN, D.B., JANEK, J.F. AND WALLIS, J.R. 1972. Some ecological consequences of a computer model of forest growth. Journal of Ecology 60:849-872.

CHRISTENSEN, N. L., BARTUSKA, A.M., BROWN, J. H., CARPENTER, S., D'ANTONIO, C., FRANCIS, R., FRANKLIN, J. F., MACMAHON, J. A., NOSS, R. F., PARSONS, D. J., PETERSON, C. H., TURNER, M. G., AND WOODMANSEE, R. G. 1996. The report of the Ecological Society of America committee on the scientific basis for ecosystem management. Ecological Applications 6:665-691.

CLEMENTS, F.E. 1916. Plant Succession: An Analysis of the Development of Vegetation. Publication 242. Washington, D.C.Carnegie Institute of Washington.

DALE, V. H., HOUGHTON, R. A. AND HALL, C. A. S. 1991. Estimating the effects of land-use change on global atmospheric carbon dioxide concentrations. Canadian Journal of Forest Research 21: 87-90.

DAVIS, L. S., JOHNSON, K. N. AND DAVIS, K. P. 1987. Forest management. McGraw-Hill, New York.

GLENN-LEWIN, D.C. AND VAN DER MAAREL, E. 1992. Patterns and Processes of Vegetation Dynamics. New York: Chapman and Hall.

HENDERSON-SELLERS, A., DICKINSON, R. E., DURBIDGE, T. B., KENNEDY, P. J., MCGUFFIE, K. AND PITMAN, A. J. 1993. Tropical Deforestation - Modelling Local-Scale to Regional-Scale Climate Change. Journal of Geophysical Research-Atmospheres 98:7289-7315.

HOF, J. G. AND BEVERS, M. 1998. Spatial optimisation for managed ecosystems. Columbia University Press, New York.

KESSLER, W. B., SALWASSER, H., CARTWRIGHT, C. W. JR., AND CAPLAN, J.A. 1992. New perspectives for sustainable natural resources management. Ecological Applications 2:221-225.

LEE, K.N. 1993. Compass and Gyroscope: Integrating Science and Politics for the Environment. Washington,D.C., Island Press.

MCNULTY, S. G., VOSE, J. M., SWANK, W. T., ABER, J.D. AND FEDERER, C. A. 1994. Regional-scale forest ecosystem modelling: Database development, model predictions and validation using a Geographic Information System. Climate Research 4:223-231.

MUNRO 1974. Forest growth models - a prognosis. In: Growth Models for Tree and Stand Simulation, Royal College, Stockholm, Forest Research Notes 30.

NRC 1990. Forestry research: A mandate for change. National Academy Press, Seattle, Washington, USA.

PASTOR, J. AND POST, W. M. 1988. Response of northern forest to CO2 induced climate change. Nature 334:55-58.

PEET, R.K. AND CHRISTENSEN, N. L. 1980. Succession: A population process. Vegetatio 43:131-40.

PICKETT, S.T.A. AND WHITE, P. S. 1985 The Ecology of Natural Disturbance and Patch Dynamics. Orlando, FL: Academic Press.

PICKETT, S.T.A., KOLASA, J. AND JONES, C. G. 1994. Ecological Understanding. Orlando, FL: Academic Press.

QI, Y. AND GONG, P. 1996. Metabolic and phenological response of vegetation to temperature gradient: evidence derived from AVHRR data, Geographical Information Sciences 2(1-2):64-72.

ROBERTSON, F.D. 1991. Charter: New Perspectives for managing the National Forest System. USDA Forest Service, Washington, D.C., March 29, 1991. Cited by Salwasser, H. 1992. From New Perspective to Ecosystem Management: Responses to Frissell et al. and Lawrence and Murphy. Conservation Biology 6(3):469-472.

ROBERTSON, F. D. 1992. Ecosystem management of the national forest and grasslands. Memo to regional foresters and station directors. USDA Forest Service, Washington, D.C., June 4, 1992. Cited by Salwasser, H. 1992. From New Perspective to Ecosystem Management: Responses to Frissell et al. and Lawrence and Murphy. Conservation Biology 6(3):469-472.

SHUGART, H. H. 1998. Terrestrial ecosystems in changing environments. Cambridge University Press, Cambridge ; New York.

SIERRA NEVADA ECOSYSTEM PROJECT (SNEP). 1996. The Status of Sierra Nevada, Vol.1. Wildland Resource Center. Report No.39, University of California, Davis.

SPIES, T. 1997. In: Kohm, K.A. and Franklin, J.F. (eds.). Creating a Forestry for the 21st Century: The Science of Ecosystem Management. Washington, DC. Island Press:7-30.

TIAN, H., HALL, C. A. S. AND QI, Y. 1998. Modelling primary productivity of the terrestrial biosphere in changing environments: Toward a dynamic biosphere model. Critical Reviews in Plant Sciences 17:541-557.

USDA Forest Services 1989. RPA assessment of the forest and rangeland situation in the United States, 1989. United States Department of Agriculture, Forest Services Resource Report Number 26.

WU, J. G. AND LEVIN, S. A. 1994. A Spatial Patch Dynamic Modelling Approach to Pattern and Process in an Annual Grassland. Ecological Monographs 64:447-464.

YAFFEE, S. L., PHILLIPPS, A. F., FRENTZ, I. C., HARDY, P. W., MALEKI, S. M. AND THORPE, B. E. 1996. Ecosystem management in the United States: An assessment of current experience.

YEH, HUI-YI 1997. The Relationship Between Tree Diameter Growth and Climate for Coniferous Species in Northern California. Ph.D. Dissertation, University of California, Berkeley.

Descriptive Indicators of Sustainable Forest Management

Tove Enggrob Boon and Finn Helles

Abstract

The changing role of forests in society provides new challenges to forest management planners, as conceptualised in 'sustainable forestry'. Using an end-means rationality, there is a need to develop criteria for sustainable development and indicators to measure whether development goes in the desired direction. Recent European efforts of operationalising sustainable forest management have focused on forest output and ecological processes, whereas socioeconomic processes and structures, how they affect output, and what processes and structures are considered legitimate by the different actors, have been ignored at large. However, operationalisation of sustainable forest management implies policy formulation, and the question appears: whose sustainability are we striving at? With the 1998 Lisbon resolution L1 on socioeconomic conditions, the issue has been highlighted again, however, without providing means to monitor the social sustainability of forestry. The present paper aims at filling a methodological gap by developing a framework for descriptive indicators of socially sustainable forest management. The framework is based on the Helsinki and Lisbon resolutions, but the overriding idea of providing valid and reliable descriptive (qualitative) indicators could be transferred to operational level planning with only slight modifications.

Keywords: Social sustainability, rationality, qualitative indicators, institutions, Lisbon resolutions

1. Introduction

The role of European forestry is undergoing structural change. Until recently, the core objective was to meet the economic demands of forest owners. Today, society demands multiple uses of forests, including biodiversity, recreation and environmental protection. In addition to the changing values related to forests, society is increasingly interested in the production processes and the principles of forest management, reflecting the varying perceptions of efficiency (Scott 1992). Thus, planning has to include the changeable, diverse and sometimes contradictory demands on forest output.

Forest planning and management is gradually evolving to reflect the changing role of forests in society. Multiple-use forest management is being replaced by process-oriented management, as we improve our knowledge about the links between output of forest goods and benefits and the underlying ecological and organisational, or other social processes. This is reflected in the concepts of, e.g. ecosystem management and, on the social side, participatory management. In addition, there is a recognition at a national level of the link between social structures (institutional conditions), processes and the resulting forest output. Such structures are, e.g. professional skills, level of technology, financial resources, organisational structures, vertical or horisontal co-ordination with nurseries, wood industries etc.

Recent management planning has been based on an assumption of end-means rationality, where the processes and structures are ultimately the means to reach the objective of an optimal output of multiple goods and benefits. The current challenges of planners are then to disclose the causal links between structures, processes and output in terms of operational indicators to be integrated in management planning. One example could be that we need to know how to design public participation structures in order to provide an efficient process without conflicts, with optimal input of local and professional knowledge on the forest and local preferences, all in order to have an optimal output of forest goods and benefits.

The concept of sustainable forestry describes the new role of forestry. Sustainable forestry in a European context is conceptualised in the Helsinki resolutions 1993, adopted by the European Ministers of Forestry. Since then, the concept has been operationalised in terms of a set of six criteria and a number of most suitable quantitative indicators (MCPFE 1994; MCPFE 1998a) alongside with a range of similar national and international efforts, e.g. CSCE (1994a;1994b;1995), CCFM (1994), FSC (1994), IWGF (1994a;1994b). These criteria of sustainable forest management have further been operationalised into forest management planning at forest property level, e.g. (MCPFE 1998a; Nepenthes Consult 1996).

The Helsinki resolutions, criteria and quantitative indicators for sustainable forest management primarily focused on the ecological functions and processes and to a less extent on the socioeconomic functions of forests, whereas socioeconomic processes and structures were barely mentioned, comparing it to the Rio Forest Declaration and the NGO-driven certification process of the Forest Stewardship Council (FSC 1994). A set of European indicators to describe the socioeconomic structures, descriptive indicators, had been developed already in 1995 (MCPFE 1995) but was left out in the continued c & i process. These indicators should document the 'existence and effective implementation of a related policy framework..' (MCPFE 1994), in terms of legislation, institutional framework, economic/financial framework and information. In this paper, 'descriptive' is interpreted as synonymous with 'qualitative'. However, the socioeconomic processes and structures in relation to forestry are now being highlighted in European forestry with Lisbon resolution L1 on socioeconomic aspects of sustainable forest management (MCPFE 1998b) as well as with L2 on criteria and indicators for sustainable forest management (MCPFE1998a) where the aim is to work for further development of the existing quantitative and descriptive indicators.

Considering the many possible and conflicting needs, operationalisation of sustainable forestry cannot be based on an assumption of consensus between utilisation and protection of forests. Rather, the concept of sustainable forestry implies forest policy formulation. Operationalisation provides a common framework for discussing the distribution of scarce resources within forestry and in relation to society. And in planning for sustainable forestry, and when developing criteria and indicators, we need to ask ourselves whose sustainability we are striving at: who chooses what criteria and indicators, and what are their possible distributional consequences?

There is no 'one best way' to ensure sustainable forestry. There is likely to be disagreement not only on what the output should be, but also on what processes and structures are agreeable and appropriate to rely on. This basically means that the end-means rationality is not sufficient when planning for sustainable forest management. We need to broaden our definition of efficiency to also include institutional efficiency, based on value rationality (Beck Jørgensen and Melander 1992), where survival of the organisation (or, broader, the forest sector) is a value in itself, and where the keyword is legitimacy in relation to the surrounding society. The question on output is not only: 'are we producing the right amount and quality', but: 'are we producing the right things?'. The question on processes is not only: 'are we having an optimal process in relation to producing output', but: 'are we using acceptable processes, based on acceptable values?'. The question to structures is not only: 'are we having optimal structures, considering our surroundings and the aimed output', but: 'are our structures based on acceptable norms and values?'

Reconsidering the example of public participation: From an institutional perspective, public participation in management planning now becomes (i) an end in itself, when participatory structures provides legitimacy to society; (ii) public participation can be a means to reach legitimate processes (e.g. discussing whether the use of pesticides, and size of clear-cuts, etc. is acceptable), and (iii) public participation serves to provide a legitimate and maybe optimal output of goods and benefits. Clearly, the choice of indicators to reflect the role of participation differs significantly from indicators derived from the technical efficiency perspective illustrated above.

The challenge to forest management planning increases drastically. Maintaining the end-means approach, using a criterion-indicator model, how do we take into account these conflicting values and how do we disclose the assumed causal links between output, process and socioeconomic structures?

Apparently there is a need for developing indicators of socioeconomic conditions (structures and processes) that take this into account. There is a need to overcome what may have been the reasons for descriptive indicators not being adopted in the Helsinki process: (i) Descriptive indicators do not fulfil scientific standards as applied to quantitative indicators. Consequently, there is a need to develop a theoretical framework for descriptive indicators based on a social science interpretation of requirements of data validity and reliability; (ii) Descriptive indicators tend to determine the content of the other criteria for sustainable forestry, as they provide knowledge and values by which 'sustainable' is defined. Descriptive indicators assume some objectives that should be stated as criteria (e.g. that there is a need for legal framework to set minimum levels of income for forest workers) or the reporting of descriptive indicators should be discussed or judged by an appointed panel of politicians, experts and/or officials as they cannot be interpreted objectively in their own right. Therefore, descriptive indicators should not measure to what degree they support or reflect the requirements of sustainable forestry as set in other criteria. Rather, indicators of socioeconomic conditions should reflect the extent to which interests, values and knowledge are recognised and passed on to forest policy formulation, leading to revised criteria and indicators of 'sustainable forestry'.

Accordingly, by using the Helsinki and Lissabon processes, the aim of this paper is to formulate a framework for descriptive indicators of socioeconomic conditions related to sustainable forest management.

The remainder of the paper will use the term 'institutional conditions' as a unifying concept encompassing the four issues of the Helsinki descriptive indicators, i.e. legislation and regulatory framework, institutional framework, economic/financial framework, and information.

2. Descriptive Indicators

2.1 Defining an indicator

The criterion-indicator model is basically that the concept of sustainable forest management is specified into a set of criteria, each of which 'describes the different sides of sustainability on a conceptual level'. Each criterion 'is a distinguishing element or set of conditions or processes by which a forest characteristic or management is judged' (MCPFE 1994). Indicators are then *measures* of an aspect of each criterion for sustainable forest management. The indicators should measure achievements built in criteria, *showing changes over time* and demonstrating how well each criterion reaches the objectives set for it. The indicators should *support the reporting process* and make it internationally *credible*. The aim of developing (quantitative) indicators is to 'provide a choice of indicators that are *scientifically valid, technically feasible and cost-effective*' (MCPFE 1994).

The criteria cannot stand on their own, as they do not include information regarding to what extent these elements must be present to provide sustainable forest management. A criterion must be combined with a threshold value or an objective to provide an operational management information (Egestad 1995). This information is partly provided by the Helsinki resolutions H1 and H2. The relevance of indicators can therefore not be determined by looking at the criteria only. Rather they should be developed on the basis of the resolutions.

With respect to descriptive indicators, it is stated that 'some aspects of the criteria can only be judged through the existence and effective implementation of a related policy framework. These are considered as descriptive indicators'(MCPFE 1994).

Indicators serve two main functions (OECD 1993):

- They reduce the number of parameters and measurements needed to give an exact description of a situation. To maintain a comprehensive overview the number of indicators should be limited and yet provide sufficient information.
- They simplify the communication process by which the information of results of measurement is provided to the user. This simplification may be at the expense of the scientific demands to demonstrate causal relations.

Thus, indicators should fulfil the following requirements: Provide *adequate, exact* information *relevant* to the criterion and policy behind it. The information should provide international *credibility*, being scientifically *valid.* And, generation of data should be technically and financially feasible. In order to provide credibility, information should be valid as well as reliable. Reliability of information refers to the consistency of the research findings, i.e., a judgement of the monitoring process, the gathering and the interpretation of data. That indicators should provide a valid description of a situation means, in ordinary language, that the statements should be true and correct (Kvale 1996).

2.2 Material reality and social constructs

The way we perceive our criteria and underlying objectives and, thus, what is relevant knowledge and how it should be generated, is determined by our *ontology*, i.e., how we basically look at the world, and how we can obtain knowledge about the world, our *epistemology* (Maaløe 1996).

The basic ontological question is: Does there exist one true reality, independent of our ways of perceiving this reality? In a positivist philosophy, knowledge is a reflection of the real world. Reality exists independent of our perceptions and is of a material character. This means that there is a singular *truth* about this reality which can be described by means of *objective* data, namely by using quantitative methods. Subsequently, we should try to disclose this truth by means of social structures and social facts, e.g. by testing hypotheses and using quantitative methods (Maaløe 1996).

In a post-modern perspective, the conception of knowledge as a mirror of reality is replaced by knowledge as a social construction of reality (Kvale 1996). The extreme assumption held by social constructivists is that reality only exists through our perception. Therefore, there is not one true reality, but several competing 'realities'. In relation to research, all findings are subjective and scientific knowledge is *constructed* by the scientist. Subsequently, the researcher aims at disclosing these social constructions and deriving meanings from them, namely by means of qualitative methods (Maaløe 1996). The consequence is that truth becomes relative and what is considered a 'fact' changes over time and space. Truth is constituted through dialogue. Validity of knowledge claims is discussed and negotiated among the members of a community, with the norms and debate being the court to evaluate knowledge claims. In other words, validity becomes the issue of choosing among competing and falsifiable interpretations, of examining and providing arguments for the relative credibility of alternative knowledge claims, 'intersubjectivity' (Polkinghorne 1983), involving three issues: (i) Focus on the process of producing facts rather than on the produced facts; (ii) Communication of knowledge; (iii) Pragmatic proof through action, as the positivist justification of knowledge is replaced by the ability of knowledge to perform effective and ethically right action (Kvale 1996).

Choosing an intermediate point of view, we acknowledge that part of the observed reality mirrors a material reality, whereas part of it is a social construct. Hereby the purpose of descriptive indicators becomes twofold: to describe the 'factual' content, and to disclose the social constructions embedded in *legal and regulatory framework, institutional framework, financial instruments and informational means.*

2.3 Validity of indicators

Validity refers to the extent to which our measures correspond to the concepts they are intended to reflect (Mannheim and Rich 1995). A valid measure, i.e. the set of indicators, should therefore be both appropriate and complete seen in relation to the criterion they are to describe, i.e., they should be relevant to as well as provide an exact and adequate measure of the particular criterion and policy.

Considering the social constructivist notion of competing realities rather than one truth, the relevance to criterion and policy becomes a question of: (i) Documenting the process; (ii) Arguing for the relevance of data in relation to indicator, criterion and policy; and (iii) Considering the consequences of the chosen indicators. This could be specified into a set of actions, based on Kvale (1996), Yin (1994) and Mannheim and Rich (1995):

Construct validity - Document the process

One should argue for the choice of indicator on the basis of formally stated criteria and policy objectives (in this case the Helsinki and Lisbon resolutions H1, H2, L1 and L2) and the following criteria-indicator process. Data generated should relate to the indicator. This means that the questions asked to the source of information (documents, interviews etc.) should be considered relevant to the indicator, as should the indicator to the criterion, and as should the criterion to the policy. This relevance is documented in the indicator report by specifying:

- The propositions/hypotheses underlying the choice and interpretation of indicators.
- Sources of evidence that are being used.
- Data generated for a given (set of) indicator(s) to reflect a picture of the competing realities. This is done by using multiple sources of evidence, eventually by means of different methods. Hereby converging lines of inquiry can be developed by triangulating data.
- Different perspectives on the same data set and discussing credibility and strength of the different interpretations.
- What variables are focused at when analysing evidence in relation to each indicator (i.e. the words of each indicator and their meaning).

Communicative validity

One should argue for the relevance of data in relation to indicator, criterion and policy, i.e.

- Specify what scale the indicator and data relate to (local, national, regional).
- Have the draft descriptive indicator report reviewed by key informants known to have differing viewpoints.

Pragmatic validity

One should consider the consequences of using the particular indicators:

- Check results obtained from use of the indicator against results obtained from use of another indicator that is known to be a valid measure of the concept, or test the *predictive validity* of the indicator by using it to predict events that reflect the concept being measured (Mannheim and Rich 1995).

Pragmatic validity also raises a question of power and truth. Using the words of Kvale (1996 p.283): 'Who has the power to decide what are the desired results or directions of change'? The aim of a repeated reporting process is to show progress in some way. Having a set of criteria and indicators as a measure of this, it seems likely that the mere presence and wording of an indicator can have an effect on the national forest policy objectives. And for the reporting of it, who should decide what is to the better and what is to the worse?

For instance, the proposed descriptive indicators of the Helsinki process includes *'Legal / regulatory framework - existence and extent to which it provides for legal instruments for securing income levels in forest sector'* (indicator no. 9 for criterion no. 6). Would it be regarded as progress if minimum levels of forest workers´ income change from being directed by negotiations between the unions and the employers´ organisation, as it is in Denmark, to being settled by legislation, i.e. 'complying with' or at least filling this indicator?

The suggestion here is, that, as stated above, the national reports are being revised on the basis of comments from key informants known to hold differing views of the state of forestry and forest politics. Additionally, a paneuropean panel of experts could be appointed to give their intersubjective, nonauthoritative but guiding judgement of the development based on the national reports on descriptive indicators.

2.4 Reliability of indicators

Reliability refers to the consistency of the research findings (Kvale 1996). A measure (of an indicator) is reliable when a later investigator, following exactly the same procedures as described by an earlier investigator and conducting the same study all over again, arrives at the same findings and conclusions as the earlier investigator. This creates a need of documenting the procedures, the steps of the qualitative monitoring project, so that the description of a given indicator can be traced back to its original sources and the circumstances under which the data were generated. Therefore, the reporting of qualitative indicators should fulfil the following requirements (Yin 1994):

- Indicator and data can be interpreted unambiguously.
- Data should be traceable and reproducable.

In its consequence, this is done by establishing chains of evidence so that in principle any observer can follow the derivation of any evidence from initial question to ultimate conclusions. The observer should also be able to trace the steps in either direction. This is assured by sufficient citation in the report of the relevant evidence (judicial documents, interviews, etc.) used and indicating under what circumstances this evidence is collected:

- Specify propositions regarding interpretation of the specific indicator.
- Specify the sources of evidence relied on.
- Specify what questions are asked/what issues/variables are focused on when analysing evidence.

2.5 Summing up: Framework for choosing/evaluating descriptive indicators

Summing up on the above discussion of how to ensure the validity and reliability of qualitative data, the list below shows the requirements to the choice of descriptive indicators and their reporting. The meaning of the requirements is illustrated in section 4.

- Specify the propositions/hypotheses underlying the choice and interpretation of indicators.
- Specify what variables are focused on when analysing evidence in relation to each indicator.
- Specify sources of evidence that are being used.
- Have data generated for a given (set of) indicator(s) to reflect a picture of the competing realities.

- Take different perspectives on the same data set and discuss credibility and strength of the different interpretations.
- Specify what scale the indicator and data relate to.
- Have the draft descriptive indicator report reviewed by key informants.
- Check results obtained from use of the indicator against results obtained from use of another indicator that is known to be a valid measure of the concept, or test the *predictive validity* of the indicator.
- Have the national descriptive indicator reports discussed by a paneuropean panel of experts.

Left is to consider the financial, technical and temporal feasibility of reporting. Following the above requirements, the reporting process is expected to be rather costly in terms of salaries for making the report. The 'technical' feasibility depends on whether sources of evidence are present and in what form. If extensive interviewing is needed, financial requirements will increase. With respect to time, it seems as a reasonable judgement that major reporting is only done, say each five years, whereas subsidiary reporting is done yearly.

3. Illustrating the Framework in Terms of the European Descriptive Indicators

In 1995, the Liaison Unit in Helsinki issued a draft set of qualitative indicators for sustainable management of forests in Europe (MCPFE 1995). They are briefly discussed in order to illustrate the above requirements to descriptive indicators. A draft Danish report of descriptive indicators of sustainable forestry has been made, using this framework (Boon 1998) which provide the examples below.

3.1 Relevance: Specify the propositions and variables underlying the choice and interpretation of indicators

As discussed in section 2.3, one should argue for the choice of indicator on the basis of formally stated criteria and policy objectives. In this case, the relevance of each indicator should be considered in relation to the contents of the Helsinki and Lisbon resolutions H1, H2, L1 and L2, as well as in relation to national policies and binding international policies, like the EU forest strategy.

Helsinki and Lisbon resolutions

The proposal for descriptive indicators (MCPFE 1995) mentions for each concept area of a criterion the relevant parts of the Helsinki Resolutions H1 and H2, The UNCED Forest Declaration, Biodiversity Convention, and Agenda 21, chapter 11 'combating deforestation'.

The adoption of Lisbon resolution L1 seems to change the way we interpret the existing indicators. Taking a glance at L1, the following elements are emphasised in comparison to H1:

- The role of forestry in rural development (L1.1, L1.4, L1.5, L1.7) This is specifically mentioned only in relation to employment and income opportunities in indicators 6.7 and 6.8.
- Multiple-use forestry assessment, management, planning for and marketing of forest products and services other than just wood (as compared to former emphasis on wood).
- The need to enable and motivate private small and/or absentee forest owners for sustainable forest management (L1.2). (Could be considered under indicator 6.9).
- Furthering co-operation, namely between private organisations.
- A pro active market oriented approach (L1.3).

- Quality of work, in terms of working conditions, safety and health, job security and remuneration (L1.7) as well as consideration of equal opportunities for women in employment income, training etc. (L1.8).

It appears that these new elements could fit into existing indicators, only providing additional meaning to the variables, e.g. 'forest resources' not being wood only, but all forest goods and services.

Relevance in relation to national policies and practices

Also, the relevance of indicators should be discussed in relation to national policies, practices and natural conditions. In Denmark for instance, there is no tradition of designating forest areas as protection forests. As a general rule, almost all forests should remain as such (~ 90 % is designated as forest reserves) and the Forest Act's rules of 'good and multiple use forest management' pertain to these areas. Therefore, the indicators 5.1-5.8 related to protection of soil are more or less irrelevant in a Danish context.

Other indicators may be relevant but reveal a lack of possibility to describe 'another possible way of doing things the right way'. An example is indicator 6.9 about the extent and capacity of legal framework to secure income levels in forestry. In Denmark, there is no such legislation. Rather, minimum salaries are determined by negotiations between the labour unions and the employer associations. The labour union would probably be against legislation, fearing a loss of support and power and, in the end, a lower minimum wage relative to the one negotiated.

Variables

The data relevant to an indicator is determined by the propositions and variables with often implicit meanings, meanings that may vary from country to country, from context to context, from time to time. Some are given meaning by reading the background documents (Helsinki Resolutions) whereas others are open to interpretation. These variables should be identified and defined. As a common example, take the following indicator (criterion 1, indicator no. 5):

1.5 Legal / regulatory framework - existence and extent to which it maintains forest resources and prevents forest degradation; clarifies property rights and provides for appropriate land tenure arrangements.

This indicator contains a set of variables that ideally need to be defined:

- What is meant by *legal/regulatory framework?*
- What sort of *forest resources* are included, what is the minimum area under tree cover to regard as forest?
- What is the threshold value for *maintaining* versus *degrading?*
- What rights are included in *property rights*?
- When are land tenure arrangements *appropriate* and in relation to what?

The only definitions given in direct relation to the descriptive indicators are the four main issues upon which all indicators are based: the legal/regulatory framework; the institutional framework; the economic policy framework and financial instruments; informational means to implement the policy framework (MCPFE 1995).

3.2 Specify sources of evidence that are being used

The reporting of each indicator should be documented by references, e.g. partial reporting for criterion 1.5, about whether legal or regulatory framework exists and maintains forest resources, and clarifies property rights, could sound like: '...The general regulations of the Forest Act ensure, that forest areas remain under forest cover. It is estimated that minimally 90 % of the Danish forest area are encompassed by the rules of permanent forest cover (Swainson 1995)...'

And the list of references would include:

- 'Forest Act, no.959, 2. november 1996.
- Swainson, A.G. 1995. Loven og Skoven I: Miljø- og Energiministeriet. *Retligt set - synsvinkler på arealanvendelseslovgivningen*. Miljø- og Energiministeriet, København. Pp. 315-327...'

3.3 Have data generated for a given (set of) indicator(s) to reflect a picture of the competing realities

Use multiple sources of evidence and use varying methods, e.g. document analysis and interviews. Hereby, converging lines of inquiry can be developed by triangulating data. In the reporting process, conflicting views should be illustrated, if any. For instance, for criterion 1.5 mentioned above it could be relevant to draw on legislation and instructions, research reports evaluating legislative measures as well as grassroots organisations´ statements relating to the issue, either in documents or/and by interview.

It is acknowledged that some of the official documents have been commented, if not validated, by the community already, such as is the case for the Danish Strategy for Sustainable Forest Management (Miljøministeriet, Skov- og Naturstyrelsen 1994) that was written by a board comprising main stakeholders in Danish Forestry.

3.4 Take different perspectives on the same data set and discuss credibility and strength of the different interpretations.

The findings should be analysed and discussed from different perspectives. This should be done for each indicator, for the joint set of indicators related to each criterion and for the indicators as a whole. This requirement is linked to the above requirement to analyse and reflect a picture of competing views.

3.5 Specify what scale the indicator and data relates to

The Helsinki descriptive indicators all relate to the national level. However, data may be relevant only to parts of a country, specific stakeholders etc. This should be explicitly stated.
Taking indicator 1.1 as an example: the existence of legal or regulatory framework to provide an overall policy framework for conservation and sustainable management of forests, the Danish report would include the following notions on scale:

'Legislation pertains to all forests, from local to national scale. At some points it differentiates between state forestry and other forests: the state forestry has a special responsibility to pay regard to recreational utilisation, nature, environment and cultural heritage in forests (Forest Act). Moreover, public access to forests is wider in state forestry as compared to other forests (Nature Conservation Act)'.

3.6 Have the draft descriptive indicator report reviewed by key informants known to have differing viewpoints

This includes two different issues:

Key informants that have contributed with information should read the relevant parts of the report in order to validate the meanings drawn by the author from informants' own contributions.

A group of experts, NGOs, politicians and/or officials could be asked to read the report and give their general comments and viewpoints on the contents.

3.7 Check results obtained from use of the indicator against results obtained from use of another indicator

In order to find the minimum set of indicators providing an 'exact and adequate, relevant measure', results from using different (sets of) indicators should be held against each other and considered in relation to what (sets of) indicators provide the broadest understanding of the criterion and underlying policy. This could be done at the national and paneuropean level after the first reporting with a given set of indicators has taken place.

3.8 Test the *predictive validity* of the indicator

Another way of determining the use of indicators is to use them to predict events that reflect the concept being measured (Mannheim and Rich 1995). This is, however, considered to require a rather comprehensive reporting format. An example could be that if, for the second reporting, the situation of biodiversity is worse off than predicted by the reporting on indicators, the indicators and/or the reporting (sources of evidence and their analysis) have not been appropriate.

3.9 Have the national descriptive indicator reports discussed by Pan-European panel of experts (and national key informants as considered above)

The need for such a panel is illustrated by the indicator 6.8 about the existence of legal or regulatory framework to provide for legal instruments for securing income levels in the forest sector: Would it be regarded as progress if minimum levels of forest workers´ income change from being directed by negotiations between the unions and the employer organisations, as is the case in Denmark, to their being settled by legislation, i.e. 'fulfilling' or at least filling this indicator?

4. Concluding Remarks

The role of forestry is changing. The focus shifts from demands of the forest owner towards demands of society, from output towards processes and structures, and from producing an optimal output towards producing a legitimate output in a legitimate way. Technical rationality is succeeded by a value-based rationality that has to be taken into account in planning and monitoring and, as here, by developing descriptive indicators that take the relative nature of 'truth' into account.

The present framework has been based on national planning and monitoring of sustainable forest management. Nevertheless, the framework could be applied to participatory forest management planning at a property level as well, acknowledging that many structures would be considered fixed factors beyond the jurisdisction of the forest manager or fixed in relation to the time scale.

References

BECK JØRGENSEN, T. AND MELANDER, P. (eds.) 1992. Livet i offentlige organisationer [Life in Public Organisations]. DJØF Forlag, Copenhagen.

BOON, T.E. 1998. Descriptive Indicators for Sustainable Forest Management. The Royal Veterinary and Agricultural University, Unit of Forestry, Frederiksberg.

CCFM 1994. Criteria and Indicators for the Sustainable Management of Forests: The Canadian Process.

Canadian Council of Forest Ministers. Information folder. C/I Secretariat, c/o Canadian Forest Service.

CSCE 1995. Working Group on Criteria and Indicators for the Conservation and Sustainable Management of Temperate and Boreal Forests. Statement on Criteria and Indicators for the Conservation and Sustainable Management of Temperate and Boreal Forests. Natural Resources Canada. Canadian Forest Service, Ottawa.

CSCE 1994a. Seminar of Experts on Sustainable Development of Boreal and Temperate Forests. September 27 to October 1, 1993 Montreal, Quebec, Technical Report. Natural Resources Canada. Canadian Forest Service, Ottawa.

CSCE 1994b. Working Group on Criteria and Indicators for the Conservation and Sustainable Management of Temperate and Boreal Forests. September 9, Olympia, Washington. Discussion Draft. Natural Resources Canada. Canadian Forest Service, Ottawa.

EGESTAD, P. 1995. Setting Standards for Sustainable Forest Management. The Royal Veterinary and Agricultural University, Unit of Forestry, Frederiksberg.

FSC (FOREST STEWARDSHIP COUNCIL) 1994. Forest Stewardship Principles and Criteria for Natural Forest Management. Board Approved Version, June 1994. Forest Stewardship Council. Oaxaca, Mexico.

IWGF (INTERGOVERNMENTAL WORKING GROUP ON FORESTS) 1994a. Intergovernmental Working Group on Forests. Criteria and Indicators for Sustainable Forest Management. Ottawa/Hull, Canada, October 10-14 1994. International Working Group on Forests.

IWGF (INTERGOVERNMENTAL WORKING GROUP ON FORESTS) 1994b. Intergovernmental Working Group on Forests. Report, Second Meeting, Ottawa/Hull, Canada, October 10-14 1994. International Working Group on Forests.

KVALE, S. 1996. Interview. An Introduction to Qualitative Research Interviewing. Sage, London.

MAALØE, E. 1996. Case-studier af og om mennesker i organisationer. [Case Studies by and about People in Organisations]. Akademisk Forlag, Copenhagen.

MANNHEIM, J.B. AND RICH, R. C. 1995. Empirical Political Analysis. Research Methods in Political Science. Longman Publisher, White Plains.

MCPFE (MINISTERIAL CONFERENCE ON THE PROTECTION OF FORESTS IN EUROPE) 1994. European List of Criteria and Most Suitable Quantitative Indicators. Ministerial Conference on the Protection of Forests in Europe.

MCPFE (MINISTERIAL CONFERENCE ON THE PROTECTION OF FORESTS IN EUROPE) 1995. Descriptive indicators. Ministerial Conference on the Protection of Forests in Europe, Liaison Unit in Helsinki.

MCPFE (MINISTERIAL CONFERENCE ON THE PROTECTION OF FORESTS IN EUROPE) 1998a. Resolution L2. Pan European Criteria, Indicators and Operational Level Guidelines for Sustainable Forest Management. The Lisbon Conference on the Protection of Forests in Europe, June 1998, Lisbon, Portugal.

MCPFE (MINISTERIAL CONFERENCE ON THE PROTECTION OF FORESTS IN EUROPE) 1998b. Resolution L1. People, Forests and Forestry Enhancement of Socio-Economic Aspects of Sustainable Forest Management. The Lisbon Conference on the Protection of Forests in Europe, June 1998, Lisbon, Portugal.

MILJØMINISTERIET, SKOV- OG NATURSTYRELSEN 1994a. Strategi for bæredygtig skovdrift. Betænkning nr. 1267. Miljøministeriet, København. [A partial translation can be found in: Ministry of the Environment 1994. Strategy for sustainable forest management. Ministry of the Environment, Copenhagen.]

NEPENTHES CONSULT 1996. Projekt bæredygtig skov. Projektrapport [Project Sustainable Forest. Project Report]. Nepenthes Consult, Aarhus.

OECD 1993. OECD Core Set of Indicators for Environmental Performance. Synthesis report by the group on the state of the environment ENV/EPOC/GEP(93)5/ADD. OECD, Paris.

POLKINGHORNE, D.E. 1983. Methodology for the Human Sciences. SUNY Press, Albany.

SCOTT, W. R. 1992. Organizations. Rational, Natural and Open Systems. Third edition. Prentice Hall, Englewood Cliffs.

YIN, R.K. 1989. Case Study Research, Design and Methods. Sage, London.

Discounting in the Context of Multiple Objectives

Colin Price

Abstract

The inclusion of many objectives other than profit maximisation should raise questions about using a financially based rate to discount all costs and benefits. Psychological time preferences is best explained as a preference for consumption in the immediate time period, not for consumption earlier in time: this gives no democratic basis for discounting. Personal mortality risk is not well represented by a discount function. The debate about the entitlements of future generations on balance suggest their interests should be included in resource decisions. Because of physical and psychological changes through time, an action affecting attainment of aesthetic objectives may have monotonically declining importance: but it may also have increased, or cyclical importance. Diminishing marginal utility has widely differing applicability to technology-rich products and to those intensive in requirement of natural resources. Efficiency in allocating investment funds is only attained by discounting if revenues are wholly reinvested at a rate of return equal to the discount rate. Other, more sophisticated means should be used to select investments efficiently.

Keywords: Discounting, environment, sustainability

1. Introduction

Should there be anything *particular* to say about discounting in the context of multiple objectives? An answer might be attempted from three different perspectives.

1 The summary objective of conservation has often been described in terms such as achieving 'the greatest good of the greatest numbers for the longest time'. However, applying this definition requires an impossible simultaneous maximisation. Discounting is one means of making the necessary policy trade-offs between *duration* of good against *intensity* and *kind* of good, and against the *number of beneficiaries* of good.

2 The increasing recent tendency to monetise non-market benefits and costs makes it possible to bring environmental and social factors into cost–benefit analysis. In this framework discounting is the normal way to treat future values. But new questions are raised as to whether discounting's philosophy, developed in relation to *actual monetary* values, is also applicable to values given only in terms of *monetary equivalents.*

3 Since the Rio Earth Summit of 1992, sustainability has become a focus of attention, the treatment of costs and benefits arising in the long term being a central issue. Discounting is a means – a very particular one – of expressing the importance, or otherwise, of such long term effects.

Actually one of the problems with the recent obsession about sustainability is that it has often been treated as a free-standing desirable product, having the same character as profit, landscape and biodiversity, rather than as an expression of the time period over which the profit, landscape and biodiversity values are aggregated.

However, the high importance ascribed, politically, to sustainability leads to a rather obvious conflict. Sustainable development is 'development that meets the needs of the present without compromising the ability of future generations to meet their own needs' (Brundtland 1987). Yet if an organisation *actively sought* to compromise the ability of future generations to meet their needs, it could hardly have chosen a more effective method than discounting – the technique adopted world-wide by government and international agencies in the appraisal of projects! Any system which reduces the importance of the United Kingdom's GDP, 500 years hence, to the price of a bag of potato crisps has clear potential conflict with sustainability. Yet this is what discounting, at the UK's advised 6% discount rate (Department of the Environment 1991), would do.

The purpose of this paper is to examine the justifications put forward for this apparently malevolent technique, and to show that they are – particularly in relation to the multiple objectives of forestry – massively unconvincing. The arguments are selected from those detailed in Price (1993), from which examples are also drawn.

Because of the widespread acceptance of discounting among investment appraisers, and the enormous effect which discounting exerts on long-term values, the implications of the argument are immense. If there are errors in it, they should be identified and exposed. But if not, practising economists – and all those whose well-being depends on the outcome of their analyses – ought to prepare themselves for very fundamental changes.

2. The Meaning of Time Preference

Let us begin with human impatience as an argument for discounting. A few years ago, David Heigham, then Chief Economist of the Department of the Environment, posed the following question to a conference audience, composed largely of vociferously anti-economic ecologists and geographers.

Which of you, faced with the choice of getting your pet project, your new bird reserve, your cleaned up trout stream, or whatever, this year rather than in five years' time, would not take it this year? (Heigham 1992).

And those listening had to agree, that the sooner, the better. 'There!' he concluded triumphantly. 'You are all discounting.'

Of course, consumer sovereignty is a pillar of democracy: if it appears that the population of a country discounts, then a democratic government should take on board its preference for early consumption. But is preference for early rather than late actually what time preference means? The choice posed by Heigham involved no trade-off whatsoever between early and late consumption: only an additional five-year period of consumption, starting now, with no penalty in the future. That *more* consumption of good things is preferred to *less* is a central axiom of utilitarian economics, but that tells us nothing about relative preferences for early and late consumption.

2.1 Inconsistent time preference

To establish such preferences, choices are needed in which some amount of a product can be consumed early only by sacrificing some other amount of that product later. Such choices, real and hypothetical, have been extensively explored by economists and other social scientists. One disquieting application has shown, repeatedly over many years, that forestry students typically discount future consumption of beer at rates of around 400% per annum. Fortunately for the world's forests, when these students graduate they are generally investing other people's money in forest enterprises....

Examination of more detailed evidence, however, shows that preference for earliness, apparently so consistently displayed in such choices, turns out to be anything but consistent. Firstly, the rate and functional form of discounting differ remarkably from the uniform negative exponential function by which most economists undertake discounting.

Large sums of money, and important life decisions, are subject to lower rates of discount than small sums and trivial decisions.

Financial losses and 'bad things' are discounted less heavily than positive events (Benzion et al. 1989).

Lower rates of discount are considered appropriate for public decisions than for personal ones (Pope and Perry 1989).

Long periods are – completely contrary to what most financial economists have believed – discounted at a lower rate than short periods (Thaler 1981). More precisely, lapse of a given time period incurs less discount if it is far in the future: tomorrow may be considerably discounted in relation to today, but 366 days in future is hardly discounted at all in relation to 365 days in future.

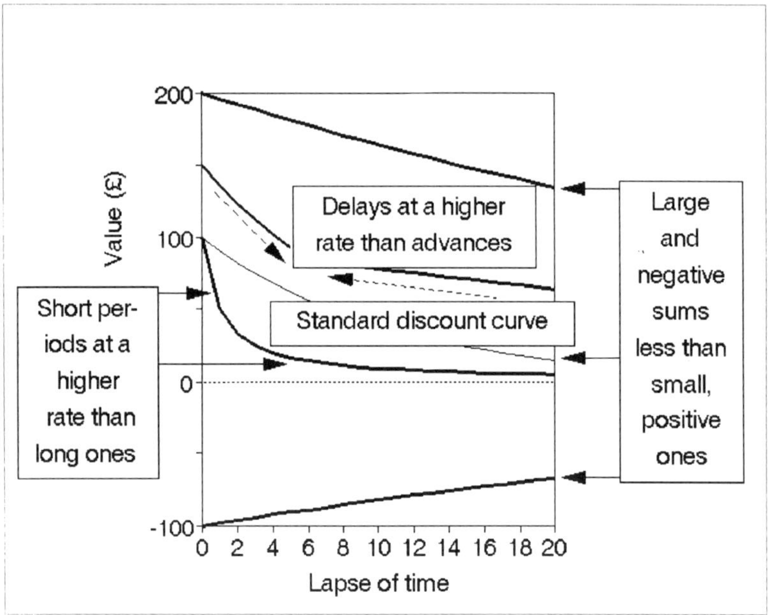

Figure 1 Inconsistency in time preference functions

Postponement of pleasure causes a greater proportional change in value than advancing pleasure by the same period (Loewenstein 1988) – experience with young children shows conclusively that the gratification imparted when treats are brought forward does not compare with the fury unleashed when treats are postponed.

These results cast doubt both on the algebraic formulation of discounting, and on the validity of applying rates derived from one context to a different context. The particular importance for long-term environmental values should be evident. Moreover, they create a dynamic problem, recognised (partly) long ago by Strotz (1956).

- Product A expected 10 years hence may be preferred, when discounted at a low rate, over product B expected at the same time, which happens to be discounted at a higher rate. But when 10 years have passed discounting no longer affects the relative value of the products and product B may be preferred over product A.
  The choice of optimal forest rotation (80 years or 90 years) will appear in one light at the time of planting: little discounting takes place over the additional 10 years, and a 90-year rotation may be preferred. But when the crop is 80 years old, high rates of short-term discount may result in preference switching to the 80-year rotation.

2.2 Negative time preference

Both experiment and personal experience reveal cases where greater weight is given to future benefits and costs than to present ones. In one North American experiment, postgraduates were asked for their willingness to pay for 'a kiss' from their nominated film star (Loewenstein 1987). As expected, willingness to pay for the experience in the distant future was less than for the same experience immediately. However, the highest willingness to pay was for the experience three days in the future, implying negative discounting over a short period.

The explanation for such preferences lies in the pleasure of anticipation, which over short periods may outweigh impatience for fulfilment. Similarly, apprehension about bad future events adds to their disutility.

2.3 Reversing time preference

Typically, preference for early consumption of goods things – or for postponement of bad things; or for having good things immediately even though bad consequences follow – changes to regret as time passes. We wish we had saved the box of chocolates for later pleasure; we wish we had undertaken the unpleasant task earlier; we regret the over-indulgence of the night before, in the cold light of the morning afterwards.

2.4 Now-preference

The evidence against regarding time preference as *preference for earliness* is seemingly overwhelming. What hypothesis explains the data more consistently? Some clues may be gathered by looking at our response to the past. Two years ago, while glancing at an hotel brochure, I found the following suggested destination for a day's outing:

Fotheringay Castle, where Mary, Queen of Scots, was executed!

Executed, exclamation mark! Yet this is no exclamation in the normal grammatical sense. The effect here is jocular, reducing the violent death of a sentient human being to the focal point of a jolly historical theme park experience.

The same era of British history features in a curiously insensitive advertisement for a postage stamp. This depicts Anne Boleyn, the second wife of King Henry VIII 'perfectly executed by Royal Mail'. (For those not familiar with heavy British humour, it should be pointed out that 'executed' has the two meanings 'carried out' and 'decapitated', while 'Royal Mail' alludes to both the British postal service and King Henry VIII (= royal *male*).) Nor is the joke laid to rest there. In the following paragraphs we find:

'Some women really lost their heads over him [King Henry].'
'Some, like Anne, are given the chop.'
'So, be decisive, like Henry VIII, take a sharp object and … Off with the coupon.'

Regrettably, whenever I mention these cases, there is merriment in the audience, not sadness or anger. Would we respond to the judicial assassination of a contemporary political figure in this way? I think not: it is not the gravity of the event as such, but its position in time, which makes it socially acceptable to trivialise it. We view the dim and distant *past* in the same ungenerous light as that in which we see the dim and distant *future*. Yet what changes through time is not the *importance* of such events, but our *perception* of their importance: it is an absolute error to understand, from that perception, that earliness in history confers entitlement to a greater weight.

Even our improper perceptions do not imply such a weight: rather, it is '*nowness*' or immediacy in time, that is perceived as important, in relation to '*other-timeness*' – whether that is past time or future time.

As well as by this *reinterpretation* of evidence, it is possible to gather evidence directly which supports the nowness hypothesis. When asked directly about reasons underlying their preference to have something good (in this case, a luxury holiday) early, respondents to two pilot surveys gave the distribution of answers shown in Table 1.

Note that a significant number of respondents did not favour early consumption at all. Secondly, the hypothesis almost-universally-accepted by economists – that earliness is itself better – finds little favour. A number of reasons to do with changing personal circumstance, rather than time preference as such, gathered some support. Over the two surveys, preference for immediacy was the most frequent response.

Table 1 Stated reasons for time preference

Why is it better to have good things earlier in historical time?	Pilot I	Pilot II
(a) Earliness just *is* better.	3	1
(b) I may not live until 'later'.	4	5
(c) I can enjoy things more when I'm young.	3	2
(d) Good things have more impact if I expect to receive them immediately.	7	8
(e) My future tastes may change.	0	0
(f) I will be rich later on – good things now have scarcity value.	2	1
(g) It's difficult to predict my future availability.	0	9
(h) I actually prefer later.	3	7
(i) I don't care about timing.	0	3

Evidence of a more anecdotal kind is enshrined in the form of the student protest chants of the 1960s:

> 'What do we want?'
> [Name some politically favoured outcome.]
> 'When do we want it?'
> '*Now!*'

Once pure time preference is understood as *preference for now*, other problematical issues are resolved. Naturally, preference for good things earlier turns to regret with the passage of time, because earlier in time then comes to mean 'in the [less-preferred] past' rather than 'at the [more-preferred] now'.

Moreover, inconsistencies no longer pose a problem for the evaluation of preference. If a product is preferred for its 'nowness', changing preferences are *expected*, not inconsistent. The relevant point in time at which to value it, is the time at which it is consumed.

To summarise:

- our perceptions of relative importance are unstable, and time preferences are contingent upon the point in history at which we make our evaluation;
- even a single transient perception, expressing as it certainly does a preference for the present over both past and future, is no warrant for giving more weight to early in history.

The significance of these points can hardly be over-estimated.

If *now* is what people weight heavily, what are the implications for a democratic government? Simply, that it *should not discount* – at least, not on this basis. Barring accidents which bring human history to a close, all future times will, equally, become *now* in due course.

The particular importance of this reinterpretation of time preference for multiple objectives is this: ideas of return on investment, which have tended to dominate economists' thinking in relation to discounting, have no relevance to non-financial objectives. Thus time preference has been *seen as* a key justification for discounting the future attainment of such objectives. Clearly, that is an invalid perception – time preference does not justify discounting *anything*.

3. Mortality and Future Generations

Row 2 of Table 1 shows that, for some people, human mortality is an explicit reason for putting less weight on their future. It is rational to want experiences, including those of the natural world, before death deletes them from personal possibility. It is not selfish to have such a preference, since, by enjoying such experiences ourselves, we do not preclude later generations from enjoying them.

However, actuarial statistics show clearly that the negative exponential of conventional discounting is a very poor representative of the impact of mortality. Lifetime survival prospects show a reversed sigmoid relationship with duration of survival. Discounting does not even promote the efficient pursuit of self-interest.

Moreover, in some cases early consumption by ourselves is gained at the expense of late consumption by future generations. This is obvious for non-renewable physical resources, such as fossil fuels. The argument also applies to ecologically damaging over-use of the environment, as on popular footpaths or at sites where wildlife is vulnerable to disturbance. Where a constraint limits the use of such resources, it is relevant that passing time brings the prospect of birth as well as the risk of death. Resources or experiences, which we may not survive to enjoy, are thereby made available to those who come after us: death does not eliminate utility, but only changes the beneficiary.

3.1 Ethical positions

Nonetheless, there are philosophical arguments to confront, about whether future generations have any claim on our consideration. Heilbroner (1981), for example, asks satirically:

> What has posterity ever done for me?

Marglin (1963) asserts that

> a democratic view of the state does not countenance governmental intervention on behalf of future generations,

because, he claims, it is 'authoritarian' to assume a set of preferences for future generations: (perhaps they would *like* to live in a world poor in resources but rich in pollution?).

Pearce (et al. 1990) considers that the claims of future generations are represented in the pleasure *their* prospective well-being gives to *us*, rather than in the pleasure *our* benevolence might offer *them*.

Beckerman (1983) argues that even the existence of future generations does not matter, since

> if one is not born, one cannot miss the happiness one might otherwise have had.

(This is an extraordinary argument: if the potential happiness imparted by a possibility which does not become actuality is to be ignored, the whole notion of opportunity cost is vapid, and the prospect of rational choice between alternatives is eliminated.)

Against these rather casual views are arrayed serious arguments. At first glance the libertarian philosophy – that one may do as one pleases with what one has justly acquired – seems to repudiate the claims of the future: if what one pleases is to bequeath an undamaged Earth, then future generations are the fortunate recipients of intertemporal benevolence; if not, they have no just cause to complain.

But there is ambiguity even in this position. It has, for example, been declared that:

> No generation has a freehold on this earth. All we have is a life tenancy with a full repairing lease.

The words, not of Gro Harlem Brundtland, but of that other formidable lady prime minister, the fully-paid-up libertarian, Margaret Thatcher (1989), addressing the Conservative Party Conference. And the seminal libertarian, Locke (1690) had stipulated earlier that 'just acquisition' did not confer entitlement without limits. Natural resources should remain available

> enough and as good ...; and more than the yet unprovided could use

The 'yet unprovided' surely includes future generations.

The contractarian view is that individuals will voluntarily grant rights to others in exchange for recognition of their own rights. At first this seems to offer no comfort to future generations. They have done nothing for us, they *can* do nothing for us, except refrain from cursing our memory. But in determining rules of justice, there is a mental game which indicates how mutual self-interest might be worked out. Rawls (1972) envisages a situation where representatives of all generations meet to determine how resources will be distributed through time. The rules of the debate are that the representatives are *almost* perfectly knowledgeable, but that a 'veil of ignorance' prevents them from knowing the generation to which they will be born. Under this arrangement, nobody knows at what point in time their personal future begins, so to penalise the future risks penalising oneself.

Utilitarian philosophers are often, though unjustly, reviled for a supposedly short-termist outlook. And yet Sidgwick (1874), concluded that

> It seems clear that the time at which a man exists cannot affect the value of his happiness...; and that the interests of posterity must concern a Utilitarian.

And the early utilitarian economists, Hume (1739) and Bentham (1789) were explicitly opposed to political incorporation of those the psychological propensities of humans which led them to devalue the interests of future generations.

4. The Changing Conditions of Consumption

The passing of time *in itself* is not sufficient cause for discounting. If a case is to be made, it must be cast in terms of those processes which through time might be expected to affect the value of consumption. The further we look into the future, the dimmer is our vision, the lesser our certainty that physical things will be as they are today.

The physical processes of time bring attrition of good things and amelioration of bad. For example, asbestos barns, initially garish intrusions in the pastel tones of the British landscape, tend to weather and become colonised by sympathetically-coloured lichens and mosses. But time does not treat all materials so kindly. Throughout Europe, the 1960s brought a tide of brutalist architecture, to universities as much as to other types of building. These buildings have not *mellowed*, but become indescribably *drab*; and exposed reinforcing rods weep rusty tears of chagrin down grim, disintegrating concrete slabs.

Tastes change also. On Fylingdales Moor in north-east England there is a scene extraordinarily rich in the symbols of time's passing. An early Anglo-Scandinavian enterprise, a stone Christian cross, stands on a Bronze Age burial mound some 3500 years old. In the background, until recently, there were arrayed three giant white spheres, part of the United Kingdom's air attack early warning system. At first the landscape effect of these twentieth century intrusions was almost universally condemned. However, new generations grew up not knowing the moor different. Through time some people became more accustomed to them, while others grew to hate them more. Yet others contracted an affection for them, and when changing technological and military circumstances made them redundant, a society for their preservation sprang up. It is inconceivable that the net effect of these different taste changes could produced a negative exponential path of value through time.

But for positive changes to landscape an initial favourable reaction may improve with time, as was the case with the landscaped estates of eighteenth century England. There are also cycles of taste: the fashionable style of church architecture between thirteenth and sixteenth centuries came to be known as 'Gothic', as a term of abuse, in the following period; yet it was revived as the 'only true' church architecture in the nineteenth century. Before it, after it, and between its phases, the Classical architectural style had held sway.

Moreover, rates of change of taste differ markedly. Over a period of weeks or months in the worlds of fashion and popular music, the amazingly *chic* becomes the unspeakably *passé*. In the British landscape, however, the taste of the eighteenth century has persisted to the end of the millennium. Perhaps its constancy has been particularly valued precisely *because* so much else has been changing in our lives.

So, in neither rate, nor shape, nor even in direction of change can we expect discounting (usually at a rate derived from financial markets) to represent the way in which meeting aesthetic objectives would change through time.

5. Marginal tility, Diminishing at a Variety of Rates

In addition to taste-driven *shifts of* the demand curve, there is *movement down* the demand curve as a result of technological advance and increasing supply.

5.1 Technology-rich products

The uses and prices of *marginal* computing power over 30 years illustrates the point convincingly. In 1968 computers were the preserve of research establishments and high-level financial institutions. Even a modest machine – capable of nothing more complex than computing a discount factor table – cost a sum equivalent to a university lecturer's salary for a year. Nowadays such power is available for a university lecturer's salary for an hour (and this is not because of vast increase in university lecturers' salaries). Commensurately, marginal computing power has lapsed from a frontier-defining role, to use for trivial entertainment.

5.2 Environmental purposes

But how would this argument apply to environmental services? In 1972 a prediction was made (Treasury 1972) that recreation in the UK Forestry Commission's forests would increase at 10% per year for 20 years, then at 3% per year indefinitely thereafter. Use of the 10% discount rate, then current for state enterprises, meant that every year beyond 1992 the *present* value of recreation declined by about 7%. But how so? It has been shown that the passing of time *in itself* is not an argument for discounting. No explicit prediction is made of a shift of taste away from forest recreation. Why, then, would a population choose to make *more* visits to a forest, if they imparted *less* value, at *greater* cost? The fatal error, repeated elsewhere in the literature, is to use diminishing *marginal* utility of increasing consumption as an argument for discounting it; then applying the process to *total* utility.

5.3 Substitution among sites

It is true that the power of technology to diminish the value even of whole recreation sites can be demonstrated historically. In the mid-nineteenth century, municipal parks were a vital recreation resource for the industrially employed populations of the UK. Then the advent of cheap public transport put successively the rural fringes, the seaside and mountain areas, continental Europe, and the world at the recreational disposal of the average working family. In some cases the municipal parks have fallen into dereliction: 150 years ago their founders would have been justified in discounting their value for the present day.

There are, however, limits on the efficacy of this transportation process. The ability of the transporting medium to absorb pollution from cars and aeroplanes is a more serious limitation than the technology or the affluence required for mass transport. Moreover, the frenetic search for 'new leisure destinations' has extended to the remotest parts of the Earth. There are always the Moon and the nearer planets, of course.... But perhaps the urge for the new will meet a limits of inherent human conservatism, at the moment when the resource demands of such space tourism come to be seen as ultimately unreasonable. Perhaps, and perhaps not. There is no absolute guarantee that technology can overcome such natural resource limits.

5.4 Substitution of technology for resources

A more purely technological solution cuts out natural resources completely. My particular pleasure in the countryside is, at the end of the academic year, to walk one of the UK's long distance footpaths – this year the Thames Path from the source of the river to the sea east of London. But, conceptually at least, technology can displace the need for resource-hungry travel to south-east England and ecologically damaging traverse of the physical route. Instead, I take myself to the local multi-media exercise centre. I select a compact disk titled 'The Thames Path Experience'. I put on the virtual reality helmet, and switch on. The exercise belt begins to move, carrying me successively across well-manured pastures; along gravelly quarry bunds; through rampant biodiversity in riverside nature reserves. Through the morning hologrammatic waterfowl glide over a simulated river; in the afternoon rain-type precipitation envelopes me; and in the evening I select from the pull-down night-menu whether I shall walk on until dawn, spent the night in a Thames-effect pub module, or pitch my tent on polypropylene turf by the light of a neon moon.

It is said that human beings can achieve anything that they can imagine, and I find that thought terrifying. For I know that in some software house a team of machine-code freaks is labouring mightily to realise my nightmare. I go to rural areas to *escape* the oppression of excessive technology, and it seems a perverse kind of homeopathy to seek the cure of my aversion, by burying me under a truck-load of hairs of the dog that bit me.

5.5 Lives at a discount?

The most grotesque misapplication of diminishing marginal utility as a justification for discounting is to the value of future human lives. Projects in transportation, safety at work and health investment, and policy appraisal of climate change have all done so. And the outcome of the process is, that the entire world population in 450 years' time is given the same importance as one person existing now.

5.6 Implications

Clearly marginal utility diminishes at rates which are highly product-specific. Similarly, the marginal utility of income changes differently for different nations, groups and individuals. No mean figure for growth rate of income can correctly reflect these differences. This is so, even if the affected group is a representative cross-section of the population for which the mean is derived (Price and Nair 1985).

The significance for multiple objectives is very clear: the process of aggregating across objectives and groups can only be undertaken *after* the aggregation of time periods has been accomplished, probably by use of some function other than a negative exponential, and certainly at a variety of rates. Aggregation across objectives and groups, followed by mass discounting is the routine procedure of cost–benefit analysis: but it is a wholly unreasonable custom.

6. Discounting for Opportunity Cost

The most common remaining reason for mass discounting is to assure efficiency of investment, for the benefit of future generations. Seen in this light, sustainable development is the *very justification* for discounting. Two different – and seeming contradictory – mechanisms may be invoked.

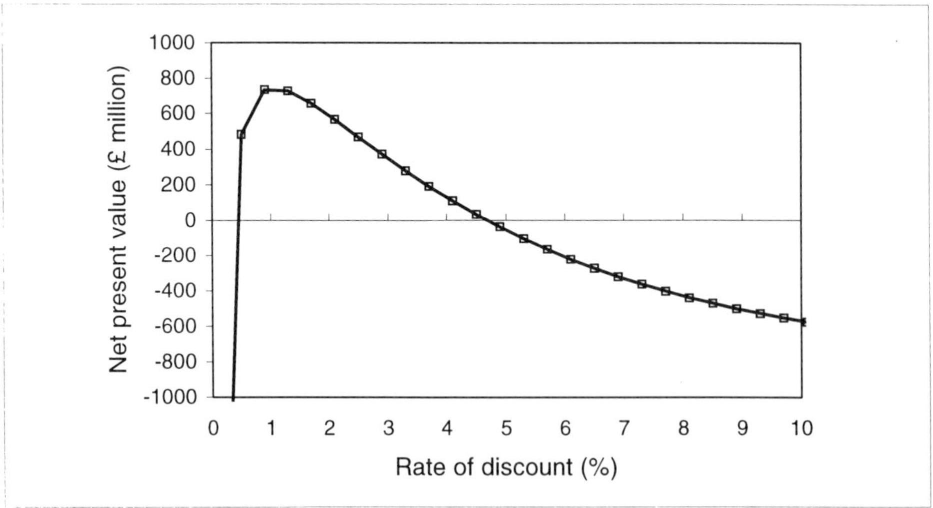

Figure 2 Discounted cost–benefit flows for the Sizewell B Nuclear Power Station. Official figures, with £5 million per year long-term environmental cost

6.1 Commercial discounting reduces the probability of adopting investments which may cause long-term environmental damage

Figure 2 illustrates the point. At a commercial 10% discount rate, the focus is on the short-term construction costs, and the project is seen to be inviable (as was demonstrated during the pre-construction public inquiry). At a lower 3% rate (often deemed to be 'more environmentally friendly') the focus shifts to the medium-term benefits of fossil fuel saving, without giving much weight to the long-term damage. However, if environmental damage is *not* considered less important (merely because it occurs in the long term) there is no reason for discounting it at all. As the discount rate drops below 1%, long-term environmental damage comes to have overwhelming importance, so that eventually net present value becomes negative again.

The best way to protect the environment in the long term is to give proper weight to it, not to hope that the chance effects of high discount rates on commercial cash flows will do the job. Down that path of argument lies global deforestation, elimination of the non-renewable resource base, and trivialisation of the centuries-long impacts of climate change.

6.2 Discounting increases the probability that the most productive investments will be selected

Suppose that one is confronted with the following options in a commercial logging operation in the tropics:

(a) log all commercial trees, with immediate earning of $5.000 per hectare, and little prospect of a valuable second cut in 50 years' time;

(b) retain and protect a few seed trees of valuable species, with immediate earning of $4.000 per hectare, and the likely prospect of a second cut worth $3.000 in 50 years' time.

Option (b) offers long-term benefit of $3.000 per hectare. But option (a) has greater potential: the extra $1.000 per hectare could be invested at, say, the 10% rate of return which the World Bank seeks, to offer a compounded sum of $117.000 after 50 years – a much better deal for the beneficiaries in future generations. Discounted at 10%, the $3.000 offered by sustainable forest management in 50 years yields a present value of only $26 for a present opportunity cost of $1.000. This decisively rules out the 'unproductive' forest investment.

However, for those familiar with the economic growth debate of the 1950s and 1960s (Rostow 1955; Westoby 1962), the idea that pillaging natural resources invariably leads to growth of productive capacity, at a compound rate equal to the rate of return on investment, may have a hollow sound. In practice, even if income was not repatriated, it was all too often spent on luxury consumption items for the few, on competitive growth of military hardware (for neighbouring countries were also pillaging their resources for this purpose), and on coping with the environmental and social crises that were precipitated by deforestation.

Nor can those developed countries which used the windfall of North Sea Oil revenue in lieu of current taxation claim such justifications for discounting the proceeds of investment.

As for the achievement of environmental objectives, it is at first sight difficult to understand by what mechanism landscape values, say, are invested: so why should aesthetic delight early in time be more valuable to the national economy than aesthetic delight later? Potentially, of course, revenues from eco-tourism *could* be invested, but there is no reason to suppose that projects undertaken in the era of sustainable development will, in fact plough back all their profits into development of long-term productive capacity. The following quotation gives little hope that this is the official view:

> the setting aside of part of the proceeds [of natural resource exploitation] in reinvestment is only a metaphor (El Serafy 1989)

Only a metaphor! But metaphorical reinvestment brings benefits only to metaphorical future generations. It has no consequence for real future people. While governments and multi-national agencies are still swayed by short-term considerations, ironically a major justification for discounting – that it selects those investments which promise maximum long-term good – is unlikely to be valid. If there is not *complete* and *perpetual* reinvestment of net benefits from a project, then the rate of return on investment does not offer the means of transforming values between time periods. Hence it is no proper basis for discounting.

7. Conclusions

Arguing against conventional discounting does not solve the problem of allocating investment funds for the benefit of future generations. But there are solutions to it which do not rely on airy theories, pretty metaphors and the beguiling algebra of conventional discounting. They are reviewed in Price (1996; 1997).

Such solutions, which consider the real processes of investment and of value change through time, do not always favour environmental against commercial values, when multiple-objective options are being compared. But favour to one class or other of benefits is not the purpose of multiple-objective decision making. Rather, it is to give appropriate emphasis to benefits (and costs), whatever their nature, whoever the beneficiaries, and whenever they may accrue. This, conventional discounting dismally fails to do.

References

BECKERMAN, W. 1983. Human resources: are they worth preserving? In: P. Streeten and Maier, H. (eds), Human Resources, Employment and Development, Vol. 2: Concepts, Measurement and Long-run Perspective. Macmillan. London.

BENTHAM, J. 1789. An Introduction to the Principles of Morals and Public Legislation. Athlone Press, 1970.

BENZION, U., RAPOPORT, A. AND YAGIL, J. 1989. Discount rates inferred from decisions - an experimental study. Management Science 35:270-84.

BRUNDTLAND, G.H. (World Commission on Environment and Development) 1987. Our Common Future. Oxford University Press.

DEPARTMENT OF THE ENVIRONMENT 1991. Policy Appraisal and the Environment. HMSO. London.

EL SERAFY, S. 1989: The proper calculation of income from depletable natural resources. In: Ahmad, Y.J., El Serafy, S. and Lutz, E. 1989. Environmental Accounting for Sustainable Development. The World Bank. Washington.

HEIGHAM, D. 1992. Some economic terms and ideas explained. In: Talbot, H. (ed.): Our Priceless Countryside: Should it be Priced? Countryside Recreation Research Advisory Group, Bristol.

HEILBRONER, R. 1981. What has posterity ever done for me? In Partridge, E. 1981: Responsibilities to Future Generations. Prometheus Books.

HUME, D. 1739. A Treatise of Human Nature. Oxford University Press.

LOCKE, J. 1690. The Second Treatise of Civil Government. Blackwell, 1946. London.

LOEWENSTEIN, G. 1987. Anticipation and the valuation of delayed consumption. Economic Journal 97:666-84.

LOEWENSTEIN, G. 1988. Frames of mind in intertemporal choice. Management Science 34:200-214.

MARGLIN, S.A. 1963. The social rate of discount and the optimal rate of investment. Quarterly Journal of Economics 77:95-111.

PEARCE, D.W., MARKANDYA, A. AND BARBIER, E.B. 1990. Sustainable Development. Earthscan. London.

POPE, C.A. III AND PERRY, G. 1989. Individual versus social discount rates in allocating depletable natural resources over time. Economics Letters 29:257-264.

PRICE, C. 1993. Time, Discounting and Value. Blackwell. Oxford.

PRICE, C. 1996. Long time horizons, low discount rates and moderate investment criteria, Project Appraisal 11:157-168.

PRICE, C. 1997. Evaluating forest investments when rates of return are high and discount rates are low, Scandinavian Forest Economics 36:121-133.

PRICE, C. AND NAIR, C.T.S. 1985. Social discounting and the distribution of project benefits. Journal of Development Studies 21:525-532.

RAWLS, J. 1972. A Theory of Justice. Harvard University Press.

ROSTOW, W.W. 1956. The take-off into self-sustained growth. Economic Journal 66:26-48.

SIDGWICK, H. 1874. The Methods of Ethics. Macmillan. London.

STROTZ, R.H. 1956. Myopia and inconsistency in dynamic utility maximization. Review of Economic Studies 23:165-180.

THALER, R. 1981. Some empirical evidence on dynamic inconsistency. Economics Letters 8: 201-207.

THATCHER, M. 1989. The Revival of Britain, edited by A. Cooke. Aurum Press. London.

TREASURY 1972. Forestry in Great Britain: an Interdepartmental Cost/Benefit Study. HMSO. London.

WESTOBY, J.C. 1962. The role of forest industries in the attack on economic underdevelopment. Unasylva 16:168-201.

Stated and Revealed Preference Analysis: Cost–Benefit Analysis, Democracy and Multiple-Objective Decision-Making

Colin Price

Abstract

Compared with multi-criteria techniques, cost–benefit analysis values a restricted set of options and gives minimal scope for interaction with decision-makers. Its utility maximising assumption is refuted by actual behaviour in public and private arenas. Yet utility maximisation for those who bear the outcomes of decisions is a worthy objective, and alternative decision techniques arguably over-emphasise the unimportant short-term values and the perspectives of those who construct decisions. Numerous cases in UK forestry demonstrate their shortcomings. The need is to filter such values from cost–benefit analysis.

Keywords: Cost–benefit analysis, democracy, decision-making, preference

1. Introduction

Cost–benefit analysis (CBA) has become widely accepted as a 'preferred technique' for evaluating natural resource decisions, in which there are normally multiple objectives, and by which multiple groups are affected. Government departments and agencies prescribe it; financial institutions advocate it. CBA purports to reveal 'what society prefers' in an overall sense. But its conclusions are constructed from stated or revealed preferences *of* individual consumers *for* individual goods and services *in* particular time periods *under* particular scenarios. It is in its approach to aggregation of these atomised preferences that it has drawn most criticism.

This paper explores why the CBA protocol should have come to be preferred (in public pronouncements if not in practice) over less restrictive protocols, and more open to direct democratic influences. It reviews critiques of its atomistic, individualistic and money-based approach to valuing preferences; considers implied alternatives; and constructs counter-critiques. By examples drawn from UK forestry it shows that when CBA incorporates holistic and democratic elements, it produces the least useful measures of preference.

2. Cost–Benefit Analysis and Alternatives

2.1 An outline

Many people *believe* that they know what CBA is, and offer a trenchant critique of something which is usually *not* CBA. It is therefore helpful to establish precisely what is involved. CBA is, in intention, an economic appraisal of all the costs and all the benefits, whether marketed or not, to whomsoever accruing, both present and future, in so far as possible in a common unit of account, of alternative courses of action (Price 1989). The common unit of account is invariably related to money (under some forms the base value of the money unit is tightly defined by reference to who possesses it, and for what purpose).

Crucially, CBA assumes

- that maximisation of utility *to* everyone, *from* all objectives, *in* all time, *under* all scenarios is the appropriate goal for multiple-objective decision-making;

- that greater *personal* utility is preferred to less personal utility in individual choices (people act as utility maximisers).

Intensity of preferences for individual products is measured from willingness to pay for them, preference against bad things, by willingness to accept compensation for tolerating them. People trade off one thing against another, and are willing to pay more for things they value more. *Given that people try to maximise their own utility*, prices willingly paid should be proportional to marginal values.

Traditionally, economists have favoured willingness to pay *revealed* by actual expenditure of consumers on preferred products. However, the rising importance of non-market environmental and social effects has increased emphasis on willingness to pay

- indirectly *imputed* from related decisions of consumers, and
- directly *stated* by individuals in response to questionnaires.

Individuals' preferences are aggregated by summing their willingnesses to pay, on the reasonable assumption that a good thing is twice as good if it provides the same benefit to twice as many people. Similarly aggregated costs are deducted from total willingness to pay to give net benefit. Everyone affected by a decision is in principle included. If ability to pay constrains willingness to pay, either politicians can make appropriate adjustment (economic CBA) or willingness to pay can be weighted according to the greater presumed value of money to low income individuals (social CBA).

Benefits and costs at different times are aggregated by discounting according to futurity. The discount rate used generally reflects possibilities of transforming values through time by investment (economic CBA), or preferences for substituting between values at different times (social CBA). These processes are not uncontroversial (Price, this volume). In particular, are *possibilities* of transformation in practice taken up to the optimal extent? does substitution account only for diminishing marginal utility (generally agreed to be proper), or also incorporate debatable psychological propensities.

Aggregation across possible future states of nature is usually done by weighting the outcome under each state by the probability of that state arising. The outcome might be judged, in social CBA, in terms under which gains have less weighting, DKK for DKK, than losses. Psychological attitudes on the risk-aversion–risk-seeking axis are not commonly incorporated.

2.2 Multi-criteria decision-making (MDCM) versus CBA

The ascendancy of CBA seems at first surprising, given the capabilities of other multi-criteria decision techniques, which offer both flexible combination of sets of activities and decision-maker participation in weighting objectives.

Although combination of CBA and programming approaches is possible (Mgeni and Price (1993) in practice CBA usually:

- evaluates *given* options, rather than being concerned with the *design* of options;
- views projects in isolation, not as part of an interdependent activity set;
- aims at choosing a single option, not a combination of activities.

Given that we live in an interactive world, offering an indefinite set of activity combinations, why should a technique generally confine itself to considering only a small set of discrete and independent options?

The political reality

In practice most decisions do have limited scope: decision-makers choose within a restricted choice set. Sometimes there are only a few discrete technical options, between which there is not a continuum of combinations. There are also institutional limitations: usually when there is a change of land use, it occurs on one ownership unit at a time, without regard to the optimal distribution of uses across all units, if all could be changed simultaneously. (Contrast this with the decision-making context modelled in Meilby and Strange (this volume)).

Even when a wider range of options is feasible, political circumstances may present the analyst with a small subset.

Independence of activities

It may be reasonable to assume that the available options are independent.

No significant interactions might exist between options – no other contemporary or subsequent option is pre-empted by adopting a particular one.

- Interaction might occur only in that resources required, or products made, are traded at the margin at a price which fully reflects normal costs or benefits (But see Price (1990)).
- The most significant interactions may be specified in the options set, e.g., the future effect of establishing plantations on the viability of a sawmill designed to saw wood from existing forests.
- Options included in CBA may be generated as local optima.

Hardware and software limits of programming solutions

Infeasibility of solutions may result from the following.

- There may exist non-linearities, discontinuities and intractabilities in mapping from activities to products to values, especially for localised environmental values. For example, aesthetic values tend to increase at first with more trees in the landscape, then to decline (Schroeder 1986).
- The preference structure for one product may depend on that for another, violating the assumption of separability embedded in simple optimisation procedures (Bogetoft, this volume).
- Activities themselves may be interdependent, especially where complex environmental spillovers exist between land units (e.g., mutual effects of CO_2 and plant growth).
- No recognisable homogeneous classes of activities may exist. (It sometimes seems that every hectare in Wales has a unique combination of characteristics, so no activity can be replicated precisely).

These problems mean that, in the real world, incorporating interaction at the *activity level* squeezes out interaction within *design of options*.

There is always a limit...

Whatever the ideal, optimisation procedures always abstract from all possible interactions. For example, the highly developed US harvesting scheduling models assume that the rest of the world follows a pre-determined path. Yet, as Sedjo (1996) notes, environmental constraints on harvest have impacts via increased harvests in other countries supplying world markets: these too affect environmental quality in the producing nations, or even globally.

Disenfranchising politicians?

CBA's formal rules of aggregation into a single criterion of performance may effectively trivialise decision-makers' input, confronting them with a clearly denoted 'best' option. The arcane language and unfamiliar mathematics of CBA excludes decision-makers. This contrasts with MCDM protocols under which decision-makers choose directly from possibilities offered on an efficient frontier, or test the effect of different weights on objectives.

Cost–benefit analysts might well answer that sensitivity analysis also allows decision-makers to change debatable assumptions underlying CBA; that its complexities reflect a complex real world, so decision-makers *ought* to understand its basic characteristics; that CBA forces decision-makers to confront the particular consequences of general political goals; that allowing decision-makers free choice of positions on an outcome surface underwrites political prejudice and self-interest with spurious appearances of objectivity.

2.3 The Sagoffian and allied critiques

A well-known critique of CBA, particularly relevant to environmental values, is that of Sagoff (1988), which develops ideas expressed earlier by Adams (1974) and Self (1976). Sagoff argues that when people take decisions about important matters, like environment or social justice, they act not as self-interested consumers attempting to maximise personal utility, but as responsible citizens who regard the common good. This perspective is deployed both in approving the legislative framework of policy, and in forming attitudes to particular projects.

People do not trade-off objectives such as economic efficiency and maintaining environmental standards. They have *lexicographical preferences* in which achievement of certain goals is not negotiable (justice outweighs wealth-creation, biodiversity trumps material production). The procedures of CBA cannot reveal these preferences quantitatively. When confronted by questions about willingness to pay for environmental benefits, or willingness to accept compensation for bearing environmental costs, some respondents either refuse to answer, or register 'protest bids' – zero, or 'unreasonably large' values. To the extent that they *can* be elicited, utility-maximising consumer preferences, as revealed in willingness to pay, do not imply citizen values that conventional democratic ways might revealed.

Economics is an appropriate evaluative framework only for decisions about private goods, commercially produced: it should not be accepted as a trading mechanism in a multiple objective framework. Efficiency is not only *not superior* to justice: it is not even a worthwhile goal.

The meaning of willingness to pay is also questioned by Kahneman and Knetsch (1992). They argue that people express willingness to pay in order to derive the 'moral satisfaction' of supporting (theoretically) the provision of public goods. By so doing they 'purchase' the good opinion of the interviewer and of themselves. When asked to evaluate a *particular* environmental good their willingness to pay is symbolic of *general* concern for the environment (Blamey 1996).

Characteristically, when asked their willingness to pay for a particular environmental good, respondents offer a sum from an account mentally earmarked for environmental purposes. When extra environmental goods are added to the package, willingness to pay increases only marginally. The pre-determined environmental account constrains willingness to pay.

In short, pursuit of efficiency and achievement of maximum utility from individual goods and services is probably not a worthy goal, and is not what people seek in articulating opinions, either democratically or in response to willingness to pay questions.

2.4 Referendum democracy

Why can the democratic process not be trusted to deliver the correct verdict on 'what society prefers' overall? Voting provides a means to aggregate conflicting preferences. However, some paradoxes reveal underlying shortcomings. Consider the following options, presented to an electorate which includes approximately equal numbers of individuals with single objectives: aesthetes, conservationists, and hydrologists.

Option A offers excellent landscape, moderate biodiversity, and high pollution levels.
Option B offers spoiled landscape, maximum biodiversity, and moderate pollution levels.
Option C offers pleasant landscape, little biodiversity, and no pollution.

If A and B are compared, conservationists and hydrologists outvote aesthetes, preferring B by a two-to-one majority. However, if B and C are compared, a two-to-one majority of aesthetes and hydrologists over conservationists prefer C. Finally, comparing C and A, aesthetes and conservationists outvote hydrologists, and A is selected.

Thus A > C > B > A.

On the other hand, if all options are considered together, 'divide and rule' strategies may be accidentally or knowingly employed which split interest groups' votes across several sub-options, all of which provide their preferred product.

Nor do more complex voting systems such as single transferable vote resolve the problem. Consider the situation with option D added to A, B and C.

Option D offers very attractive landscape, a great deal of diversity, and very little pollution.

In the first round of voting, option D is eliminated, receiving no votes, because each interest group identifies an option which it prefers. In the second and third rounds two out of A, B and C are eliminated, with votes for eliminated options transferred to the next preferred option. Yet whichever of A, B or C is the final choice, the first-eliminated option, D, is preferred over it by a two-thirds majority.

These problems of non-transitivity of preference arise because ordinal preferences, revealed by voting for aggregated benefit, do not measure *intensity* of preference.

2.5 Deliberative democracy

Deliberative democracy (Sagoff 1988; Jacobs 1997) attempts to weigh intensity of preference by open and searching discussion. Those who feel strongly about an option make their views known, and others respect them, in comparison with, perhaps, their own mild preferences. It seeks an ethical outcome to conflict – 'one which those affected accept' according to Pruzan (pers. comm.). However, some considerations cast doubt on whether the result must be ideal.

Who deliberates?

Who, among affected people, joins the deliberations, and therefore is empowered to accept or reject outcomes? not future generations; not those unaware of the debate; not (usually) groups large in number but so mildly affected as individuals, that it is not worthwhile to bear the costs of making representations; not those who doubt their persuasiveness. The discussants are most likely to be self-selected individuals who, though not numerous, are much affected, and who trust their abilities to influence decision-making.

What are the objectives?

It may be unresolved whether the participants' job is to *express* their own preferences, *guess* at the group's preferences, or *estimate* objective benefits and costs offered by available options.

How to weight views?

Although intensity of preference is expressed, problems exist in interpreting the strength of other people's expressed preferences, and in knowing how wide is the constituency that participants in the debate represent.

(a) What happens when the lexicographic preferences of different individuals conflict, e.g., when an irresistible force for change ('this is intolerable') meets an immovable object of attachment to the status quo ('this is my birthright')?
(b) How does a deliberative citizen know and interpret others' minds? One person may say 'I know in my soul that...' (Buongiorno, pers.comm.). The problem for social aggregation arises when 'different souls know different answers' – either because outcomes have different physical impacts on different individuals or because they weight objectives differently. I may then personally wish to include Buongiorno's soul-felt solution in my own decision, but I have no means to weigh his preference against my own.
(c) How reliable even is citizens' *self*-understanding?
(d) Biases may be created, according to the information presented.
(e) Representative citizens have limited competence in valuing complex processes, such as biodiversity conservation.
(f) Without formal quantification, weighing many marginal preferences against a few intense ones is difficult.
(g) Within temporary colloquia of unrelated people, incentives exist to exaggerate one's case, to attain the desired outcome. Compare these circumstances with those of a permanent and trusting relationship (in which one truly makes the statement 'I don't feel well enough to wash the dishes tonight').
(h) While issues may be global, communities making decisions are, increasingly in the modern world, transient. The costs of dishonesty in such communities are less: there is less chance of detection, and communal sanctions have less effect.
(i) The suspicion that other, little-known people have an incentive to dishonesty itself makes honesty a less attractive policy: this threatens to become a self-fulfilling prophecy.

***Cui bono*?**

Perhaps there is reason to be wary of processes which are passionately supported by middle-class intellectuals who have comparative advantage in silver-tongued persuasiveness. For them, deliberative democracy is a process relatively easy to capture and subvert, by outgunning the inarticulate, swaying the impressionable, and browbeating the unconfident into 'accepting the outcome'. It may be a very efficient way to obtain 'public good' for the special benefit of one interest group, at the community's expense.

This critique is not merely theoretical, but has arisen from long observation of deliberative decision-making. This is not to say that referenda or deliberative democracy can never be effective, but only that they, like CBA, have potential drawbacks. The drawbacks of CBA do not themselves establish the case for also-defective alternatives.

2.6 The critique from prospect theory

An earlier critique of utility maximisation was initiated by Allais (1953), and developed by Thaler (1980) and Kahneman and Tversky (1982), under the title *prospect theory*. This posits that even *private* decisions are not made according to utility maximisation.

Risk and the framing of decisions

Consider the following example, adapted from Thaler (1980).

First version: In addition to whatever you own, you are given DKK10,000. You must now choose between two *further* prospects:

EITHER

a gamble offering 50% probability of winning DKK10,000 more plus 50% probability of winning nothing

OR

the certainty of winning a further DKK5,000

Second version: In addition to whatever you own, you are given DKK20,000. You must now choose the 'less bad' among the two *further* prospects:

EITHER

a gamble offering 50% probability of losing DKK10,000 plus 50% probability of losing nothing

OR

the certainty of losing DKK5,000

In both versions the sets of outcomes are the same: the overall result of the 'certainty prospect' is DKK15,000; that of the 'gamble prospect' is 50% chance of DKK20,000 plus 50% chance of DKK10,000. Utility maximisation predicts that, whatever the attitude of respondents to risk, in either version the same proportions of respondents should favour the gamble and the certainty. In practice, however, the proportions usually differ significantly, a much smaller proportion preferring certainty when it is presented as *certainty of loss* than when it is presented as *certainty of gain*. It is not just the outcome that is valued, but the *way it is reached*.

Same present choice, different history

I purchase a good seat at the ballet for DKK500. I arrive in the foyer to discover that I have lost the ticket. I realise that I can either pay another DKK500 for another seat, or return to my hotel. I do not have a cash flow problem (VISA is accepted) and the extra expenditure would not materially alter the marginal utility of money to me. The decision I face now is essentially the one I faced when I made the first purchase. If I pay DKK500 I see the ballet: if I don't pay, I don't see it.

And yet my intuition is that if I go in I will see the ballet, and I will pay DKK1,000 for the privilege. I may, intellectually, be aware that bygones are forever bygones, that irretrievable past costs should have no influence on present decisions, that the most important question in economics is *what changes* between one present course of action and another. Yet I am also a creature of instinct: even when intellect wins I am unhappy with the decision.

Proportional saving counts

I am about to buy a bottle of fine wine in central København for DKK200. A friend tells me that the identical vintage may be bought at a sale in Frederiksberg for DKK50. I leave immediately for Frederiksberg. But suppose that I am about to buy a car in central København for DKK150,000. A friend tells me that the identical model may be bought in a reputable Frederiksberg showroom for DKK149,850. I reply 'it's not worth going to Frederiksberg, just to save DKK150.'

My intuition works on proportional reductions: 75% is a massively appealing saving on DKK200: 0.1% on DKK150,000 is trivially irrelevant, in that defective part of my gut where decisions are actually made.

But the opportunity cost of making the more expensive purchase, in terms of other purchases forgone, is identical in the two cases.

Choosing not to choose

Frequently, when offered a choice about a technical matter, especially a medical one, lay-persons return the decision to the professional – 'what option do *you* think I should do?' No use for the expert to present data – even with probabilities attached – for the affected person to make informed trade-offs: the expert's job is seen as *being* expert and taking decisions.

Given only that the information offered is accurate, utility maximisation theory says that affected parties, having their own weights on objectives and riskiness, make better decisions themselves. Referring the decision to someone else at best produces utility no higher, and at worst much lower. Choosing not to choose is a choice of worse over better outcomes.

This ignores the matter of where responsibility will lie for a wrong decision. 'I'll never forgive *myself* for …' is a less comfortable state of mind than 'I'll never forgive *him* for …' especially where momentous decisions may involve permanent disability, or bereavement for one's family. Passing the buck is a means of alleviating heavy responsibilities.

For less potentially traumatic decisions, public participation exercises have an element of two-way buck-passing. The ideal outcome is a situation where following a bad outcome the public say 'the planners took the final decision' while planners say 'well, we gave them what they asked for'. Obversely, if the outcome is good, the public congratulate themselves on averting a (probably) disastrous decision, and the planners (at the least) on their openness to voices of protest.

Pre-commitment to restrict future choices

The choice set may also be voluntarily restricted by precommitments, made earlier than necessary, that oblige or proscribe some course of action. This recognises that decisions may change as time moves on, and that sometimes we can anticipate that not acting in a utility-maximising way (Elster 1984). In particular, when confronted with immediate self-indulgence which we will later regret, discounting between the moment of gratification and the moment of regret occurs at a peculiarly high rate. Shackle (1958) asserts that

There is for us a moment-in-being, which is the locus of every actual sense-experience, every thought, feeling, decision and action.

That moment in being, now, is confirmed as the perceived focus of time preference by many people (Price, this volume).

If, however, we step backwards in time, the moment of gratification and the moment of regret are, alike, *not-now*, and as such compared according to more 'normal' and rational discounting. This 'excessive' discounting over the immediate future – hereafter loosely referred to as 'hyperbolic' discounting – has profound consequences for the interpretation of decision-making.

Mental accounts

People borrow money at 20% interest to make debt purchases, while continuing to invest in an account that earns only 5% (Shefrin and Thaler 1988). Reduction in net expenditure could be achieved by paying off the debt, instead of augmenting the savings. But there are psychological factors too. The capital accounts are mentally separated. There is no trading between them, and this may be rational: the savings habit may be painfully acquired: once the account is seen as liquid, it may be drawn down in pursuit of ephemeral pleasure.

The irrelevance of not-adopted alternatives

The utility of strategies that were not chosen, or the outcome of states of nature that did not materialise, have no bearing on the outcome that I actually enjoy (or suffer). In the logic of utility maximisation, *not having something good* is identical to *not having something bad*. And yet it is a commonplace that our evaluations of well-being are relative: we compare ourselves with our neighbours (too often, through our human penchant for self-torment, with our more affluent neighbours); we construct ideas of our present happiness in a framework of what else we might have had. We experience relief at our lucky – or clever – escapes (not bonding ourselves to unsuitable partners, not investing in enterprises which subsequently collapsed). Conversely, we grieve not only at the outcome of bad decisions, but about the wilfully abandoned better possibilities that went a-begging.

A perceptive insight by Savage (1951), the minimax regret criterion, provides a mechanical protocol for taking decisions under risk. It focuses on potential gaps between actuality and the might-have-been, for all the strategies that could be adopted. It selects the strategy for which the *greatest* possible gap under any state of nature is the least among all strategies. It is a masterly summary of our human irrationality.

Near misses

I leave after the ballet (having decided that it *is* worth another DKK500 to see it), and remember that I need milk for my early morning tea. At 22.02h I arrive at the NETTO supermarket hopefully, seeing people still inside. However, the entrance is locked, and I discover that it closed at 22.00h. Is my utility level the same as if I had found the shop in darkness, with a closing time of 17.30h? Utility maximisation says that the outcome is the same: I have no milk for my tea. Prospect theory says that in the realms of the might-have-been, the oh-so-nearly-might-have-been is less satisfactory than there-never-was-any-chance. My well-being is diminished by this feeling, *however unreasonable I know the feeling to be.*

In summary

The context of the decision and the process by which it is made may have comparable status with the actual outcome, in determining how choices are made. So to behave may be irrational, judged by the utility maximisation model. Yet clearly such considerations do affect people.

3. The Counter-Critique

Sagoff considers that publicly expressed preferences and values cannot be understood as self-centred utility maximisation. Prospect theory suggests that even private decisions cannot be interpreted in these terms. One possible conclusion is that the utility-maximisation concept embodied in CBA is a bad model of democratic processes, and therefore not an appropriate basis for decision-making.

There is, however, an alternative perspective: that the maximisation of utility *ought* to result from decision-making. Thus democratic processes are not the most appropriate means of decision-making. To the extent that democratic processes do not follow utility-maximising behaviour, CBA is corrupted by democratic consultation. However, alternative decision processes, which rely entirely on the preference-revealing behaviour of democratic debate or choice by elected representatives, are even further removed from utility maximisation. People take decisions with a very unbalanced time perspective, located as they are at one point in time when *outcomes* may be less psychologically influential than *processes* and *contexts* of decision-making – in due course when they face the actual outcomes they might have wished another decision to have been made.

Moreover, the interests of decision-makers are not always those of the people on whose behalf they make decisions: blatantly self-serving objectives (maximising promotion prospects) and perhaps self-focused sentiments (worry about outcomes) should be filtered out, rather than reproduced, in decision processes intended to improve the wider good.

Seen from this perspective, the problem is not that CBA fails to replicate democratic decision-making processes, but that it is contaminated by them. The need is for CBA to circumvent the distorted instincts of decision-makers and the ill-advised choices of the people.

Thus to doubt the ability of democratic processes to fulfil the 'real needs' of the people is to invoke the justification of every dictator. Before one could venture down this path it is desirable to see precisely where the decision process fails. The following sections analyse the perspectives of the actors in the process, then the nature of the values accruing to them.

3.1 Decision constructors and outcome bearers

Decision constructors are a wider group than decision-makers classically understood. They include not only those who choose positions on an options frontier or insert weights on goals or formally speak for or vote in favour of a particular proposal, but also those who exert political pressure or whose views are solicited during decision-making, together with those who respond to evaluative questionnaires in willingness to pay or other terms.

Outcome bearers are those who benefit – or suffer – from implementation of a decision. They take no part in making the decision, through lack of information, lack of interest, lack of time, lack of money, lack of luck (e.g., in being born into a later generation). They are, however, the group on which CBA focuses. The utility consequences for outcome bearers are precisely what CBA's evaluative process seeks to establish.

For outcome bearers, the importance of things is measured by how much they themselves are actually affected, *post hoc facto*, rather than how great the effect is perceived to be, *ex ante*, by decision constructors.

The sets of values accruing to or perceived by these groups are different. The circumstances of decision constructors emphasise the context and process of decision, which the Sagoffian view and prospect theory also focus on. By contrast, outcome bearers may be unaware that any formal decision has taken place. For them the route to the decision has no importance. What matters is resultant costs and benefits: a good process is one that tends to give the greatest preponderance of benefit over cost – in fact, that leads to utility maximisation.

The groups also have different perspectives on values. Since context and process signify in the immediate present they are given even greater importance, for decision constructors, by hyperbolic discounting. By contrast, for outcome bearers the time perspective is that of the moment in which benefits accrue or costs occur: pure time preference – hyperbolic or other – has no relevance, only the circumstances of the moment (how wealthy they are, what their tastes have become).

Depending on the culture of the organisation, promotion may come more readily to professional decision constructors with 'a safe pair of hands' – who do not take moderate risks even when so doing offers large potential rewards to society. Not only promotion prospects, but personal peace of mind may arise from taking – or evading – decisions which involve personal responsibility for an outcome. Risk matters to outcome bearers too, but only in the objective terms of probability distributions of outcomes, and what effect extreme outcomes have on the marginal utility of consumption. Because they have no responsibility for the decision, they are less likely to suffer personal anxiety about outcomes before the event.

One individual may play both roles, especially in intensely personal decisions. Even so, within an individual there may exist different persons, e.g. what Thaler and Shefrin (1981) term the doer and the planner, who have different perspectives on immediate and long-term gratification, and would make different choices.

3.2 Values: what kind and to whom?

What values influence decision-making? *which* ones embody long-term outcomes for individuals? *which* are genuine values arising in decision-making? and which are transient or illusory values, that might be termed 'clutter'? And to whom and how widely do they accrue?

The value of a specific outcome actually happening

These are the values, accruing to outcome bearers, that conventional utility maximisation assumes will dominate decision-making. They are what CBA's enquiries into stated and revealed preference aim to quantify. They would normally be subjected to whatever discounting and allowance for risk is deemed appropriate in utility maximisation.

Some such values are likely to be ignored or played down in decision construction. Brodie (this volume/pers.comm.) asks 'Why should we [foresters] be bothered about efficiency in making plans? No-one criticises us for inefficiency: they just don't like the plans.' However, they don't like the increase in taxes or reduced personal consumption that *results* from inefficiency. Responses to presented scenarios will not incorporate consequences which are not evident.

Values arising in the process of making a decision

As decision constructors perceive the decision, the issues of 'same present choice, different history' and 'proportional saving' may be influential. But even for them, these values are transient, carrying weight in relation to outcome values only through hyperbolic discounting. For outcome bearers, knowing nothing of the process but paying the opportunity costs, they are irrelevant.

Worry about what the outcome will be between times of decision and of the outcome's materialising

Anxiety exists, which is itself a disutility independent of the actual outcome. This disutility, however, affects only those who know. Psychological and corporate response to risk and responsibility come into play. The gamble with DKK10,000 has serious counterparts in medical and military decision-making, where choosing certain death for a given number is less acceptable than choosing certain life, although the outcome may be the same. Choosing not to choose is similarly a responsibility-evading strategy, in which the consequence of not-acting are perceived to result from 'fate', 'nature' or 'forces outside my control'.

A different aspect of risk response relates to small changes in risk. A decision constructor puts less weight on a programme which reduces death rate of a disease from 50% to 49%, than one which reduces it from 1% to 0%, because the latter programme removes any vestige of anxiety. Yet the benefits to outcome bearers are identical: out of every 100 contracting the disease, one life is saved by either programme.

Once again, hyperbolic discounting means that worry antecedent to outcome has undue weight, compared with the outcome. From personal experience I know I frequently take bad decisions, earlier than I should, just to avoid the short-term cost of uncertainty.

The effect of *being aware* that a specific outcome has happened, rather than not-happened, as a result of a deliberate decision

Even when an outcome becomes known, only decision constructors may be aware that there was ever any alternative (and perhaps also outcome bearers who trouble to inform themselves of the decision process). They are the ones who make comparisons with other things that might have happened, had they behaved differently; they who grieve over the nearness of misses (when an heroic gamble just fails to pay off). These sensations register strongly in the immediate aftermath of the decision, and also, by anticipation, during the period of decision constructing. Soon, however, their significance fades, even to decision constructors who failed to pave themselves a prior path out of accepting responsibility.

Citizen, symbolic and associated values

When decision constructors assign symbolic values in a particular decision, they hang the whole weight of a class of values (environment, justice) onto an individual case. Yet as a class those values are only marginally affected in the outcome of each decision. Arithmetic absurdities arise from any attempt to reduce the sum of good by those general values, every time the environment loses something particular. There is, admittedly, a non-trivial sense in which no-one is an island; that the world groans when one of its parts diminishes; that environmental integrity is always, repeatedly and cumulatively damaged when one fragment is destroyed. Yet this is a sensed or else a metaphysical value whose meaning at least bears searching inquiry.

When decision constructors act as citizens, they hazard at values which other people may, or may not hold, and which ought, in a democracy, to be estimated and inserted in some other way. They may, from sense of duty, make judgements about the results of processes which they understand little, and in so doing displace or dilute the judgements of those with better understanding.

CBA is concerned with values which individual citizens experience as a result of a decision. It is imperative to remove citizen values that incorporate and thus double-count values already included in valuation by other, more accurate and specific means.

Identifying with 'right' sentiments (apple pie values)

Decision constructors are human. They may on the whole act with good intent. They wish also to be seen to act with good intent. Whoever declares willingness to pay to conserve this or that species or habitat not only expresses symbolic rather than specific value, but also aligns (at no actual cost to themselves) with the angels in the eyes of those who hear the declaration. Motherhood and apple pie may have been displaced by sustainability and multipurpose forestry as emblems of unequivocal good, but the social rites of approval remain the same.

But perceived 'right' sentiments may be wrong sentiments. Signatories to the UK Forestry Accord (Forestry Authority 1998) declared in favour of 'every form of sustainable forestry'. But sustainable includes sustainably bad – doing harm in perpetuity – as logically as sustainably good.

The question, however, is not so much 'is this a statement of unequivocal good?' but 'how much good comes from embracing this statement?' Personal benefit, to the numerically small decision constructors, may indeed flow in the induced regard of others and in self-regard. For the outcome bearers, such self-gratification is irrelevant.

The obverse case is 'bad' sentiments to be rejected, monetarily, symbolically, or however-how. Asking rhetorically 'how can you put a value on a child's life (or on a beautiful sunset in the forest)' might make the speaker feel good, but it does not save children's lives, nor increase the frequency of sunsets.

No government spends its entire budget on saving a child's life: if it did it would rightly draw censure for thereby not funding other programmes for saving lives with greater economic efficiency (a term used by Sagoff dismissively and pejoratively). Thus a protest bid is not a vote for putting indefinite value on a child's life, but in effect authority to value a child's life by intuition and political expediency. Lexicographic preference may express a highly principled refusal to compromise on ultimate values, or a self-indulgent choosing not to choose and declining to accept responsibility for choice.

Value of the system of decision-making, e.g. democratic consultation, acting as a 'good citizen'

Many aspects of democracy are rightly viewed as healthy in themselves, irrespective of the outcome of particular decisions. But we should not confuse attachment to the process (meaningful to decision constructors) with the value any particular outcome. Moreover, the process ceases to have imbued virtue, if it routinely produces bad outcomes. From this viewpoint what matters is the maintenance of a meta-democracy which is competent to change the rules under which decisions are made.

3.3 An alternative viewpoint

We might make the following assertions.

- Atomistic consequences of decisions for individual outcome bearers matter more than instant gut feeling (resulting from the framing of the decision) and good vibes (resulting from 'acting as a good citizen') which affect decision constructors.
- Context and process of decision-making are immediate and, as a result of hyperbolic discounting, have excessive influence compared with long-term actual well-being resulting from the chosen alternative.
- Publicly expressed preferences may be conscience-easing and status-enhancing rather than consequence-regarding.
- Encouraging people to act as citizens in decision processes obscures their own value system and substitutes a value system of uncertain meaning.
- CBA has better potential to filter out irrelevant and overweighted 'clutter' values than either intuitive responses by politicians and professional decision-makers or participatory decision-making processes. It should be made *less* like the democratic process, not more, so that it can reach over the heads of the decision constructors to the outcome bearers whom democracy is meant to serve.

4. Danger! Decision Constructors at Work

The following sections review UK examples in which CBA and democratic processes have failed to emphasise outcomes. Where possible they also comment on how the result might be improved.

4.1 Windthrow

In the 1960s windthrow of the UK's plantation forests was first perceived as a major problem: windthrown timber is sometimes damaged and invariably more costly to harvest. The (intuitive) management response showed the following features:

- dominating worry about the crop's future, causing
- desire for a minimal-risk strategy, rather than deliberation on how much risk was worth taking;
- the active decision to fell crops early, a process of 'imposing a management outcome', in contrast to passively 'responding to an imposed outcome' (clearing windthrown timber);
- with early felling the norm, doing nothing becoming a conscious decision, for whose bad results responsibility rested with the manager.

In the 1970s options were reviewed by financial appraisal, with empirical probabilities. Premature felling imposed a heavy penalty, because trees that might have yielded mostly sawlogs were felled at low-valued pulpwood size. Objectively, it was worth accepting considerable risk. Rotations were markedly extended (Price 1981). Intuitive decisions of managers had alleviated anxiety and burden of responsibility, but by sacrificing better outcomes.

4.2 Forest recreation

Until 1958 UK state forestry had the primary objective of creating a strategic reserve for times of war. Since then social and economic objectives have dominated. As it became clear that afforestation was not conventionally profitable, foresters turned increasingly to environmental justifications. Working from preferences revealed by willingness to pay to travel to forests (Clawson 1959) forest economists claimed that recreation values would eventually exceed those of timber production in state forests (Treasury 1972). These and other environmental and social benefits justified finance for afforestation (MAFF 1972).

But what change in outcome would accrue from more afforestation? At that time, most afforestation was occurring in areas remote from population centres, and where in any case substantial little-used forest area was already available for recreation. Without afforestation, recreation could take place existing forest destinations, or on the land (to which there was usually free access anyway) where afforestation was proposed.

Here professional decision constructors could be seen as self-interestedly protecting territory and job prospects within the profession, rather than acting for the outcome bearers. Remarkably, in subsequent debate these false assumptions have rarely been challenged, and often repeated, to the present day (e.g. Pearce 1994).

Only recently has emphasis moved towards creating forests in areas where few outdoor recreation opportunities exist. But here evaluation has been undertaken, not by travel cost analysis but by the contingent valuation method (Whiteman and Sinclair 1994) whose potential for introducing clutter is discussed in section 5.

4.3 Stated design preferences

Recently the UK produced a Forestry Standard (Forestry Authority1998), detailing social and environmental desiderata for forest management. An early draft specified that large scale felling should only be accepted if rarely seen by the public. Yet this general conception of aesthetic preference is not supported by the Forestry Commission's own publications on landscape design – 'The scale of woodland and forest should reflect the scale of the landscape' (Lucas, 1991). Kielder Forest on the England–Scotland border covers a topography having huge horizontal scale. Twenty years ago, local officers favoured clear-felling areas of around 5 ha, and were apologetic about larger areas. However, small coups fussily contradict the topographic scale. A bolder policy has ensued, and recently fellings of 200 ha have been made – but only in plateau areas 'out of the public eye'. Yet if there is aesthetic merit in large-scale land use operations in large-scale topography, the case is *even stronger* when they lie in the public eye.

Still, along the shores of a large-scale artificial reservoir, coups of 5 ha continue to be made, on the grounds that 'this is what the public want', and public questionnaires have indeed elicited preference for small scale. It is asserted that this preference would also be found in interviews on this site. However the questions arise:

- Are people responding for this specific situation, favouring apple-pie prescriptions, like 'small is beautiful'?
- Might implementing a genuinely case-centred preference for small scale diminish diversity of experience, because all places are treated according to one 'ideal' prescription?
- What would be the response, if the question were not 'do you think clear-felling should be done on a small scale', but 'do you think clear-felling should match the scale of the landscape'? This is apple pie at a higher level, but might more usefully inform design procedures.

4.4 Stated preferences: landscape values

Asking citizens so emotional about their national landscape as those of the UK what they are willing to pay for landscape, induces much symbolic clutter not relevant to the outcomes under debate. Opposition in principle may be expressed to *any* landscape change: but no-one can offer a package of unchanging landscape.

Next, the process of evaluating landscape may be attacked, with assertions like 'beauty is in the eye of the beholder'. A valuation system *which compares like with like as far as possible* helps to avoid this clutter. This might involve either preferences between pairs of photographs or scoring landscapes on a pure aesthetic scale, as devised by Fines (1968) and later modified.

Table 1 Landscape evaluation scales

Descriptive category	Fines's original scale	Harding/Thomas
Unsightly	0–1	0–5
Undistinguished	1–2	5–10
Pleasant	2–4	10–15
Distinguished; attractive	4–8	15–20
Superb; excellent	8–16	20–25
Spectacular; exceptional	16–32	25–30

While participants routinely protest that 'landscape values are wholly subjective', landscape evaluations on this scale show statistically impressive agreement. In one exercise requiring individuals to score 16 landscapes, all 136 comparisons of scores had positive R values, and more than half the R^2 values exceeded 50%.

For such aesthetic preferences to be combined with other, economically quantified preferences, a means is needed to map the aesthetic scale to a monetary one. Voluntary contributions towards the cost of maintaining aesthetic objects may indicate their importance. Cathedrals provide an interesting example of revealed preference: but donations here may have mixed motives (including being seen to side, in a rather specific sense, with the angels), yet suffer from the free rider problem. Ely Cathedral, which had received a mean donation of £0.38 per visitor, was eventually forced by financial exigency to charge £2.50 for 'tourist' entry. This seven-fold increase in the 'admission price' affected visitor numbers little. Clearly voluntary contributions under-revealed people's true preferences.

Contingent valuation has also been applied in cathedrals, revealing an even more curious result at the World Heritage Site of Durham (Willis et al. 1993). Contrary to expectation, *stated* willingness to pay was less than *revealed* voluntary contribution. Perhaps people value the experience more highly as something for which they make a voluntary offering *post hoc* than as something for which they are obliged to pay *ex ante*. Or they may be registering protest at the concept of market-researching such issues. The stated value is partly an artefact of the method of constructing a decision. It says nothing reliable about values experienced by visitors.

The most promising source of uncluttered preferences utilise preferences revealed when people are seeking outcomes by making choices, rather than engaging in conscious decision construction. For example, tourists seemed to travel further to find better landscapes in Wales (Bergin and Price 1994).

5. The Flow Country Controversy

During the 1980s, tax-avoidance private forestry expanded in north-east Scotland. The unproductive land, whose low price attracted afforestation, had great nature conservation interest:

in particular, it was said to support half the European breeding population of greenshank (*Tringa nebularia*). The fact that tax concessions encouraged already wealthy people to commit environmentally damaging acts was politically contentious, and eventually the furore led to withdrawal of the tax arrangements. Thus from the outset general political clutter influenced valuation of specific environmental effects.

The controversy attracted environmental economists interested in estimating *passive use values*, accruing to consumers who do not actually visit a site of environmental or cultural importance. Such people may value:

- the *option* to visit a site in future (even if they never subsequently take the option up);
- the *vicarious* experience of a site by others who do visit (the much-mentioned *bequest* value is part, but not all of this value);
- the knowledge of the *existence* of this site (and conversely, they are unhappy if they discover that the site has been destroyed).

Since enjoying these values needs no direct or indirect monetary transaction, questionnaire approaches have been widely used to evaluate them.

5.1 The questionnaire

My own survey of passive use values from the Flow Country was made with the objective of discovering to what extent stated values might be clutter, induced by the *process of trying to elicit them*. A fuller account appears in Price (1999). The questionnaire was completed by 55 first year students. This was not intended as a valid cross-section of the population, nor was the sample size as large as normally recommended. However, it illustrated problems that would be found in a more representative survey.

After brief description of the site, willingness to pay to protect the Flow Country against damaging commercial afforestation was elicited. The mean response was £76. It is often asserted that respondents need scientific information before they can make a rational evaluation of the site (Samples et al. 1986), so further detail about the Flow Country was given. The new mean willingness to pay almost doubled, to £136.

5.2 Informing inappropriate citizen values

Respondents were asked to apportion willingness to pay among several categories of benefit. For example, 30% of willingness to pay was attributed to maintaining the known and unknown ecosystem functions of the Flow Country. But what does such willingness to pay reveal? 'Citizen values' given on behalf of the community are not appropriate. How can respondents know how many people are affected by these functions? Is it a small local population, or are the ecological effects of land use change global? Even 'consumer values' formed self-interestedly are not appropriate. How do respondents understand the magnitude of the ecological effects (especially the unknown quasi-option ones)? It is deeply insulting to scientific and economic experts (whose judgements are selectively scorned in the present political context) to suggest that questionnaire respondents can judge the importance to themselves of carbon fluxes in soil and vegetation, on the basis of a five-minute information-bite. Experts are the most appropriate people to compile a valuation of these effects. This does not exclude consumers from the decision. Their values are incorporated via *revealed* willingness to pay for the *outcomes* of ecosystem processes, not *stated* willingness to pay for ecosystem processes which deliver the outcomes.

Giving information misleads and flatters individuals, and promotes them to decision constructors in arenas where they are not competent.

5.3 Information and bias

Thirty-four per cent of willingness to pay was for genuine passive use values. But the information-loaded process gives a biased assessment of their value.

- It invites the declaration of symbolic and apple pie values – yes, nature conservation is good – which say virtually nothing about *this* site's. In fact, once this particular controversy has been resolved, the symbolic willingness to pay can be transported to the site of the next controversy, and be deployed against development all over again.
- Even if respondents distribute their entire willingness to pay for conservation across individual sites or species, the very act of asking about the conservation of *this* site hints at its importance and rearranges perceived priorities towards it. The higher the profile created by more information, the greater the rearrangement.
- Information distorts the intensity and even direction of passive use values which people already (legitimately) hold on the basis of their own psychological response to particular creatures. I am not proud of myself for suffering from arachnophobia, but I *do* suffer from it, and it is inappropriate for arachnologists to tell me that I undervalue the psychological benefits to myself of spiders' existence. People may genuinely not like wetlands and their flora and fauna, but be apple-pied into pretending that they do.
- Ideally, contingent valuation compares absolute values in alternative states. Yet, even if poor question construction does not gives cues about *change* of state, hyperbolic discounting is likely to focus valuation on the *process of losing* and on the *decision to not-have the alternative.* These values are temporary. New conservation values may emerge in the altered state of the land, but respondents, even if informed of them, may consider them remote compared with the immediacy of losing what exists.
- Not only does information aggrandise the clutter, it also makes the respondents atypical of the vast majority of still-uninformed outcome bearers: decision constructors become, through their role, definitionally unrepresentative people.

Good information is an intrinsic part of democracy. But information of equal quality should be given, simultaneously and to everyone, on all sites which are threatened by development, and even on all those which are not. Only so can distortions be avoided which arise from focusing symbolic values onto a particular issue. It is *anti-democratic* to supply partial information.

5.4 Passive use values and human outcome bearers

Passive use values are experienced, or lost, by non-visitors who actually know, not only about the site's existence, *but whether the proposed land use change is implemented.* Only 15% of respondents actually knew even of the Flow Country's existence before the questionnaire. Probably fewer still remember the outcome of the controversy. But 15% of what base population? Table 2 shows how the data should be deployed.

Table 2 Geographical distribution of awareness of the Flow Country

Region of origin	Number of respondents	Aware of Flow Country	Percentage awareness	Approximate population (million)	Population times % awareness
Scotland	1	1	100%	5	5 000 000
Rest of UK	42	7	17%	50	8 500 000
Rest of Europe	4	0	0%	500	0
Rest of World	8	0	0%	5000	0

A larger survey is needed to impart narrow confidence limits. But, amazingly, even in principle this simple expedient for quantifying outcome bearers does not seem to have been attempted by the many investigators who have puzzled over the problem of the relevant base population.

Irrespective of outcome, the *possibility* of site development created anxiety costs for 73% of respondents. But these are experienced only by those who know, not only of the site's existence, but of the existence of a threat (the questionnaire did not establish this).

5.5 Intrinsic wildlife values –wildlife as an outcome bearer

Intrinsic values are 'the value to wildlife of *being* wildlife' the creature's preference to be alive rather than not alive. Human attitudes to intrinsic values may be dismissive: one respondent declared himself 'totally indifferent' to the well-being of other species. And economists are particularly prone to argue that 'what matters, is what matters to people'. Yet intrinsic value was highly valued by respondents in general, attracting 25% of stated willingness to pay. Neither the moral reprobates, however, nor the enthusiasts give a correct picture of intrinsic value, rightly understood. Intrinsic value is not a function of human preferences: it exists, if at all, outside the human value framework. Declared willingness to pay reflects a different, human value of including intrinsic value in the valuation system, or an apple-pie value of embracing – or being seen to embrace – an ethical position in which other species count. This benefit is achieved merely by being considered. It should not be attached to one or other outcome of the issue, so is irrelevant to choice of outcomes.

Moreover, ethical judgement may be misplaced, for example, that creatures have a greater right to live if they belong to a rare species (60% of respondents thought so). Afforestation of the Flow Country would reduce numbers of rare greenshank, but increase those of common woodland species. The judgement of human decision constructors promises little for these *potential* beneficiaries.

5.6 So, what?

When improper values are excluded, the genuine passive use values amounted to only 0.4% of the original total. But even this overstated, because for a short and atypical period of the site's life, the site becomes a symbol for general environmental concern, clutter-value which says hardly anything specific about the site and what its loss would mean in the long term. And its former state could pass into oblivion, not missed, not mourned by those phantom outcome bearers who *did not even know that it had been there.*

6. Goodbye expert, hello citizen

In 1996 Environmental Resource Management published its studies of public perception of conservation value in British public forests (Environmental Resource Management 1996). Everything currently perceived as 'proper' were in place: focus groups identified concerns and interpretations, a contingent valuation study established willingness to pay, a post-contingent uncovered the motivations of respondents. In forming conclusions on the appropriate level of biodiversity conservation, respondents chose compromise levels which included timber production value. In the context of finding a *biodiversity* value, this is obstructive and distorting clutter.

a) If they included in CBA, timber values are double-counted: once by participatory judgement, and once by economists' valuation.
b) Randomly selected citizens know little about UK timber production – how self-sufficient the country is, what the timber supply position is within the European Union. Applicants to study

forestry at university rarely know such things, and even professional foresters may be hazy about them.

c) If the public redefine the task to answer the wrong question in contingent valuation, why should they not to do so when asked directly for a management preference through deliberative democracy?

7. Conclusion

The critiques of utility maximisation by Sagoff, Allais and others do in practice appear to undermine CBA. But the problems arise only when CBA becomes more like democratic processes. The apparent need is to *purge* clutter values from CBA, not to emphasise them. Viewed in this way, the critiques bear far more destructively on participatory and interactive decision-making than on CBA.

So, given the presence of so much clutter, what criteria, if any, can be found for establishing valid and relevant preferences in multiple-objective decision-making?

a) People's preferences *reveal* more of uncluttered outcome value, the *further* they are removed from formal decision-making: revealed preferences in unfocused purchases say more than *stated* ones, which are usually elicited in artificial situations, plainly intended to feed into decision-making.
b) Routine preferences, to which people are habituated in the day-to-day matter of creating and bearing outcomes for themselves say more than special preferences, where they must make up their mind in unusual (and often decision-oriented) conditions.
c) Stated preferences are more reliable if defined so that respondents don't incorporate extraneous values: for example preferences might allow trade-off of biodiversity level and management cost, *holding forest production constant.*
d) Preferences are less likely to incorporate clutter when respondents are not *choosing something for money*, but *choosing between things*. Important steps have been taken in developing the stated preference method (Sievanen et al. 1992; Adamowicz 1995) in which respondents make trade-offs among baskets of commodities rather than between a specified basket (which is laden with symbolic values) and money.

The tone of this paper should not be construed as anti-democratic: democracy is a good thing and we should have more of it. But we should be wary of identifying democracy with certain institutions and processes which indulge the whims of decision constructors at the expense of outcome bearers. By contrast to its arcane and elite image, CBA is arguably the *truly* democratic means of aggregating interests in multiple products. Here weighting is done, not by decision constructors, nor by self-selecting participants in discussion, but by *outcome bearers* – all those whose lives are affected by the decision.

References

ADAMOWICZ,W.L., BOXALL, P.C., LUCKERT, M.K., PHILLIPS, W.E. AND WHITE, W.A. (EDS) 1996. Forestry, economics and the environment. CAB International, Wallingford.

ADAMOWICZ, W.L. 1995. Alternative valuation techniques: a comparison and movement to a synthesis. In: Willis, K.G. and Corkindale, J. (eds). Environmental valuation: new directions.CAB International, Wallingford.

ADAMS, J.G.U. 1974. ...and how much for your grandmother? Environment and Planning A 6:619-26.

ALLAIS,M. 1953. Le comportement de l'homme rationnel devant le risque, critique des postulats et axiomes de l'École Americaine. Econometrica 21:503-46.

BERGIN, J. AND PRICE, C. 1994. The travel cost method and landscape quality. Landscape Research19:21-3.

BLAMEY,R.K. 1996. Citizens, consumers and contingent valuation: clarification and the expression of citizen values and issue-opinions. In: Adamowicz et al., q.v.

CLAWSON, M. 1959. Methods of measuring the demand for and value of outdoor recreation. Resources for the Future, Washington. Reprint 10.

ELSTER, J. 1984. Ulysses and the Sirens. Cambridge University Press.

ENVIRONMENTAL RESOURCES MANAGEMENT 1996. Valuing management for biodiversity in British forests: Final report. Environmental Resources Management. Edinburgh.

FINES, K D. 1968. Landscape evaluation: a research project in East Sussex. Regional Studies 2:41-55.

FORESTRY AUTHORITY. 1998. The UK Forestry Standard. Forestry Commission, Edinburgh.

JACOBS,M. 1997. Environmental valuation, deliberative democracy and public decision-making institutions. In: Foster, J. Valuing nature? economics, ethics and environment. Routledge,London.

KAHNEMAN, D. AND KNETSCH, J.L. 1992. Valuing public goods: the purchase of moral satisfaction.Journal of Environmental Economics and Management 22:57-70.

KAHNEMAN, D. AND TVERSKY, A. 1982. The psychology of preferences. Scientific American246:136-42.

LUCAS, O.W.R. 1991. The design of forest landscapes. Oxford University Press.

MAFF 1972. Forestry policy. HMSO, London.

MGENI, A.S.M. AND PRICE, C. 1993. Planning of forest plantation investments with the aid of linear-programming – a case study of Sao Hill Forest, Tanzania. Forest Ecology and Management 62:51-72.

PEARCE,D. 1994. The environment: assessing the social returns from investment in temperate zoneforestry. In: Layard, R. and Glaister, R.G. (eds). Cost–Benefit Analysis 2nd ed. Penguin, Harmondsworth.

PRICE, C. 1981. When does windthrow risk happen and does it matter? Y coedwigwr 33:14-25.

PRICE, C. 1989. The theory and application of forest economics. Blackwell, Oxford.

PRICE, C. 1990. Pecuniary externalities and project appraisal. Project Appraisal 5:39-48.

PRICE, C. 1999. Contingent valuation and retrograde information bias. In: Park, A. and Stewart Roper, C. (eds). The living forest: proceedings of the international symposium on the non-market benefits of forestry, Edinburgh, June 1996. HMSO. London.

SAGOFF, M. 1988. The economy of the earth. Cambridge University Press.

SAMPLES, K.C., DIXON, J.A. AND GOWEN, M.M. 1986. Information disclosure and endangeredspecies valuation. Land Economics 62:306-12.

SAVAGE, L.J. 1951. The theory of statistical decisions. American Statistical Association Journal 46:55-67.

SCHROEDER, H.W. 1986. Estimating park tree densities to maximize landscape esthetics, Journal ofEnvironmental Management 23:325-33.

SEDJO,R. A. 1996. Local timber production and global trade: the environmental implications of forestry trade. In: Adamowicz et al., q.v.

SELF, P. 1976. Econocrats and the policy process. The politics and philosophy of CBA. Macmillan,London.

SHACKLE, G.L.S. 1958. Time in economics. North Holland.

SHEFRIN, H.M. AND THALER, R. 1988. The behavioral life-cycle hypothesis. Economic Inquiry 26:609-43.

SIEVÄNEN, T., POUTA, E. AND OVASKAINEN, V. 1992. Problems of measuring recreational value. Scandinavian Forest Economics 33:231-43.

THALER, R. 1980. Towards a positive theory of consumer choice. Journal of Economic Behavior and Organization 1:39-60.

THALER, R. AND SHEFRIN, H.M. 1981. An economic theory of self-control. Journal of Political Economy 89:392-406.

TREASURY 1972. Forestry in Great Britain: an interdepartmental cost/benefit study. HMSO, London.

WHITEMAN,A. AND SINCLAIR, J. 1994. The costs and benefits of planting three community forests: Forest of Mercia, Thames Chase, and Great North Forest. Forestry Commission, Edinburgh.

WILLIS,K., BEALE, N., CALDER, N. AND FREER, D. 1993. Paying for heritage: what price Durham Cathedral? Countryside change working paper 43. University of Newcastle upon Tyne.

Chapter II

Multi-Criteria Decision-Making: Theory and Methods

Choosing Between Models of Choice: On the Selection of MCDM Planning Modes

Peter Bogetoft

Abstract

Considerable resources have been devoted to research on Multiple Criteria Decision Making (MCDM) during the last 20-30 years. Countless approaches have been suggested, many of which rather ingenious. Little work has been done, however, to synthesise the methods and to discuss how to choose between them. In this paper, we first propose a taxonomy for multiple criteria planning modes. The taxonomy depends on the overall organisation of the investigation, communication and choice activities involved. Next, the multiple costs and benefits of these activities are identified. Furthermore, the relative importance of these costs and benefits is related both to the planning mode and to the contextual characteristics. This provides the basis for a contextual theory of the choice between alternative MCDM planning modes.

Keywords: Multiple Criteria Decision Making (MCDM), taxonomy, procedural rationality

1. Introduction

In the seventies and eighties, the most expanding branch of operations research and management science was probably Multiple Criteria Decision Making (MCDM) or, to use a more recent terminology, Multiple Criteria Decision Aid (MCDA). For an interesting review, see Korhonen et al. (1992). The overall theme of the MCDM literature is how to cope with conflicting objectives in a planning context. Countless approaches have been proposed, and the thinking embedded in these approaches has gradually disseminated from the purely technical journals to the more applied journals, including journals on natural resource planning.

Unfortunately, little work has been done to synthesise the exiting body of methods and only little guidance has been offered for choosing between them. The focus has been on the development of new methods rather than on the relative merits of these. This is troublesome for at least two reasons. First, from a theoretical point of view, a maturing scientific discipline should offer an overview and a critical perspective on its own techniques. Secondly, from a practical point of view, it is necessary to know the relative merits of alternative methods in alternative settings to make a good choice of method.

In this paper we discuss how to choose between multiple criteria approaches. Our answer is straightforward: by using a multiple criteria approach. Of course, this leads directly to a problem of infinite regress and we do not want to embark upon such a fruitless excursion. Therefore our approach is pragmatic. We suggest a series of criteria for the choice of approach, we tentatively evaluate how alternative approaches perform according to these criteria, and we demonstrate how these evaluations may provide a basis for the choice of an approach in a given context. What really counts here is not our particular choice of criteria nor the evaluations of the approaches based upon these, but the overall theme of employing a multiple criteria approach to evaluate multiple criteria approaches. We aspire to provide a rational and logical background for the choice of a MCDM approach as well as some food for thoughts as to the various costs and benefits of decision making procedures. We want to discuss how to choose a choice procedure, i.e. procedural rationality or at least procedural efficiency, as opposed to what the optimal choice looks like given a choice procedure, i.e. substantive rationality or efficiency.

The outline of the paper is as follows. First, in sections 2-4, we introduce a simple taxonomy for MCDM methods. It is based on the overall organisation of the investigation, communication and choice activities they involve. Four organisations or modes are identified, and it is the choice between these modes that we address in this paper. Hence, we operate at the top of a three-tiered choice hierarchy whose levels are: 1) choice of an overall organisation of the decision making activities, i.e. a mode, 2) choice of a MCDM technique within the mode chosen, and 3) application of the MCDM technique to the decision problem at hand to make a choice of alternative. We are confident that the methodology presented could be further refined so as to be applicable also to the integration of the first two levels, i.e. to the problem of choosing both a mode as well as a particular MCDM technique within this mode.

In section 5, we consider the criteria which can be employed to characterise the investigation, communication and choice activities in a decision making process. Also, on a rather abstract level, we evaluate how the multiple criteria planning modes perform with respect to these criteria, independent of the specific planning context.

Finally, in section 6, we use a hypothetical problem to demonstrate how this approach can be used to choose a particular mode within a given context. The specific context, including the particular problem being faced as well as the particular organisation and its traditions, customers, competitors and community, determines the relative importance of the various criteria. Combining this 'weight'-information with the performance profiles of the four modes gives a logical basis for the choice of mode in a given context. We choose a mode that performs well according to the important criteria.

2. The MCDM Setting

We begin by introducing a bit of the traditional MCDM terminology and by offering a conceptual model of the fundamental issues in planning with multiple criteria.

In multiple criteria planning, we assume that there exists an underlying *decision problem.* For simplicity, we assume that a choice x must be made from a set X of feasible plans. We typically refer to X as

$$X = \text{set of plans in decision space}$$

That is, X is the set of detailed production plans, blueprint design plans or similar.

Typically, it is not all aspects of the plans that are important for choosing between them. We assume therefore that there exist n aggregate properties that really matter. We refer to these as criteria

$$f_i(.): X \to R, \quad i = 1, \ldots, n$$

Using the criteria, we may map the set of basic plans X into criteria space

$$Y = \{(f_1(x), .., f_n(x)) \mid x \in X\} = \text{set of plans in criteria space}$$

Often, we shall think of the choice problem as one of choosing from Y and ignore the underlying set X.

Even though Y gives a simplified and presumably more decision relevant picture of X, the choice from Y is still non-trivial. In multiple criteria decision problems, there are typically no optimal solutions. Even if we assume that more is better than less for all criteria (which strictly speaking requires some amount of preferential independence, cf. Bogetoft and Pruzan(1991/97)), the different criteria will usually point towards different alternatives and we have what is known as a *conflict of interest*

conflict of interest $\quad$ No single decision x* solves Max $f_i(x) \in X$ for all i

In such cases, and without further information, we cannot determine a best decision. We can, however, restrict our search for good decisions to the set of *efficient alternatives*:

x is efficient There is no $x' \in X : f_i(x') > f_i(x)$ all i and $f_j(x') > f_j(x)$ for some j

i.e. the set of alternatives for which there is no other alternative that is at least as good in all criteria and strictly better in at least one criterion.

It is obvious that to make a choice, we must, implicitly or explicitly, introduce information about the relative importance of the n criteria. Only hereby can we compromise the different concerns expressed by the n criteria. For this reason, the final decision is referred to as a *compromise solution* in MCDM.

Of course, if we assume that there exists an underlying *value function* aggregating the different criteria

$$V(.):Y \to R$$

we could define the *optimal solution* as the one solving

$$x^* \text{ optimal decision} \quad x^* \text{ solves Max } V(f_1(x), \dots, f_n(x)) \text{ st } y \in Y$$

In this case, we are back to ordinary single criterion decision making. In MCDM we assume that no such V exists - or at least that it is not readily available.

Let us comment also on the origin of the multiple criteria.

One reason why multiple criteria are often present is *modelling limitations*. Models are hardly ever all-inclusive and comprehensive. To compensate for this, we may introduce criteria that can proxy for the positive or negative effects our decision has on non-modelled areas. If, for example, our model is static in a truly dynamic context, we may choose to work with two criteria like profit and flexibility, even though profit may be our only real concern. Flexibility may give an indication about future profits.

Multiple criteria may also originate from *intra-personal conflicts*, i.e. conflicts within a single decision maker who may be uncertain as to how he feels one criteria should be traded-off against the other. This is the kind of conflicts we shall typically think of in this paper. It is also the kind of conflicts for which MCDM was primarily developed.

There are other types of conflicts leading to multiple criteria problems, however. The most important of these are *inter-personal conflicts*, i.e. conflicts between a group of decision makers, and *systemic conflicts*, i.e. conflicts between decision makers and decision receivers. Inter-personal and systemic conflicts can be thought of a multiple criteria problem where each person's interest is represented by one (or more) of the criteria. If we ignore the fact that individuals with conflicting interests may try to behave strategic to manipulate the outcome, much of the MCDM literature may be relevant to such conflicts. If, however, strategic behaviour must be reckoned with in the specific context, approaches based on social choice and game theory will be more relevant.

The fundamental difficulties of multiple criteria decision making may conceptually be thought of as one of decentralised and not immediately available information about preferences and possibilities.

We have, as our conceptual model, an *ideal decision problem* as depicted to the left in Figure 1. That is, we regard the decision maker's underlying problem as one of selecting an alternative y from the set of alternatives Y so as to best achieve his objectives as reflected by the preference function, V(y). This corresponds to the rational ideal model of human behaviour which dominate most of economics and operations research. Or to the single criterion decision problem in the terminology of MCDM, despite of its success and obvious relevance as a threshold, the rational ideal is not an adequate description of real decision problems. Things are usually complicated by *information being decentralised and not-immediately available.* Clear cut expressions of V and Y do not exist when the decision problem is first encountered. That is, the ingredients of the rational ideal model, the preference function V and the alternative set Y, are not known to begin with.

To be able to make decisions in this context of decentralised and not immediately available information, we must assume that some data collection is possible. Specifically, as illustrated in the right-hand side of Figure 1, we imagine that one 'agent', the *decision maker* (DM), is able to investigate the preference function V, while another 'agent', the *analyst* (AN), is able to investigate the set of alternatives Y. We can look at the DM and the AN as being able to investigate (i.e. answer certain questions about) the contents of the black boxes containing V and Y, respectively.

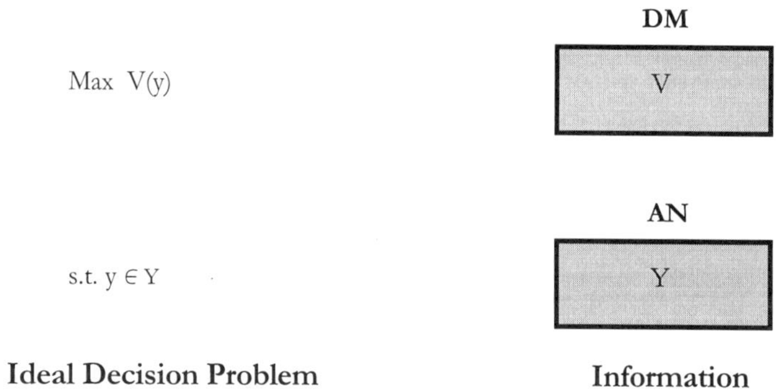

Figure 1 The MCDM setting

Within a context of decentralised and not-immediately-available information we cannot choose an alternative as described in the ideal decision model. We need first to *investigate* the preference function V and the set of alternatives Y. In addition we must integrate whatever we may learn about V and Y, which requires *communication*, so that the final *choice* of an alternative may reflect both the possibilities Y and wishes V delineated. In other words, to determine a compromise solution we have to engage in a *decision making process* involving investigation, communication and choice activities.

Decision making theories have traditionally focused on the choice moment of planning. They have been outcome oriented, stressing substantive rationality. The MCDM literature offers a somewhat broader perspective by considering how the investigation and communication activities may be supported, structured and integrated with the choice activity. MCDM is more process oriented with a view towards procedural rationality. To emphasise this, we shall talk about decision making processes or simply *planning* as opposed to decision making. We do of course realise that 'planning' is usually much more inclusive. A useful introduction to the multiple costs and benefits of planning is provided in Matsuda and Hirano (1982, 1983).

3. The MCDM Taxonomy

A large number of MCDM techniques have been proposed in the literature. One way to get an overview of these techniques is by considering them within the overall organisation of the investigation, communication, and choice activities.

A basic design issue when organising the interaction between the DM and the AN involved in multiple criteria planning is

*** When to investigate V and Y**

At least two possible organisations of the timing may be distinguished. One is a linear or *phased* arrangement, where the investigations of V and Y are performed in two distinct phases. In the first phase, and largely detached from the specific set of alternatives, an approximate representation of V is established. In the next phase we establish those aspects of Y that are relevant in view of the V modelled in the first phase. Or vice versa. The second possible arrangement is a cyclic or *iterative* approach. It consists of a sequence of alternating investigations of V and Y such that one successively examines those details of V that are relevant in view of the information established so far about Y and vice versa.

Another basic design issue when organising the interaction between the DM and the AN is

*** Who directs the search**

This is really a set of questions about who directs the investigations, accumulates the information and makes the final choice proposal. Two obvious arrangements are the *DM-directed* approach where the DM is in charge of the investigation, communication and choice tasks, and the *AN-directed* approach where the AN is in charge of these tasks. Certainly, other arrangements are possible. A third 'mediator' could for example be introduced or a task-force composed of the DM and the AN could be in charge. It turns out, however, that to categorise almost all of the existing MCDM methods, the two pure arrangements are sufficient.

Using the above design questions, we get a taxonomy distinguishing four overall organisations, or modes, of the DM-AN interaction. This taxonomy is depicted in Figure 2.

Entity in charge Timing of Investigation	DM directed	AN directed
Phased	Prior art. Of alternatives DM ↑↑↑↑↑↑↑ AN	Prior art. of preferences DM ↓↓↓↓↓↓↓ AN
Iterative	Progressive art. of alternatives DM question ↓ ↑ answer AN	Progressive art. of preferences DM answer ↓ ↑ question AN

Figure 2 The MCDM Taxonomy

In MCDM, a classification based solely on the timing of the elicited value information has sometimes been used, cf. e.g. Evans (1984). We introduced the 'who is in charge' perspective and integrated it with the 'when' perspective. This structuring of the taxonomy stresses what we consider to be a major theme in MCDM, namely that of 'letting the DM back in'.

From a *contextual point of view*, we believe the taxonomy to be relevant in the sense that: i) the different costs and benefits of the planning process significantly depends on the organisation which is chosen, and ii) the relative importance of these costs and benefits depends on the planning context. Together, these factors provide a contextual basis for choosing between planning modes. For example, in a specific environment where the DM has very limited time to participate in the planning process, it may be advantageous to employ an AN-directed approach. Alternatively, if the issues at stake involve delicate or confidential value judgments, a DM-directed arrangement may be most favourable.

4. The Four Planning Modes

Before we proceed to a more thorough discussion of such interdependencies between context and mode, some further clarification of the different organisations will be helpful. Such clarifications are provided in this section together with organisational examples of the different modes and the MCDM theory supporting them. The main points are summarised in Figure 3.

Prior articulation of alternatives	Prior articulation of preferences
Organisational: Exceptions in hierarchies Civil servant - politician Technical: Vector optimisation Filtering	Organisational: True exceptions Exceptions become routine Technical: MAUT elicitation and est. Outranking
Progr. articulation of alternatives Organisational: Capital budgeting Technical: Varying side-constraints Varying weights Varying target / goals	Progr. Articulation of preferences Organisational: Agents learn pref. Of peers Technical: Choose among trade-off possibilities Express trade-off wishes

Figure 3 Organisational and Technical Examples

One possible organisation of the DM-AN interaction involves *prior articulation of alternatives*. In the first phase, the AN undertakes rather extensive investigations to identify possible alternatives and submits a set of proposals, Y, to the DM. In the next phase, the DM inspects the set of proposals, clarifies his preferences and makes a choice. This contemplative phase is carried out 'inside' the decision maker. Specific MCDM techniques which can be classified as belonging to this organisation do not describe how this may be done. The overall idea, though, is that the set of specific and realistic alternatives may help the DM to focus attention and to sharpen his ideas about what he really wants.

In *practice*, such a structuring of decision making is often encountered in hierarchical organisations. While routine tasks in most enterprises are handled by rules, programmes and tradition, exceptions are usually handled by hierarchical referral, cf. e.g. Arrow (1974). This involves a lower level of the organisation submitting proposals to a higher level. The higher level has the authority to formulate organisational goals or it is in a better position to judge how different alternatives satisfy overall goals and how they may affect other parts of the organisation. This approach to structuring decision making is also met in the context of public planning where civil servants perform analyses for politicians. In such contexts a typical procedure is for the civil servants (AN) to generate a few alternative plans based upon some general instructions and expressions of preferences from the political leadership (DM) and for the politicians to simply choose between these alternative plans and to present the choice in the form of a legislative proposal.

In terms of *MCDM theory*, the main question addressed is how the AN can find a relevant set of alternatives. The guiding principle focused on is typically that of *efficiency*. There is a substantial literature on different ways to search for efficient alternatives, see e.g. Bogetoft and Pruzan(1991/97) or Steuer(1986). At least three different approaches may be distinguished: *the constraint approach* based on varying minimum requirements imposed on the different criteria, the weighting approach based on varying importance weights assigned to the different criteria and *the target (or goal) approach* based on varying target values for the different criteria. This spectrum of approaches provides a useful intellectual framework for structuring the search for alternatives. Also, they have nice conceptual interpretation in terms of budget, price, and target steering.

Another multiple criteria planning mode is characterised by *prior articulation of preferences*. The aim of the first phase is to construct a model of the DM's preferences. The AN accomplishes this by studying his behaviour in previous or, most notably, fictive choice contexts or by otherwise interrogating the DM or guiding his self-examination. In the next phase the AN couples this description of the DM's preferences with the actual set of feasible alternatives in an attempt to pick the most preferred alternative or at least a reduced set from which the final choice will be made by the decision maker.

If we take a liberal interpretation of this planning mode it should be clear that it is again encountered in many and quite distinct *practical* contexts. Consider again an organisation where predictable routine tasks are governed by rules and programmes while exceptions are handled by hierarchical referral. One area of application concerns referrals which are truly exceptional, one-shot, complicated issues to be resolved. This might well require the upper level to rethink its own rules and programmes, or indeed the whole organisation to clarify its objectives and its willingness to make trade-offs between them. To structure these activities, the approaches in this mode may be useful. However, the approaches may also be useful for strongly centralised organisations where 'exceptions' start becoming a routine event. To avoid overloading the organisation's communication channels and information processing capacity, this may call for the upper level to extend the rules and programmes applied at the lower level by supplying it with additional information as to its preferences. With better knowledge about the overall preferences, the lower level may improve screening and reduce the number of referrals. Finally, these approaches may also pave the way for decentralised decision making by providing the decentral decision makers with operational information on the preferences of decision makers higher up in the hierarchy.

In terms of *MCDM theory*, the first fundamental issue in the modelling of a multiple criteria preferences is of course how to construct a useful *set of criteria*. Unfortunately, this is a question for which there does not exist precise theory. The process is basically creative in nature and there exists at best rules of thumb. Most likely, the best approach is to build the criteria structure hierarchically. The next fundamental issue is how to compromise these different desires, i.e. how to incorporate the *relative importance of the different criteria* to give a comprehensive, overall preference structure. The classical theories addressing this issue are the *multiple attribute utility and assessment theories*, cf. e.g. Keeney and Raiffa(1976). A fundamental assumption here is that the DM is willing and able to make extensive comparisons and trade-offs and to behave in accord with (accept) several axioms of rationality. If so, the result will be a complete, functional description of the DM's wishes, his overall preference function V. We also classify the so-called *outranking methods* to this mode. In these, less extensive methods compromising is requested and in particular there is room for some amount of incompatibilities of alternatives. Of course, the resulting prescriptions in the second phase are then more modest. Typically, these models only allow us to isolate a subset of alternatives from which the final choice must be made by whatever means available. Still, as a first cut into the preference structure, this may be very useful, cf. e.g. Bogetoft and Pruzan(1991/97) and Vincke(1982).

A third organisation of the DM-AN interaction is based on *progressive articulation of alternatives*. This is an iterative arrangement directed by the DM. In each round, the DM asks the AN about the set of alternatives, the AN answers and the DM evaluates the answer. He then decides either to continue the search by posing new questions or to stop the search and choose one of the alternatives identified so far. In this mode, the DM's tasks are again somewhat ill-defined. He has to investigate his preferences, to accumulate information, and to direct the search for better alternatives. How all of this is carried out is usually left entirely to the decision maker to figure out, i.e. it is not an explicit part of existing models.

Practical planning often proceeds along these lines. In a hierarchical organisation, planning typically involves several rounds, where the higher level iteratively asks the lower level to propose uses of a given budget or to propose activities in accord with some tentative priorities. Either way, the higher level seeks information about the lower level's alternatives. Several rounds of communication and modifications of the tentative plans may be needed to develop a consistent overall plan, let alone a good compromise between the multiple conflicting objectives of the organisation. From an organisational point of view, an interesting aspect of such procedures is how the search for a desirable alternative is adversely affected by the different levels using different weightings of the criteria, see Bogetoft et al. (1994).

In terms of *MCDM theory*, the approaches within this mode are basically interactive versions of the prior articulation of alternative methods discussed above. Instead of the AN generating all efficient alternatives by parametric variations of constraints, weights or reference points, the DM now successively decides for which values of these he wants to see an efficient proposal. Additionally, there are some variations in exactly which information is submitted to the DM and in which form. In most procedures, the AN is supposed to submit an efficient alternative and some additional information that may guide the DM in his subsequent search decision. In general terms, the *additional information* concerns the neighbouring possibilities and it may be supplied in terms of so-called ideal and nadir points, cross-effect matrices, sets of non-dominated solutions, visual illustration of curves from the efficient frontier, efficient trade-offs etc. Again, exactly how the DM should use this additional information is usually not discussed. To illustrate, we would roughly classify techniques like those proposed by Benayoun et al (1971), Kok and Lootsma (1985), Korhonen (1987), Korhonen and Laakso (1986), Korhonen and Wallenius (1988), Lewandowski and Grauer (1982), Nakayama and Sawaragi (1984), Nijkamp and Spronk (1980), Rietveld (1980, ch.9), Soland (1979), and Wierzbicki (1979) as belonging to this mode, cf. also Bogetoft (1986) and Bogetoft et al. (1988).

Finally, a fourth possible organisation of the DM-AN interaction involves *progressive articulation of preferences*. Again, this is an iterative approach. However, this time the AN is in charge. In each round, he poses a question to the decision maker about his preferences and the DM answers. If the AN now knows sufficient about the DM's preferences to make a choice from the specific set of alternatives available, he proposes a final choice. Otherwise, the questioning continues.

In *practice*, this mode is seldomly found within the formal decision making procedures of a hierarchical organisation since it gives the lower level - the AN - the duties of questioning the higher level - the DM - as well as the power to pick the final alternative. If, however, we interpret the mode more liberally and consider the organisation's informal procedures as well, it may be relevant. During repeated hierarchical referrals, it is conceivable that the lower level creates a picture of the DM's preferences which may be useful in its search and design activities. Furthermore, during a budgeting process, it is likely that the lower level may also pose questions about the higher level's intentions which provide it with information relevant for modelling the higher level's preference structure. Finally, when the set of opportunities is complex, as in research and development or design activities, it is often the case that the AN makes some tentative proposals which the DM comments on. In this manner the AN iteratively collects information relevant for modelling the DM's preferences.

In terms of *MCDM theory*, there is again a wide spectrum of procedures within this general organisation of decision making. They differ in the computational duties of the AN and more importantly here in the *kind of information* requested from the DM. Usually, they either require direct information about the DM's substitution wishes at a given point or they ask him to evaluate existing substitution possibilities. To illustrate, we would roughly classify techniques like those proposed in Chankong and Haimes (1983, ch.8), Geoffrion et al. (1972), Musselman and Talavage (1980), Oppenheimer (1978), White (1980), Zionts (1981), and Zionts and Wallenius (1976, 1983) as belonging to this mode, cf. also Bogetoft (1986).

5. Evaluate the Planning Modes

According to the perspective offered above, different multiple criteria planning modes correspond to different ways to organise the investigation, communication and choice activities. A natural approach to the choice of mode is therefore to consider how the different modes perform according to the criteria we may choose to evaluate the investigation, communication and choice activities.

5.1 Investigation activities

We suggest that the following criteria could be used to evaluate the investigation activities:

1. Is the information generated as to V and Y context specific or can it be employed as well to analyze other (classes of) problems?

2. Which risks are involved for the DM and the organisation? In particular as regards: i) misuse of information on the DM's preferences, ii misuse of information on alternatives, and iii) inappropriate investigation leading to opportunity costs of not choosing the best compromise solution.

3. What are the direct costs (demands on the DM and the AN) involved in performing the investigations of V and Y?

The first part of Figure 4 provides a rough synthesis of our subjective evaluations of the four modes with respect to these criteria. The interpretation of the numerical entries in this and the following evaluation are: A '0' signifies an evaluation that is neither positive nor negative. For 'benefit-criteria', '1' signifies a positive evaluation and '2' represents a rather positive evaluation. For 'cost-criteria', '-1' and '-2' are used to signify a negative and a rather negative evaluation. We underscore that the numerical entries reflect a priori evaluations without reference to an organisation's present way of planning decisions or any other such specific contextual information.

Let us comment on these evaluations. With respect to the first criterion, *contributes to future investigations*, we note that the prior articulation modes provides explicit information about Y and V, respectively. In many cases, however, information about Y will be rather context specific as the set of alternatives X will change or the performance evaluation of the alternatives, i.e. the criteria transforming X to Y, will depend on past decisions. The situation is not quite so context specific as regards information on preferences. Neither of the iterative approaches provide explicit information which can be relied upon in future decision situations as they do not result in explicit delineations of V or Y. Furthermore, whatever information is collected is only partial and depends upon when the investigations terminate.

With respect to the second class of criteria, 'risks involved', we distinguish between three types. The first of these is the risk of *misuse of the information on preferences.* In particular the prior articulation of preferences mode is problematic here. The DM can be criticised if he behaves in a way which does not optimise V and this may lead to a stifling of his creative behaviour and to an unhealthy focus on formalistic consistency rather than on innovative leadership. Also, detailed explicit information as to V can lead to manipulations as others may be tempted to tinker with the presentation of alternatives. Aside from the prior articulation of preferences, and to a lesser extent the progressive articulation of preferences mode, none of the other approa-ches are considered to be risky here. The second risk concerns the potential *misuse of information on alternatives.* The more the set of alternatives is described and characterised, the more information is potentially available to competitors. Seen from this perspective, the prior articulation of alternatives approach may be harmful. The third risk concerns the opportunity costs associated with *inappropriate investigations*, i.e. the risk that the investigation process may not provide relevant and/or correct information and therefore may lead to poorer decisions than would otherwise have been the case. We consider the prior articulation of preferences mode to be risky is this respect because it decouples the DM from the choice process and therefore prevents him from introducing his subjective, intuitive and creative insights such that the decisions made may not be consistent with the DM's true, unarticulated and perhaps unstable preferences. Also, the prior articulation of alternatives mode is problematic. Although the DM is strongly involved in the choice activity, he is more or less on his own. In particular, when the subset of alternatives presented to him is extensive and the criteria are many and overlapping, we expect that the complexity of the choice situation may lead to decisions which are inconsistent with these underlying preferences. The iterative approaches, which provide for an on-going communication between the DM and the AN as to alternatives and preferences are considered to be the least risky with respect to the inappropriate investigations criterion.

	Prior art. alternat.	Prog. art. alternat.	Prior art. prefer.	Prog. art prefer.	Regulat. Context
		Investigation			
Contribution to future inves.	1	0	2	0	+
Risks involved: A) prefer. info. B) altern. info. C) inapp. invest.	 0 -2 -1	 0 0 0	 -2 0 -2	 -1 0 0	 + + +
Direct inves. costs A) DM time etc. B) AN time etc.	 -1 -2	 -2 0	 0 -1	 -2 0	 0 0
		Communication			
Info to org.	0	0	2	0	0
Int. AN & DM	0	1	0	1	+
Contr. to learn	0	1	0	1	+
Direct costs	0	-2	-1	-2	0
		Choice:			
Integrates org. and personal valies	1	2	0	1	0
Contr. to well defined choice	1	2	0	1	0
Contr. to insight in stakeholder	0	1	0	1	0

Figure 4 Evaluation of modes and importance criteria

The third class of investigation criteria concerns 'direct costs'. The first perspective here is the *DM's time and effort.* In the prior articulation of alternatives mode, the demands on the DM may be quite severe since he is more or less on his own, with no search technology to support his thinking. Such support is provided in the progressive articulation of alternatives modes. Also, in both cases, doubts ('Am I making the best choice?') may place severe burdens on the DM. In the prior articulation of preferences modes, the AN is in charge of these duties. However, the development of a complete preference structure V may still require the DM to spend substantial effort in the initial phase. Less complete preference information is required in the progressive articulation of preferences mode. The second perspective here is the *AN's time and effort*, including computer time and the like. It is argued that the prior articulation of alternatives mode in all but the simplest cases is by far the most demanding approach because is requires extensive data collection and analysis in order to delineate the efficient set. In the progressive articulation of alternatives approach, only part of the efficient frontier will be explored. Also, the prior articulation of preferences mode may be demanding due to the large 'sunk' costs involved in developing V without any reference to Y. If we assume that in most cases the iterative approaches do not require a large number of iterations before terminating in a choice, then they should be less costly than the phased approaches. Truly, they require the application of interactive search algorithms but such algorithms most likely will be available as standard software packages and it is anticipated that the savings which result from not having to examine all of Y and/or V will, in general, compensate for the possible extra demands of the iterative approaches.

5.2 Communication criteria

Four communication criteria are selected to evaluate the multiple criteria planning modes in the taxonomy. These are:

1) Does the mode provide information which can serve as a guide to behaviour elsewhere in the organisation?

2) Does the mode contribute to the integration of analysis and decision making?

3) Does the mode contribute to the (DM's) learning process?

4) What are the direct costs of communication?

Figure 4 provides a summary of our context independent evaluations of the four modes with respect to these criteria as well.

With respect to the first communication criterion, *provides information to the organisation*, the prior articulation of preferences approach may be attractive. By providing a mapping of the DM's preferences, it can for example serve as the basis for decision rules in other parts of the organisation where the decisions to be made are of a more routine and repetitive type. We note that we previously considered the potential negative aspects of this approach under the heading 'risks of misuse of information on preferences'. The prior articulation of alternatives approach produces information about Y. However, this information is only relevant for similar problems with the same alternatives and criteria and usually, other parts of the organisation face different alternatives, or they emphasise different criteria. Neither of the interactive approaches are able to generate information which is sufficiently general and valid to be particularly useful elsewhere in the organisation.

The second criterion, the *integration of AN and DM*, is introduced because we believe that a major barrier to decision making which truly reflects a DM's preferences, is the separation of the analytical and political aspects of planning. Roughly speaking, there are two possible results of that separation. Either bureaucratic planning, where the DM has to react to detailed proposals of which he has little intrinsic knowledge since he has not actively participated in generating and analysing the proposals. Or ad hoc, charismatic decision making, where power, charm and intuition may appear convincing, but are very risky as they are not sufficiently supported by rational analysis. Only the iterative modes can contribute to the integration of the analytical and political aspects. The gradual discovering and expression of information in these modes can provide both participants with insights into their own values and roles as well as their mutual dependency. When well performed, this should also result in increased commitment to the planning approach, to each other and to the organisation as a whole. In general we do not expect these benefits to arise in the two phased modes. In these, the DM and AN work rather independently of each other and have but limited communication.

With respect to the third criterion, *contributes to the (DM's) learning process*, the iterative methods provide the DM with the opportunity to gain insight, self-understanding and improved capacity to participate in the planning process. The progressive articulation of preferences mode compels the DM to make trade-offs between his various criteria and hereby promotes considerable introspection and reflection as to what he wants and what he can do. The progressive articulation of alternatives may be regarded as a communication device or data search technology. By permitting the DM to take non-modelled issues into account, it provides a laboratory for the balancing of softer issues and more innate complexities of real-world planning on the one hand, and more traditional, quantitative and objective measures of performance on the other hand. Therefore, this approach too may stimulate reflection, insight and learning. The two phased modes will not in general contribute to learning. The DM is not actively involved in the generation of Y, and has very little to do with the generation of V, when he essentially answers hypothetical questions as to trade-offs.

The evaluations with respect to the last communication criterion, *direct costs of communication*, are quite obvious. Most likely, the prior articulation of alternatives requires least communication; the DM simply receives a list of efficient alternatives and aside from the preliminary identification of his relevant criteria there is no need for communication between the DM and the AN. The amount of communication in the prior articulation of preferences mode may also be quite limited, especially in repeated problems, where already established preference information can be utilised by the AN to determine which course of action is to be taken. Still, some amount of questioning is necessary in order to determine V, and therefore some time and effort must be spent on communication. Finally, a major part of the time and effort spent in the interactive modes is related to communication.

5.3. Choice criteria

The choice attributes of the four modes are tightly related to our proceeding analysis. Appropriate investigation and communication are the primary factors that will

0) Contribute to good choices, i.e. decisions which are in accord with the DM's preferences.

Here we want to characterise the four modes as to some additional, potential contributions to the DM actually making good choices. We propose the following choice criteria:

1) Does the mode support the integration and balancing of organisational and personal values?

2) Does the mode facilitate the implementation process? In particular, does it i) lead to well defined choices, and ii) contribute to insight as to possible reactions by 'decision-takers' or 'stakeholders', i.e. those who are affected by the decision - and whose reactions affect the success of the decision.

Our context independent evaluations of the four modes with respect to these criteria are also depicted in Figure 4.

The first criterion, *integrates organisational and personal values*, concerns how to balance the more organisation-oriented aspects of planning (e.g. will the decision lead to improved profits and market share?), with the more personal aspects (e.g. will the decision improve my prestige, power, position and income? and will it be in harmony with my moral, political and aesthetic values?). Neither of the AN directed modes directly stimulates the DM's balancy of personal and organisational values. Perhaps just the opposite is the case; both of these approaches focus exclusively on the explicit criteria and these will most often belong to the category of organisational values. On the other hand, the DM directed modes permit the DM to consider how to balance his personal values and those of the organisation. The prior articulation of alternatives mode may be well suited if the subset of alternatives presented to the DM to consider is relatively small so that he can mentally perform such evaluations. This is also true for the progressive articulation of alternatives mode in which the DM steers the iterative approach in accord with his, at no time explicitly expressed, values. A DM's behaviour will always reflect a balance between his desires to forward the organisation's expressed goals and his individual, personal goals and this balancing will be implicitly manifest in the way he arrives at his best compromise solution. Note, however, that the likely omission of explicit, personal criteria in both DM directed approaches may imply that the best compromise between personal and organisational values may not be attainable in this mode. The 'efficient' solutions presented to the DM may omit alternatives which are in fact efficient with respect to the more inclusive set of criteria.

The second class of choice criteria, the facilitation of the implementation process, is relevant in that 'good' decisions which are not carried out according to the DM's wishes and expectations, may turn out to be 'poor' decisions in practice. The first aspect of this concerns the development of a *well defined choice*. This is necessary if the implementation process is to result in tasks, plans, budgets and a delegation of responsibilities which further the DM's values. Approaches which directly involve the DM in the delineation and evaluation of alternatives will presumably lead to more precise definitions of the alternatives - and therefore to more operational plans for implementing the ensuing decisions. From this perspective, the least attractive mode is the prior articulation of preferences as it completely removes the DM from contact with the alternatives to be considered. Also, the progressive articulation of preferences mode only provides the DM with a rather limited ability to analyse the alternatives. In prior articulation of alternatives, the preliminary search amongst the set of alternatives and the identification of the efficient subset is performed by the AN, and when the DM gets involved, he is essentially on his own with only limited possibility to really probe into the definition of the alternatives. Still, if both the original set of alternatives and the subset presented to the DM for evaluation are 'small', he may have reasonable opportunity to perform investigations of the alternatives which can lead to successful implementation. More promising, however, is the progressive articulation of alternatives mode since it directly involves the DM in the examination of the set of alternatives; the DM is essentially involved in 'designing' the set of efficient solutions. Another aspect of implementation is the *insight as to stakeholder values*. To facilitate implementation, it may be crucial to consider such complex and hard to model factors as the aspirations and values of the decision takers, the demands on the organisation, the availability of motivated and capable employees, the attitudes of major customers, competitors, activists etc. The progressive articulation of alternatives mode may stimulate reflection as to stakeholder motivations and reactions. For example, procedures where weights or restrictions are modified may invite the DM to reevaluate demands as to customers services and the like. Also, the progressive articulation of preferences can contribute to such reflections as the DM is compelled to evaluate his trade-offs between the various attributes characterising the alternatives and in so doing must reflect, even if only indirectly, upon stakeholder preferences and possible reactions. Such reflection goes on 'internally' and the DM requires no explicit definition of the stakeholders, their criteria and their preferences. Neither of the phased approaches can be said to directly stimulate the DM to consider stakeholder preferences. Neither unguided choices nor abstract, hypothetical questioning can be expected to be able to meaningfully tackle the very complex issues involved here.

6. Choosing a Planning Mode

In the last section, we provided rough context independent evaluations of the different planning modes with respect to a series of investigation, communication and choice criteria. Clearly, none of the modes are universally better than any of the others. They are all efficient in the problem of choosing a mode. Which mode to choose therefore depends on additional information. In a specific context, the relative importance of the criteria can be delineated more precisely, and this may provide the basis for a 'rational' choice of choice mode.

We do not want to advocate a specific aggregation or compromise of the investigation, communication and choice criteria. First of all, our primary purpose has been to suggest the use of a multiple criteria approach to the choice of a multiple criteria planning mode. Secondly, the best compromise will most likely be very context specific. In a specific setting, the particular problem being faced as well as the relevant setting - the organisation and its traditions, customers, competitors and community - can be characterised more precisely, and this should all feed into the choice process. Thirdly, the AN will be aware of which particular MCDM *techniques* like algorithms and standard software packages - as distinguished from *modes* - could be relevant; certainly this will influence the choice of mode. We may add that the specific setting may not only affect the relative importance of the various criteria, but also the evaluations of modes according to these, i.e. the performance profiles in Figure 4, may be context specific.

To illustrate the last step, we consider a hypothetical planning problem. The problem formulation is arbitrary, very broad and only roughly defined. The problem to be considered here exemplifies a class of problems that involve rather unique, significant choices. The DM is the managing director of a public regulatory organisation, and he has to decide how to compensate land owners (farmers and forest owners) in a given region for providing additional amenity value to the general public. This is a complicated problem. There are several significantly different ways of compensating the land owners and the amount of amenity value to strive for depends in part on the cost of producing it. But the problem is more complicated than just choosing between alternative regulatory schemes. The choice will significantly affect the day to day operations of the regulatory organisation in the years to come. A simple scheme may for example lead to less administrative troubles - and less staff. Also, the final approval of the decision will most likely be made by the organisation's semi-political board. Therefore the DM will not only have to consider his own preferences, but also those of the board members as well as his ability to argue his position. At present, however, he wants to carry out an analysis which will make it clear to him, what he feels is the best decision.

The task at hand is to decide which of the four modes should be employed to structure the planning problem, i.e. to generate and examine the consequences of alternative regulatory schemes. To decide this, we consider in turn the importance of the investigation, communication and choice criteria. *Contributions to future investigations* may be significant. It is conceivable that similar regulatory problems will be faced in the future, and although the route taken in the present case will significantly influence what can be done later, the present case may still provide useful information about the land owners' costs of providing the services. This may improve the directors' bargaining position in the future. Hence, insight into the set of alternatives is particularly important. Also, there are several risks involved. An explicit delineation of V makes the DM very amenable to criticism as well as to manipulations. Detailed information about Y may be misused in subsequent deregulation efforts. Finally, the potential opportunity costs due to inappropriate investigations are very large. If the investigations of the possibilities are deficient, considerable damage can be done to the organisation - and the DM as the promoter of the scheme could find his prestige and position threatened. The *time and effort* needed by the DM and AN may be considerable and very dependent on the mode selected. Still, these criteria are considered too insignificant compared to the significance of the problem. The *provision of information to the organisation* is also irrelevant here because of the strategic nature of the problem faced. The *integration of analysis and decision making* is important due to the general significance of the decision, the combination of technical and political aspects involved, as well as the potential importance of qualitative and non-modelled issues. *Contribution to (DM's) learning*, especially regarding Y, is also important here, where new activities are being considered and where the DM's final choice of strategy must be discussed with the board. *Communication costs* are insignificant compared to the decision faced. The *integration of organisational and personal values* are potentially significant due to influence the decision will have on the very identity of the firm. The *contribution to a well-defined choice* may not be significant at this stage since the board will only make the overall policy decisions anyway. However, insight as to stakeholder values is definitely significant. The reactions of landowners, lobbyists, etc. will significantly influence the success of the selected strategy. Our context specific evaluations of the importance of the different investigation, communication and choice criteria are summarised in the last column of Figure 4 (with '0' meaning 'non-important' and '+' meaning 'important').

In conclusion, we suggest that because of the importance of the choice, the great pressure on the DM to be able to argue for his choice, the potential importance of non-modelled attributes and the fact that the set of alternatives is large and complex, a progressive articulation of alternatives mode is probably preferable. Indeed, by restricting attention to the significant criteria in Figure 4, we see that this mode does in fact dominate all the others (except for the first criterion).

Once again however, we underline that the procedure followed here is only suggestive and that an actual application would need to take many other matters into account, including the DM's style, the possible MCDM techniques being considered (which in turn depends on the mathematical properties of the problem - discrete or continuous alternatives, etc.), the organisational design and culture, as well as the possible AN support available.

7. Final Remarks

We have now been through quite a bit of mental gymnastics in an attempt to suggest a methodology for choosing between alternative multiple criteria planning modes. The message, however, is simple: To choose a multiple criteria planning mode - as well as a specific MCDM technique - one must use a multiple criteria perspective on decision processes.

To emphasise this message, we have developed a simple taxonomy of MCDM approaches as well as a series of criteria that we find relevant in attempts to evaluate the performance of the approaches according to the investigation, communication, and choice activities involved. As will be evident from the analysis of the hypothetical case, however, valid recommendations as to the choice of a multiple criteria planning mode can only be achieved by considering an actual problem in an actual organisation. We emphasise therefore that the main purpose of our presentation is not to provide results which are valid for prototype problems, but to suggest a process, a way of thinking about how to choose among alternative multiple criteria planning modes.

Hereby, the paper remains at the methodological side. We suggest, however, that work along these lines could eventually lead to a broader view on actual decision making, planning and organisational design. An important step in this direction would be the development of contextual taxonomies to complement the decision tools taxonomies like the simple four mode taxonomy in this paper. Relevant factors here would include the type of problem (private or public, strategic, tactical, or operational), the type of DM, (idealist, bureaucrat, manipulator, dictator or tycoon), as well as the type of organisation and its traditions, customers, competitors, and community. By coupling contextual characteristics or prototypes with appropriate multiple criteria planning modes, we could improve our understanding of organisational design. For a single criterion approach along these lines, see Burton and Obel (1984).

Acknowledgement

This paper has drawn on joint work with Professor P. Pruzan. For more detailed discussions, see Bogetoft and Pruzan(1991/97). The paper was finished while I was visiting Department of Agricultural and Resource Economics, University of California, Davis.

References

ARROW, K.J. 1974. The Limits of Organisation, W.W.Norton.

BENAYOUN, R., DEMONTGOLFIER, J., TERGNY, J. AND LARITCHER, O. 1971. Linear Programming with Multi Objective Functions, Step Method (STEM). Mathematical Programming 1:366-371.

BOGETOFT, P. 1986. General Communication Schemes for Multiobjective Decision Making, European Journal of Operational Research 26:108-122.

BOGETOFT, P., HJALLEFJORD, Å. AND KOK, M. 1988. On the Convergence of Reference Point Methods in Multi-objective Programming, European Journal of Operational Research 34:56-68.

BOGETOFT, P., MING, C. AND TIND, J. 1994. Price-Directive Decision Making in Hierarchical Systems with Conflicting Preferences, Journal of Multi-Criteria Decision Analysis 3:65-82.

BOGETOFT, P. AND PRUZAN, P. 1991/97. Planning with Multiple Criteria: Investigation, Communication and Choice. North-Holland / Copenhagen Business School Press.

BURTON, R. AND OBEL, B. 1984. Designing Efficient Organisations. Modelling and Experimentation. North-Holland.

CHANKONG, V. AND HAIMES, Y.Y. 1983. Multiobjective Decision Making: Theory and Methodology. North-Holland.

EVANS, G.W. 1984. An Overview of Techniques for Solving Multi-Objective Mathematical Programmes. Management Science 30:1268-1282.

GEOFFRION, A.M., DYER, J.S. AND FEINBERG, A. 1972. An Interactive Approach for Multicriteria Optimization. Management Science 19:357-368.

KEENEY, R.L AND RAIFFA, H. 1976. Decisions with Multiple Objectives: Preferences and Value Tradeoffs. Wiley.

KOK, M. AND LOOTSMA, F.A. 1985. Pairwise-Comparison Methods in Multiobjective Programming with Applications in a Long-Term Energy-Planning Model. European Journal of Operational Research 22:44-55.

KORHONEN, P. 1987. VIG - A Visual Interactive Support System for Multiple Criteria Decision Making. Belgian Journal of Operations Research, Statistics and Computer Science 27(1):3-15.

KORHONEN, P.J. AND LAAKSO, J. 1986. A Visual Interactive Method for Solving the Multiple Criteria Problem. European Journal of Operational Research 24:277- 287.

KORHONEN, P., MOSKOWITZ, H. AND WALLENIUS, J. 1992. Multiple Criteria Decision Support - A Review. European Journal of Operational Research 63(3):361 - 375.

KORHONEN, P. AND WALLENIUS, J. 1988. A Pareto Race. Naval Research Logistics 35(6):615-623

LEWANDOWSKI, A. AND GRAUER, M. 1982. The Reference Point Optimization Approach - Methods of Efficient Implementation. pp. 353-376 in A.P.Wierzbicki (ed.). Multiobjective and Stochastic Optimization. IIASA CP-82-S12, Laxenburg, Austria.

MATSUDA, T. AND HIRANO, M. 1982. A Study on the Structure of Planning Behaviour. European Journal of Operational Research 9:122-132.

MATSUDA, T. AND HIRANO, M. 1983. A Perspective of Planning Literature. European Journal of Operational Research 12:138-145.

MUSSELMAN, K. AND TALAVAGE 1980. A Tradeoff Cut Approach to Multiple Objective Optimization. Operations Research 28:1424-1435.

NAKAYAMA, H. AND SAWARAGI, Y. 1984. Satisficing Tradeoff Method for Multiobjective Programming and its Applications. Paper presented at the 9th IFAC World Congress, July 2-4, Budapest, Hungary.

NIJKAMP, P. AND SPRONK, J. 1980. Interactive Multiple Goal Programming: An Evaluation and Some Results. pp. 278-293 in G.Fandel and Gal, T. (eds.) Multiple Criteria Decision Making: Theory and Applications. Springer.

OPPENHEIMER, K.R. 1978. A Proxy Approach to Multi-Attribute Decision Making. Management Science 24:675-689.

RIETVELD, P. 1980. Multiple Objective Decision Methods and Regional Planning. North-Holland.

STEUER, R.E. 1986. Multiple Criteria Optimization: Theory, Computation and Application. Wiley.

SOLAND, R.M. 1979. Multicriteria Optimization: A General Characterization of Efficient Solutions. Decision Science 10:26-38.

VINCKE, PH. 1992. Multicriteria Decision-Aid. John Wiley and Sons

WHITE, D.J. 1980. Multiple-Objective Interactive Programming. Journal of Operational Research Society 31:517-523.

WIERZBICKI, A.P. 1979. The Use of Reference Objectives in Multi-Objective Optimization - Theoretical Implications and Practical Experience. IIASA WP-79-66, Laxenburg, Austria.

ZIONTS, S. 1981. A Multiple Criteria Method for Choosing among Discrete Alternatives. European Journal of Operational Research 7:143-147.

ZIONTS, S. AND WALLENIUS, J. 1976. An Interactive Programming Method for Solving the Multiple Criteria Problem. Management Science 22:652-663.

ZIONTS, S. AND WALLENIUS, J. 1983. An Interactive Multiple Objective Linear Programming Method for a Class of Underlying Nonlinear Utility Functions. Management Science 29:519-529.

Multiple Objective Linear Programming in Supporting Forest Management

Pekka Korhonen

Abstract

This paper gives a brief introduction to multiple objective linear programming and discusses its possibilities in the area of forest management. We overview basic concepts, formulations, and principles of solving multiple objective linear programming problems. As such problems always require the intervention of a decision-maker, behavioural aspects are also considered. We use an application by Chang and Buongiorno (1981) to illustrate how a reference direction approach developed for multiple objective linear programming can be used to deal with a forest management problem.

Keywords: Multiple Criteria, Multiple Objective Linear Programming, interactive approach, forest management

1. Introduction

Before taking a closer look at *Multiple Objective Linear Programming* (MOLP), it may be useful to first define the general concept *Multiple Criteria Decision Making.* Many different definitions exist about what MCDM actually is, but most researchers working in the field might accept the following general one: *Multiple Criteria Decision Making* (MCDM) refers to the solving of decision and planning problems involving multiple (generally conflicting) criteria. 'Solving' means that a decision-maker (DM) will choose one 'reasonable' alternative from among a set of available alternatives. It is also meaningful to define that the choice is irrevocable. For an MCDM problem it is typical that no unique solution to the problem exists. Therefore to find a solution to MCDM problems requires the intervention of a DM. In MCDM, the word 'reasonable' is replaced by the words 'efficient/nondominated'. They will be defined later on.

Actually the above definition is a strongly simplified description of the whole (multiple criteria) decision making process. In practice, MCDM problems are not often so well-structured that they can be considered just as a choice problem. Before a decision problem is ready to be 'solved', the following questions require a lot of preliminary work: How to structure the problem? How to find essential criteria? How to handle uncertainty? These questions are by no means outside the interest area of MCDM researchers. The Outranking Method by Roy (1973) and the Analytical Hierachy Process (AHP) developed by Saaty (1980) are examples of MCDM methods, in which a lot of effort is devoted to problem structuring. Both methods are well known and widely used in practice. In both methods, the presence of multiple criteria is an essential feature, but the structuring of a problem is an even more important part of the solution process.

When the term 'support' is used in connection with MCDM, we may adopt a broad perspective and refer with the term to all research associated with the relationship between the problem and the DM. In this paper we take a narrower perspective and focus on a very essential supporting problem in MCDM: How to assist a DM to find the 'best' solution from among a set of available 'reasonable' alternatives, when the alternatives are evaluated by using several criteria? Available alternatives are assumed to be defined implicitly by means of a mathematical model. This kind of problem is referred to as a *Multiple Criteria Design Problem* or a *Continuous Multiple Criteria Problem* and the term *Multiple Objective Programming* (MOP) is used to refer to a corresponding mathematical model. In addition, when all constraints and objective functions are linear, the model is a *Multiple Objective Linear Programming* (MOLP) model.

The paper consists of seven sections. In Section 2, we give a brief introduction to some foundations of MOLP. How to generate potential 'reasonable' solutions for a DM's evaluation is considered in Section 3, and in Section 4, we review general principles to solve a multiple objective programming problem. In Section 5, a reference direction approach and its dynamic version Pareto Race are introduced. A public forest management problem is solved in Section 6. Concluding remarks are given in Section 7.

2. Multiple Objective Linear Programming

The MOLP problem can be written in the general form as follows:

$$\begin{aligned} &\text{'max'} \quad q = Cx \\ &\text{s.t.} \\ &\qquad x \in X = \{x \mid Ax \geq b,\ x \geq 0\} \end{aligned} \tag{2.1}$$

where $x \in \Re^n$, $b \in \Re^m$, the constraint matrix $A \in \Re^{m \times n}$ is of full rank m, and the objective function matrix $C \in \Re^{k \times n}$. Set $X \subset \Re^n$ is a *feasible set* and the space $\Re^n$ is called a *variable space.* The functions q_i, $i = 1, 2, \ldots, k$ are *objective functions.* The set $Q = \{q \mid q = f(x),\ x \in X\} \subset \Re^k$ is called a *feasible region*, and it is of special interest in MOLP. The space $\Re^k$ is called a *criterion space.*

The MOLP problem seldom has a unique solution, i.e. an optimal solution that simultaneously maximises all objectives. Conceptually the MOLP problem may be regarded as a value (utility) function maximisation programme:

$$\begin{aligned} &\max \quad v(q) \\ &\text{s.t.} \quad\ q \in Q, \end{aligned} \tag{2.2}$$

where v is a real-valued function, which is strictly increasing in the criterion space and defined at least in the feasible region Q. It maps the feasible region into a one-dimensional *value* space $\Re$. Function v specifies the DM's preference structure over the feasible region. However, the key assumption in MOLP is that v is unknown.

Solutions to MOLP problems are all those alternatives which can be the solutions of some value function v: $Q \rightarrow \Re$. Those solutions are called *efficient* or *nondominated* depending on the space where the alternatives are considered. The term nondominated is used in the criterion space and efficient in the variable space. Any choice from among the set of efficient (nondominated) solutions is an acceptable and 'reasonable' solution, unless we have no additional information about the DM's preference structure.

Definition 1. In (2.1), $x^* \in X$ is *efficient* if there does not exist another $x \in X$ such that $Cx \geq Cx^*$ and $Cx \neq Cx^*$.

Definition 2. In (2.1), $x^* \in X$ is *weakly efficient* if there does not exist another $x \in X$ such that $Cx > Cx^*$.

Vectors $q \in Q$ corresponding to efficient solutions are called *nondominated* criterion vectors and vectors $q \in Q$ corresponding to weakly efficient solutions are called *weakly nondominated* criterion vectors. The set of all efficient solutions is called the *efficient set*, and the set of all nondominated criterion vectors is called the *nondominated set.*

Figure 1 can be used to illustrate the definitions. A shaded area denotes a feasible region in the two-dimensional criterion space. A nondominated set is line [*B,C*]. All other points are dominated, but a set of weakly nondominated points also consists of lines [*A, B*) and (C, D] - in addition to line [*B,C*]. For other points we can always find a better point on both criteria. For instance, for g^1 point q^1 is clearly a dominating point.

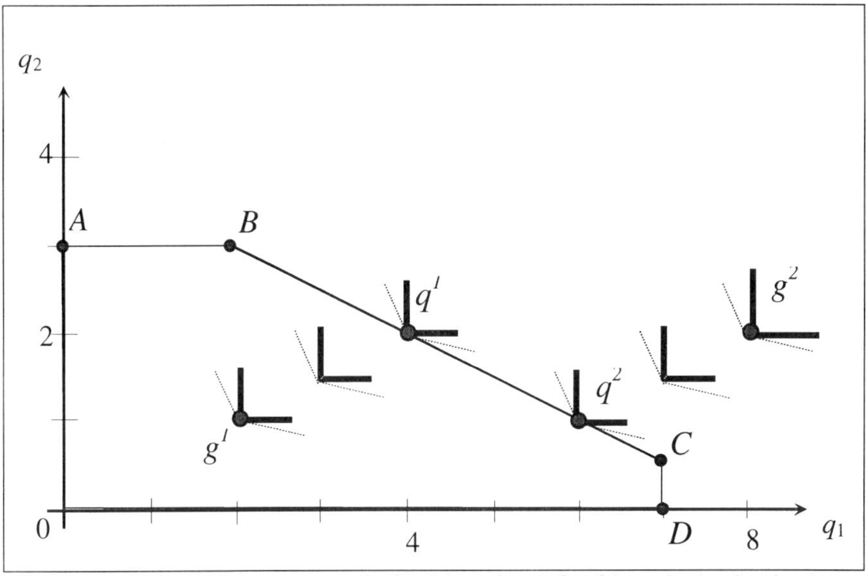

Figure 1 Illustrating the projection of a feasible and an infeasible aspiration level point onto the nondominated surface

The final ('best') solution $q \in Q$ to the problem (2.1) is called *the Most Preferred Solution*. It is a solution preferred by the DM to all other solutions. At the conceptual level, we may think it is the solution maximising an (unknown) value function (2.2). How to find this solution is a key problem in MCDM, in general.

Unfortunately, the above characterisation of the most preferred solution is not very operational, because no system allows the DM to simultaneously compare the final solution to all other solutions with an aim to check if it is really the most preferred or not. It is also just as difficult to maximise a function we do not know. For example, some properties of a good system are that it convinces the DM that the final solution is the most preferred one, does not require too much time from the DM to find the final solution in order to give sufficiently reliable information about alternatives, etc. Even if it is impossible to say which system provides the best support for a DM for his MOLP problem, all proper systems must be able to recognise, generate and operate with nondominated solutions.

3. Generating Nondominated Solutions

Despite many variations among different methods of generating nondominated solutions, the ultimate principle is the same in all methods: a single objective optimisation problem is solved to generate a new solution or solutions. The objective function of this single objective problem may be called a *scalarising function* according to Wierzbicki (1980). It typically has the original objectives and a set of parameters as its arguments. The form of the scalarising function, as well as what parameters are used, depends on the assumptions made concerning the DM's preference structure and behaviour.

Two classes of parameters are widely used in multiple objective optimisation: i) weighting coefficients for objective functions, and ii) reference/ aspiration/ reservation levels for objective function values. Based on those parameters, several ways exist to specify a scalarising function. An important requirement is that this function completely characterises the set of nondominated solutions: 'for each parameter value, all solution vectors are nondominated, and for each nondominated criterion vector, there is at least one parameter value, which produces that specific criterion vector as a solution' (for theoretical considerations, see e.g. Wierzbicki (1986)).

3.1 A linear scalarising function

A classic method to generate nondominated solutions is to use the weighted sums of objective functions, i.e. to use the following linear scalarising function:

$$\max \{\lambda' Cx \mid x \in X\}. \tag{3.1}$$

Using the parameter set $\Lambda = \{\lambda \mid \lambda > 0\}$ in the weighted sums linear programme we can completely characterise the efficient set. However, Λ is an open set, which causes difficulties in a mathematical optimisation problem. If we use $cl(\Lambda) = \{\lambda \mid \lambda \geq 0\}$ instead, the efficiency of solution x cannot be guaranteed anymore. It is surely weakly-efficient, but not necessarily efficient. (See, e.g. Steuer (1986, p. 215 and 221).) When the weighted sums are used to specify a scalarising function in MOLP problems, the optimal solution corresponding to nonextreme points of X is never unique. The set of optimal solutions always consists of at least one extreme point, or the solution is unbounded. In early methods, a common feature was to use $\lambda' Cx$ as a scalarising function, limiting considerations to efficient extreme points (see, e.g. Zionts and Wallenius (1976)).

3.2 A Chebyshev-type scalarising function

Currently, most solution methods are based on the use of a so-called Chebyshev-type scalarising function first proposed by Wierzbicki (1980). The same function was used by Steuer and Choo (1983), but in a somewhat different form. We will refer to this function by the term *achievement (scalarising) function.* The achievement (scalarising) function projects any given (feasible or infeasible) point $g \in \Re^k$ onto the set of nondominated solutions. Point g is called a *reference point*, and its components represent the desired values of the objective functions. These values are called *aspiration levels.*

The simplest form of achievement function is:

$$s(g, q, w) = \max [(g_k - q_k) / w_k, \ k \in K] \tag{3.2}$$

where $w > 0 \in \Re^k$ is a (given) vector of weights, $g \in \Re^k$, and $q \in Q$. The vector w is used to scale the objective functions (roughly) onto the same scale and/or to control the direction of projection. By minimising $s(g, q, w)$ subject to $q \in Q$, we find a weakly nondominated solution vector q^* (see, e.g. Wierzbicki (1980; 1986)). However, if the solution is unique for the problem, then q^* is nondominated. If $g \in Q$ is feasible, then $q^* \geq g$. To guarantee that only nondominated (instead of weakly nondominated) solutions are generated, more complicated forms for the achievement function have to be used, for example:

$$s(g, q, w, \rho) = \max_{k \in K} [(g_k - q_k) / w_k] + \rho \sum_{i=1}^{k} (g_i - q_i), \tag{3.3}$$

where $\rho > 0$ is small. In practice, we cannot operate with a definition 'any positive value'. We have to use a pre-specified value for ρ or to use a lexicographic formulation (see, e.g. Steuer (1986) or Korhonen and Halme (1996)).

To apply the scalarising function (3.3) is easy, because given $g \in \Re^k$, the minimum of $s(g, v, w, \rho)$ is found by solving the following LP problem:

$$\begin{aligned} & \min \quad \varepsilon + \rho \sum_{i=1}^{k} (g_i - q_i) \\ & \text{s.t.} \\ & \quad q \in Q \\ & \quad \varepsilon \geq (g_i - q_i) / w_i, \; i = 1, 2, \ldots, k, \end{aligned} \tag{3.4}$$

The problem (3.4) can be further written as:

$$\begin{aligned} & \min \quad \varepsilon + \rho \sum_{i=1}^{k} (g_i - q_i) \\ & \text{s.t.} \\ & \quad q \in Q \\ & \quad q + \varepsilon w - z = g \\ & \quad z \geq 0. \end{aligned} \tag{3.5}$$

To illustrate the use of the achievement scalarising function, consider a two-criteria problem with a feasible region having five extreme points {(0,0), (0,3), (2,3), (7,0.5), (7,0)}, as shown in Figure 1.

In Figure 1, the thick solid lines describe the indifference curves when $\rho = 0$ in the achievement scalarising function. The thin dotted lines stand for the case $\rho > 0$. Note that the line from (2,3) to (8,0) is nondominated and the lines from (0,3) to (2,3) (except point (2,3)) and (7,0.5) to (7,0) (except point (7,0.5)) are only weakly-nondominated, but dominated. Let us assume that the DM first specifies a feasible aspiration level point $g^1 = (2,1)$. Using a weight vector $w = (2,1)$, the minimum value of the achievement scalarising function (-1) is reached at a point $v^1 = (4,2)$ (cf. Figure 2). Correspondingly, if an aspiration level point is infeasible, say $g^2 = (8,2)$, then the minimum of the achievement scalarising function (+1) is reached at point $v^2 = (6,1)$. When a feasible point dominates an aspiration level point, then the value of the achievement scalarising function is always negative; otherwise it is nonnegative. It is zero, if an aspiration level point is weakly-nondominated.

4. Solving Multiple Objective Problems

Currently, the systems developed for solving multiple objective (linear) programming problems are interactive. The specifics of these procedures vary, but they have several common characteristics. For example, at each iteration a solution, or a set of solutions is generated for a DM's examination. As a result of the examination, the DM inputs information in the form of tradeoffs, pairwise comparisons, aspiration levels, etc. The responses are used to generate a presumably improved solution. The ultimate goal is to find the most preferred solution of the DM. Which search technique and termination rule is used heavily depends on the underlying assumptions postulated about the behaviour of the DM and the way in which these assumptions are implemented. In MCDM research there is a growing interest in the behavioural realism of such assumptions.

Based on the role that the value function (2.3) is supposed to play in the analysis, we can classify the assumptions into three categories:

1. Assume the existence of a value function v, and assess it explicitly. (Actually, this approach is not usually classified under the MCDM category).

2. Assume the existence of a stable value function v, but do not attempt to assess it explicitly. Make assumptions of the general functional form of the value function. (see, e.g. Geoffrion et al. (1972); Zionts and Wallenius (1976)).

3. Do not assume the existence of a stable value function v, either explicit, or implicit (see, e.g. (Wierzbicki (1980); Steuer and Choo (1983)).

For an excellent review of several interactive multiple criteria procedures, see Steuer (1986). See also for surveys Hwang and Masud (1979), Shin and Ravindran (1991), and White (1990)). Other well-known books which provide a deeper background and additional references, especially in the field of Multiple Objective Optimisation include Cohon (1978), Haimes et al. (1990), Ignizio (1976), Sawaragi et al. (1985), Yu (1985), and Zeleny (1982).

5. A Reference Direction Approach

Our purpose is to describe an interactive method called a *Reference Direction Approach* developed by Korhonen and Laakso (1986). The *Achievement Scalarising Function* is the main theoretical basis on which the method lies. By parametrising the function, it is possible to project the whole vector onto the nondominated frontier as originally proposed by Korhonen and Laakso (1986). The vector to be projected is called a *Reference Direction Vector* and the method is called Reference Direction Approach, correspondingly. When a direction is projected onto the nondominated frontier, a curve traversing across the nondominated frontier is obtained. Then an interactive line search is performed along this curve. The idea enables the DM to make a continuous search on the nondominated frontier. The corresponding mathematical model is a simple modification of the original model (3.5) developed for projecting a single point:

$$\begin{aligned} \min \quad & \varepsilon + \rho \sum_{i=1}^{k} (g_i - q_i) \\ \text{s.t.} \quad & \\ & x \in X \\ & q + \varepsilon w - z = g + tr \\ & z \geq 0, \end{aligned} \tag{5.1}$$

where $t : 0 \rightarrow \infty$ and $r \in \Re^k$ is a reference direction. In the original approach, a reference direction was specified as a vector starting from the current solution and passing through the aspiration levels. The DM was asked to give aspiration levels for the criteria. The basic idea of the original method is illustrated in Figure 2.

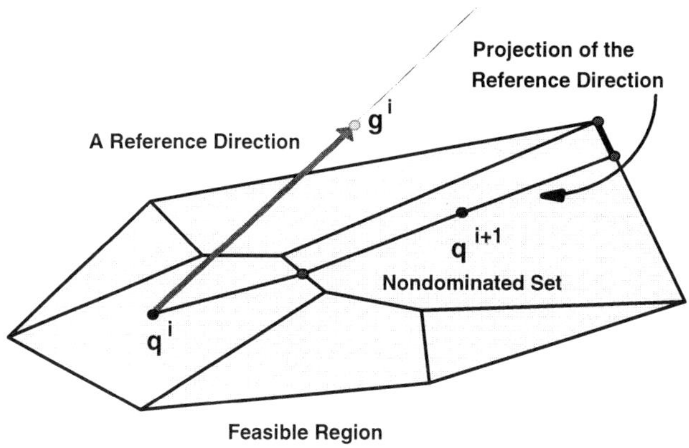

Figur 2 Illustration of the Reference Direction Approach

Korhonen and Wallenius (1988) improved upon the original procedure by making the specification of a reference direction dynamic. The dynamic version was called *Pareto Race.* In Pareto Race, the DM can freely move in any direction on the nondominated frontier he/she likes, and no restrictive assumptions concerning the DM's behaviour are made. Furthermore, the objectives and constraints are presented in a uniform manner. Thus, their role can also be changed during the search process. The method and its implementation is called Pareto Race. The whole software package consisting of Pareto Race is called VIG (Korhonen (1987)).

In Pareto Race, a reference direction r is determined by the system on the basis of preference information received from the DM. By pressing number keys corresponding to the ordinal numbers of the objectives, the DM expresses which objectives he/she would like to improve and how strongly. In this way he/she implicitly specifies a reference direction. Figure 3 shows the Pareto Race interface for the search.

Thus, Pareto Race is a visual, dynamic, search procedure for exploring the nondominated frontier of a MOLP problem. The user sees the objective function values on a display in numeric form and as bar graphs, as he/she travels along the nondominated frontier. The keyboard controls include an accelerator, gears, brakes, and a steering mechanism. The search on the nondominated frontier is like driving a car. The DM can, e.g., increase/decrease the speed, make a turn at any moment he/she likes.

To implement those features, Pareto Race uses certain control mechanisms which are controlled by the following keys:

(SPACE) BAR: An 'Accelerator'

Proceed in the current direction at constant speed.

F1: 'Gears (Backward)'

Increase speed in the backward direction.

F2: 'Gears (Forward)'

Increase speed in the forward direction.

F3: 'Fix'

Use the current value of objective i as the worst acceptable value.

F4: 'Relax'

Relax the 'bound' determined with key F3.

F5: 'Brakes'
Reduce speed.
F10: 'Exit'
num: 'Turn'

Change the direction of motion by increasing the component of the reference direction corresponding to the goal's ordinal number $i \in [1, k]$ pressed by DM.

Pareto Race does not specify restrictive behavioural assumptions for a DM. He/she is free to make a search on the nondominated surface, until he/she believes that the solution found is his/her most preferred one.

6. A Case: Model for Multiple Use Forestry

Chang and Buongiorno (1981) studied a resource allocation problem with the aim of developing a methodology for the problem of multiple use planning in public forests. They developed a mathematical model for this problem by combining goal programming and input-output analysis. Description of a public forest management problem is adopted from Chang and Buongiorno (1981).

A public forest provides society with several products and services. The management problem is to decide how much of each product and service should be provided. Since many management activities use the output of other management activities as their input, input-output tables are a convenient way to describe the relationships between various activities. First, the authors introduced an optimisation model based on input-output tables where the weighted sum of the final demand vector was maximised. The prices for outputs were proposed to be used as weights. Because it is not realistic to assume the availability of such prices, the authors proposed the use of the goal programming model, in which a desired goal was set to each component of the final demand vector. The authors were not fully satisfied with their model because of the lack of generality. To set goals, priorities, and weights is tentative in the beginning. This is why to use the model requires several trial and error iterations.

Chang and Buongiorno (1981) did not report their original model in their paper, but only a small hypothetical example. In the following, we use their example to illustrate how an interactive MOLP method (a reference direction approach) provides a flexible tool to deal with this problem. The problem consists of eight management activities (G1 ... G8) which are listed in Table 1 together with their management goals.

Table 1 Management activities and goals for hypothetical multiple use problem

#	Activity	Management Goal
G1	Timber Harvest	Supervise 5350 acres of timber harvest
G2	Timber Sale Preparation	Prepare 6200 acres of timber sale
G3	Merchantable Stand Management	Maintain 125 000 acres of merchantable Growing stock at the end of the year
G4	Non- Merchantable Stand Management	Maintain 104 000 acres of non-merchantable stand
G5	Camping Management	Provide 258 000 visitor days of camping opportunities
G6	Snowmobiling Management	Provide 90 000 visitor days of snowmobiling opportunities
G7	Road Maintenance	Maintain 24 miles of roads for general uses
G8	Road Construction	Construct 1.5 miles of roads for general uses

To produce outputs also requires resources. Seven resources are assumed to be available to achieve the various goals listed in Table 2.

Table 2 Available resources for hypothetical multiple use forest management problem

#	Type	Amount
R1	Merchantable Growing Stock	127 955 acres
R2	Non-Merchantable Growing Stock	111 657 acres
R3	Camp Ground	582 tent pads
R4	Snowmobile Trails	580 miles
R5	Roads	340 miles
R6	Manpower	7225 man-days
R7	Budget	430 000 $

The structure of the model is given in Table 3. The coefficients in each column indicate the inputs required to produce one unit of output for that particular management activity. For instance, column 5 (x_5) in Table 3 shows that to produce one camping day requires 0.93 acres of merchantable stands (G3), 12.17/1000 miles of road maintenance (G7), 1.74 tent pads (R3), 7.54 man-days (R6), and $440 (R7). All goals (G1 ... G8) are formulated as the objective functions to be maximised. In the original model, the right-hand column corresponding to those rows consisted of goal values (Table 1), and to each row a positive and negative deviation variable was attached. Negative deviations were minimised.

Table 3 The multiple objective linear programming model for the hypothetical multiple use management problem

	X_1	x_2	x_3	x_4	x_5	x_6	x_7**	x_8**		RHS
G1	1								$\rightarrow$	max
G2		1							$\rightarrow$	max
G3	-1		1		-0.93	-1.26			$\rightarrow$	max
G4			-.05	1		-1.04			$\rightarrow$	max
G5					1				$\rightarrow$	max
G6						1			$\rightarrow$	max
G7*	-1.21	-0.67	-.004	-.005	-12.17	-11.62	1		$\rightarrow$	max
G8*	-4.36							1	$\rightarrow$	max
R1			0.95					.002	$\leq$	127 955
R2				1					$\leq$	111 657
R3					1.74				$\leq$	582
R4						2.30			$\leq$	580
R5							.004		$\leq$	340
R6	.10	.31	.002	.002	7.54	6.35	.02	.045	$\leq$	7225
R7	5.73	17.59	0.11	0.11	440	476	1.21	2.55	$\leq$	430000

*) The row values are multiplied by 1000

**) The coefficients are given for 1/1000 miles

To solve the problem we use Pareto Race as illustrated in Figure 3.

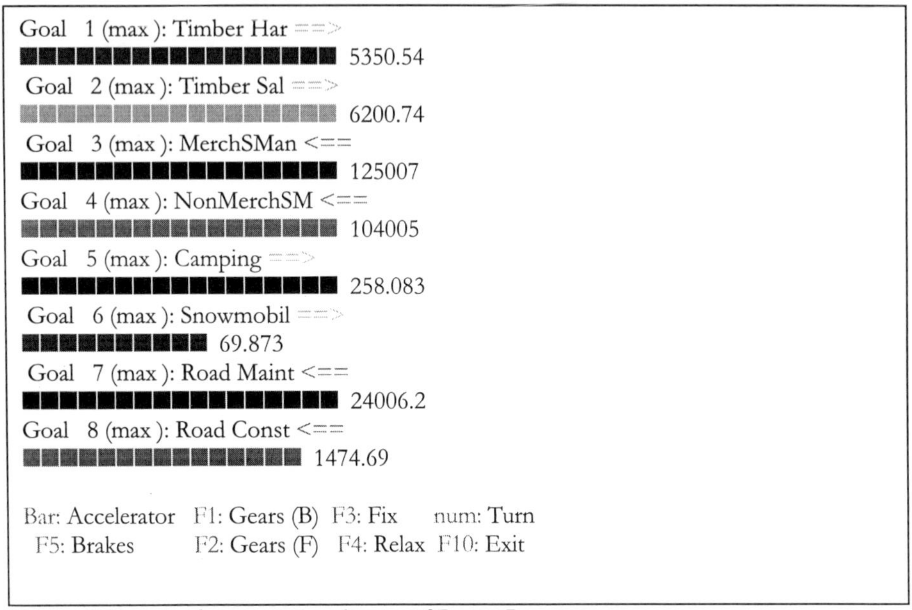

Figure 3 A screen demonstrating the use of Pareto Race

Pareto Race provides the possibility to freely search the efficient frontier. It does not require the user to give any priorities or weights. Using function keys, the user may 'walk' around the frontier. There is no need to *a priori* specify any fixed goal values. The user can also immediately observe the dependencies between different goals. The solution displayed in Figure 3 corresponds to the solution reported in the paper as the first one. However, by using Pareto Race we may search any part of the efficient frontier. We can also study which constraints are critical or which objectives are critical for trying to improve the values of the other objectives.

7. Conclusion

In this article, we have given a brief introduction to the main principles for solving multiple objective linear programming problems. We have also demonstrated with a case study picked from forest literature how a multiple objective linear programming approach can be used to support forest planning. The problem was solved by using Pareto Race (Korhonen and Wallenius (1988)) that is a dynamic version of a Reference Direction Approach originally developed by Korhonen and Laakso (1986).

References

CHANG, S. AND BUONGIORNO, J. 1981. A Programming Model for Multiple Use Forestry. Journal of Environmental Management 13:45-58.

COHON, J. 1978. Multiobjective Programming and Planning. Academic Press. New York.

GEOFFRION, A., DYER, J. AND FEINBERG, A. 1972. An Interactive Approach for Multi-Criterion Optimisation, with an Application to the Operation of an Academic Department. Management Science 19:357-368.

HAIMES, Y., TARVAINEN, K., SHIMA, T. AND THADATHIL, J. 1990. Hierarchical Multiobjective Analysis of Large-Scale Systems. Hemisphere Publishing Company. New York.

HWANG, C. AND MASUD, A. 1979. Multiple Objective Decision Making - Methods and Applications: A State-of-the-Art Survey. Springer. Berlin.

IGNIZIO, J. 1976. Goal Programming and Extensions. D. C. Heath. Lexington. Mass.

KORHONEN, P. 1987. VIG - A Visual Interactive Support System for Multiple Criteria Decision Making. Belgian Journal of OR, Statistics and Computer Science 27:3-15.

KORHONEN, P. AND HALME, M. 1996. Using Lexicographic Parametric Programming for Searching a Nondominated Set in Multiple Objective Linear Programming. Journal of Multi-Criteria Decision Analysis. 5:291-300.

KORHONEN, P., AND LAAKSO, J. 1986. A Visual Interactive Method for Solving the Multiple Criteria Problem. European Journal of Operational Research 24:277-287.

KORHONEN, P. AND WALLENIUS, J. 1988. A Pareto Race. Naval Research Logistics 35:615-623.

ROY, B. 1973. How outranking relation helps multiple criteria decision making. In: Cochrane, J. and Zeleny, M. (eds.). Multiple Criteria Decision Making. University of South Carolina Press. Columbia, SC.

SAATY, T. 1980. The Analytic Hierarchy Process. McGraw-Hill. New York.

SAWARAGI, Y., NAKAYAMA, H. AND TANINO, T. 1985. Theory of Multiobjective Optimization. Academic Press. New York.

SHIN, W. AND RAVINDRAN, A. 1991. Interactive Multiple Objective Optimization: Survey I - Continuous Case. Computers Operations Research 18:97-114.

STEUER, R.E. 1986. Multiple Criteria Optimization: Theory, Computation, and Application. Wiley. New York.

STEUER, R. AND CHOO, E.-U. 1983. An Interactive Weighted Tchebycheff Procedure for Multiple Objective Programming. Mathematical Programming 26:326-344.

WHITE, D. 1990. A Bibliography on the Applications of Mathematical Programming Multiple-Objective Methods. Journal of Operational Research Society 41:669-691.

WIERZBICKI, A. 1980. The Use of Reference Objectives in Multiobjective Optimization. In: Fandel, G. and Gal, T. (eds.). Multiple Objective Decision Making, Theory and Application. Springer-Verlag, New York.

WIERZBICKI, A. 1986. On the Completeness and Constructiveness of Parametric Characterizations to Vector Optimization Problems. Operation Research Spektrum 8:73-87.

YU, P.-L. 1985. Multiple Criteria Decision Making: Concepts, Techniques, and Extensions. Plenum. New York.

ZELENY, M. 1982. Multiple Criteria Decision Making. McGraw-Hill. New York.

ZIONTS, S. AND WALLENIUS, J. 1976. An Interactive Programming Method for Solving the Multiple Criteria Problem. Management Science 22:652-663.

The Analytic Hierachy Process (AHP): Standard Version, Forestry Application and Advances

Jyrki Kangas

Abstract

The Analytic Hierarchy Process (AHP) has attracted a lot of attention worldwide both in research and in supporting practical decision-making. It has been found especially suitable for environmental planning with multiple criteria. In natural resource management, the AHP has been most frequently applied to planning at strategic level. The AHP, as any planning method and technique, has its advantages but also drawbacks. Research related to the AHP and to its applications has been vigorous, and many methodological advances and extensions have been developed. These include, for example, statistical analyses of pairwise comparisons data and hybrid methods aiming at embody some fundamental features of the AHP together with benefits of other approaches, such as mathematical optimisation, Delphi technique, and SWOT. In Finland, a heuristic optimisation method called HERO has been developed for purposes of tactical forest planning. HERO can make use of pairwise comparisons and the eigenvalue technique in the formulation of optimisation problems in accordance with decision-makers' objectives and preferences. It can be applied to group decision support and participatory planning, too. In this article, the standard version of the AHP is first presented. After that, some experiences of the use of the method in forest management planning are discussed, including also the drawbacks. The research efforts carried out in Finland in order to further develop forestry decision support methods, which utilise some ideas and techniques as applied in the AHP, are then briefly presented. A more closer look is taken at HERO.

Keywords: Analytic Hierarchy Process, decision support, forest management, forest planning, heuristics, multi-functional forestry, optimisation

1. Introduction

The Analytic Hierarchy Process is a mathematical method for analysing complex decision problems with multiple criteria, introduced by Saaty (1977, 1980). It has been succesfully applied to many kinds of decision situations (see Zahedi 1986, Saaty 1987), for example energy policy (e.g. Hämäläinen 1992) and environmental decision analysis (e.g. Anselin et al. 1989, Varis 1989). It has also been applied to group decision making problems (e.g. Saaty and Kearns 1985, Dyer and Forman 1992). Early forestry applications include those of Mendoza and Sprouse (1989), Kangas (1991, 1992a,b) and Schmoldt et al. (1994) (see also Tarp and Helles 1995).

Due to its simplicity, effectiveness, and ability to deal with qualitative as well as with quantitative criteria, the AHP is applicable to multi-objective forest management planning. In addition, the priorities estimated using the AHP are easy to interpret. Other reasons for applying the AHP to forest planning include, e.g., that it forces decision-makers and participants to think about forestry values and objectives, and by using it the trade-offs between competing objectives and interests can easily be illustrated. The AHP is a good tool in education and learning concerning multi-purpose forestry as a whole as well as concerning trade-offs within some certain forest area. Furthermore, the AHP serves as a good introduction to the world of multiple-criteria decision support. Of course, the AHP has also its disadvantages. The drawbacks will be focused on later, as well as how they have been tried to alleviate.

2. Principles of Standard AHP

Four basic steps are involved when using the standard version of AHP to solve decision problems, in the case of a single decision maker:

Step 1. The decision hierarchy is constructed by decomposing the original decision problem into a hierarchy of interrelated decision elements (Figure 1).

Step 2. Pairwise comparisons are made at each level of the hierarchy: Which of the two factors has greater weight in decision-making, and how much greater, or which of the two decision alternatives is more preferred with regard to a certain decision attribute. Comparisons are carried out with respect to each decision element at levels 1 to n-1: those more detailed decision elements at the nearest level below the element, which refer to the element, are compared in a pairwise manner with respect to the element.

Step 3. Using the pairwise comparisons as input, the relative weights (importance/preference) of elements at each level are computed using the eigenvalue method. The resulting weights (local priorities) represent the decision-maker's perception of the relative importance or preference of the elements.

Step 4. The ratings, i.e. global priorities, for the decision alternatives are calculated based on the local priorities of the decision elements.

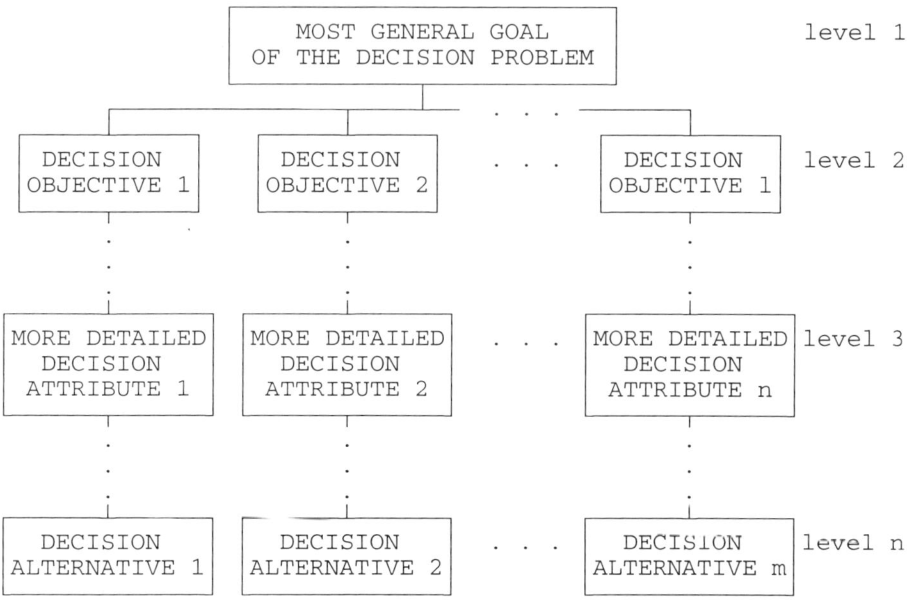

Figure 1 Standard form of the decision schema in the AHP; a hierarchy of n levels.

Pairwise comparisons give the decision-maker a basis on which to reveal his preferences by comparing two elements at a time. In the standard method presented by Saaty (1977, 1980) the decision-maker has the option of expressing preferences between two decision elements via a pairwise comparison as:

(i) equal importance or preference of both elements,
(ii) weak importance or preference of one element over another,
(iii) essential or strong importance or preference of one element over another,
(iv) demonstrated importance or preference of one element over another,
(v) absolute importance or preference of one element over another

These preference ratios are translated into pairwise weights of 1, 3, 5, 7 and 9, respectively, with 2, 4, 6 and 8 as intermediate values.

The choice of scale is based on observations of psychologists (see e.g. Miller 1956): an individual cannot simultaneously compare more than 5 - 9 objects without being confused. Using a unit difference between successive scale values, and using equal preference = 1, it follows that the scale values will range from 1 to 9. Elements with zero weights are eliminated from comparisons.

In step 3 the matrix of pairwise comparisons is determined. It is equal to the matrix of marginal rates of substitution. In the matrix the element $a_{ij} = 1/a_{ji}$. Thus, when i = j, $a_{ij} = 1$.

$$\mathbf{A} = (a_{ij}) = \begin{bmatrix} 1 & w_1/w_2 & \ldots & w_1/w_n \\ w_2/w_1 & 1 & \ldots & w_2/w_n \\ . & . & \ldots & . \\ . & . & \ldots & . \\ w_n/w_1 & w_n/w_2 & \ldots & 1 \end{bmatrix}$$

where w_i/w_j is the importance ratio, or the preference ratio, of decision elements i and j; n is the number of decision elements to be compared.

If there were no inconsistencies in judgments, matrix **A** has unit rank since every row is a constant multiple of the first row, and all the eigenvalues of the matrix are zero except one. Based on a consistent matrix **A**, relative weights could be determined by solving the equation

$$\mathbf{Aw} = \lambda \mathbf{w}$$

where λ is the only nonzero eigenvalue of a consistent matrix **A** and **w** is its right eigenvector.

The solution **w** of this problem is any column of **A**. These solutions differ by a multiplicative constant. Thus, the same relative weights are got based on any solution. In human decision making, some inconsistencies can be expected: people's feelings and preferences are not always consistent. When **A** contains inconsistencies, the estimated weights can be obtained using the eigenvector equation

$$(\mathbf{A} - \lambda_{max}\mathbf{I})\mathbf{q} = 0$$

where λ_{max} is the largest eigenvector of matrix **A**; **q** is its right eigenvector, and **I** is the identity matrix

The right eigenvector, **q**, constitutes the estimation of relative weights. It is the first principal component of the matrix of pairwise comparisons. If the matrix does not include any inconsistencies, i.e. the judgments made by a decision-maker have been consistent, **q** is an exact estimate of the priority vector. Matrix **A** has to be estimated for each decision element at levels 1 - (n-1). Each eigenvector is scaled to add up to one to get the local priorities.

Global priority of a decision element is calculated downwards from the top of the hierarchy: local priorities of the decision element are multiplied by the priorities of the corresponding decision elements in the level above and the results are added. For example, assume that the local priorities of decision objectives 1 and 2 (the only objectives) are 0.4 and 0.6, respectively. Also, the local priority of a detailed decision attribute 1 is 0.2 where objective 1 is concerned, and 0.3 where objective 2 is concerned. Then, the global priority of the detailed decision attribute is 0.4*0.2 + 0.6*0.3 = 0.26. Global priority of an element is then used to weight the local priorities of elements in the level below and so on down to the bottom level. The sum of the global priorities at each level is one.

Saaty (1977) has shown that λ_{max} of a reciprocal matrix **A** is always greater or equal to n (= number of rows = number of columns). If the pairwise comparisons do not include any inconsistencies, $\lambda_{max} = n$. The more consistent the comparisons are, the closer the value of computed λ_{max} is to n. A consistency index, CI, measures the inconsistencies of pairwise comparisons.

$$CI = (\lambda_{max} - n)/(n-1)$$

A consistency ratio, CR, also measures the coherence of the pairwise comparisons.

$$CR = 100(CI/ACI)$$

where ACI is the average consistency index of randomly generated comparisons (varies according to the size of matrix)

As mentioned above, in human decision making some inconsistencies can be expected. As a rule of thumb, a CR value of 10 % or less is considered to be acceptable. Otherwise, additional information needs to be gathered in order to resolve the inconsistencies of the pairwise comparisons: all or some of the comparisons have to be repeated or the decision hierarchy reconstructed.

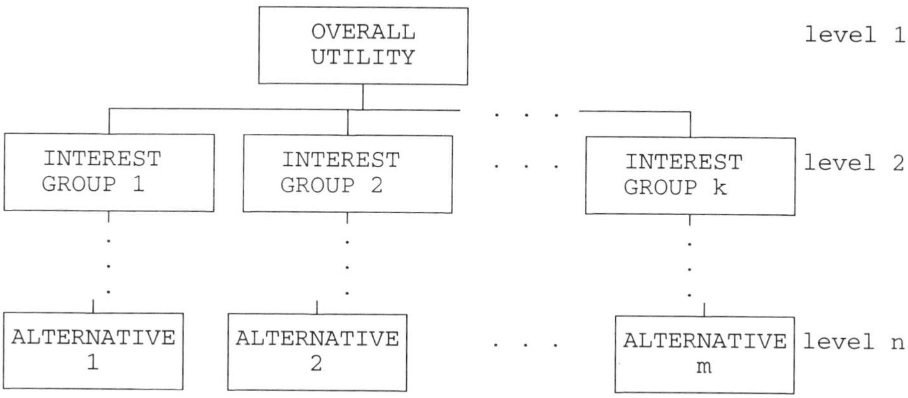

Figure 2 The decision schema when representative approach is applied.

When the AHP is applied to participatory planning or group decision making, the participants or decision makers can technically be taken as 'additional decision elements'. For example, when the representative democracy approach is applied in public participation, level 2 of the decision hierarchy can be thought to consist of interest groups (Figures 2 and 3). The weights of interest groups are determined by the decision-maker or by the organisation governing the forest area in question. Correspondingly, in a group decision making situation, the multiple decision makers form a level of 'players' in the decision hierarchy. In that case, the weights of the decision-makers can be determined according to the ownership basis, for example.

The facility for estimating priorities using the AHP depends particularly on the success in structuring and decomposing the decision problem in question; pairwise comparisons have to be meaningful and easy to understand. Simplicity of pairwise comparisons is crucial especially when public preferences have to be clarified.

For more details on the AHP theory readers are referred to Saaty (1977, 1980) and Saaty and Kearns (1985), and for forestry application of the method to Kangas (1992a, 1994).

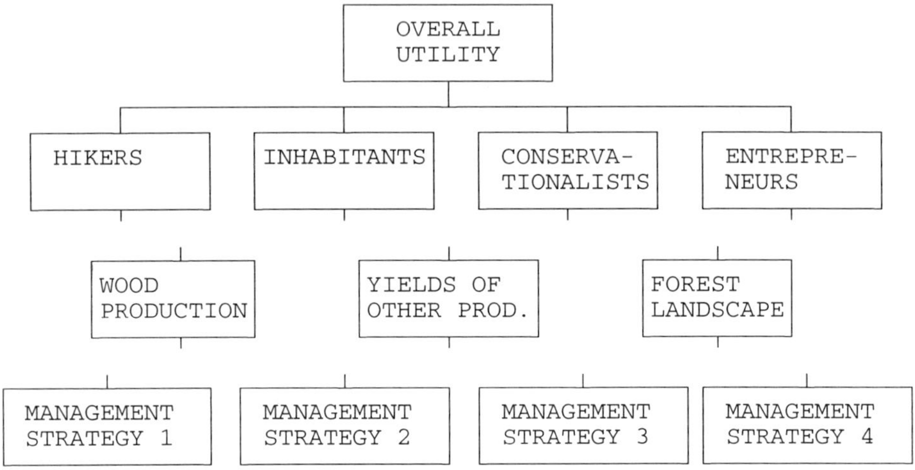

Figure 3 A figure in principle of decision schemas in participatory forest planning.

3. AHP in Finnish Forest Planning: Experiences, Drawbacks and Research Efforts

In Finland, the use of the AHP has attracted a lot of attention both in research and in supporting practical decision making. AHP has been most frequently applied to strategic forest planning with multiple objectives. In the state forestry, with state-owned forests in Finland covering one-third of all forestry land, principles of the AHP have been widely applied in participative forest planning (Heinonen 1997, Pykäläinen and Loikkanen 1997). A standard version of AHP has also been used in forest policy analysis at the province level (Kajala 1996). Pairwise comparisons and the eigenvalue technique have been utilised in modelling expert knowledge and subjective preferences in connection with planning calculations. Integrating ecological expertise with optimisation procedures has been a particular subject of study.

On grounds of the experiences obtained in the course of research, tests of the applications, and the feedback from practical forestry, AHP has been found to be a usable tool in multi-objective strategic decision situations with relatively few strategy alternatives to be compared. It has been found particularly suitable as a technique in participative forest planning and corresponding conflict management.

Although it has achieved fairly good success in strategic forest planning, the problems with standard AHP have also been noticed; especially from the viewpoint of tactical forest planning of large forest areas needing efficient calculation procedures and processes. Maybe the main drawback is the small maximum number of decision alternatives to be compared at a time. In tactical forest planning of a forest area with, e.g., hundreds of forest compartments each having several alternative treatment schedules there are far too many alternative forest plans to be evaluated. Planning of managed forests should not cost more than it benefits as decision support. Tactical forest planning must be fast, efficient, and cheap enough also in cases with many decision alternatives to be compared.

The lack of a sound statistical theory behind the AHP has also been seen as one of the major drawbacks of the method (e.g. Alho et al. 1996). Other disadvantages of the standard AHP include at least the assumption of additivity and interchangeability of decision elements, as well as problems with the comparison scale and in transforming verbal expressions into numbers. In addition, the feature of rank reversal has been reported in many case studies. Unfortunately, there is no final truth about the correct way of transforming verbal comparisons into numerical form. However, one solvable problem with the scale proposed by Saaty is that it, in fact, often forces to make inconsistent comparisons.

Forest planning research related to AHP and other decision support methods has been lively. Systematic endeavours have been made to develop hybrid methods embodying the fundamental aspects of AHP (e.g. ability to deal with multiple objectives, ease of use, comprehensibility) and efficiency of optimisation techniques applied in timber management planning (e.g. Kangas and Pukkala 1992, Pukkala and Kangas 1993) as well as the versatility of statistical methods (Alho et al. 1996, Alho and Kangas 1997, Leskinen and Kangas 1998).

Work on method development has resulted, for example, in a heuristic optimisation method called HERO, by using which pairwise comparisons and the eigenvalue technique can be utilised (Pukkala and Kangas 1993). One of the key ideas in HERO is the formulation of the optimisation problem in a manner more compliant with the objectives and preferences of the decision-maker than has been possible hitherto when using mathematical programming. HERO is more closely presented later in this paper.

Research work based on previous AHP-related studies has been carried out recently especially concerning the following topics: modelling ecological expertise (Kangas et al. 1993a, Alho et al. 1996), analysing the uncertainties involved in the resulting priorities calculated on the basis of expert judgements in a multi-objective evaluation task (Alho and Kangas 1997, Leskinen and Kangas 1998), decision-analysis techniques in participative forest planning (Kangas et al. 1996b), combined use of AHP and Delphi technique in environmental decision support (Kangas et al. 1996a, Kangas et al. 1998), taking forest landowners' preferences and cutting plans into account in determining cutting budgets at province and national level (Pesonen 1995), incorporating biodiversity assessments into planning calculations (Kangas and Kuusipalo 1993, Kangas and Pukkala 1996), and combined use of SWOT and AHP in strategic planning (Kurttila et al. 1999). The main stream of research has focused on integrating ideas, techniques and methods as applied in AHP with other theories and methods of decision support.

The forestry practise in Finland has been very interested in AHP and its 'relatives', and the interaction among researchers and foresters has been fruitful. The results of forest planning research have been applied in field tests, and researchers have received feedback on the needs of improving planning methods. At this very moment, growing interest is being shown towards combined uses of AHP and conventional optimisation techniques among forestry practitioners. Also, the possibilities to utilise ecological expertise in cases where there is a shortage of objective information based on empirical studies applicable to forest planning calculations have fascinated the forestry practice. Modelling expert knowledge is, of course, only a temporary solution to acute evaluation problems - it is used only until evaluation can rely upon information produced by empirical research. However, this 'temporary' situation most probably applies for a long time.

4. Hero Heuristic Optimisation Utilising some Features of AHP

The HERO heuristic optimisation method has been specially developed for tactical forest planning mainly at area or forest holding level. The idea is to seek - and to find - such a combination of stand-level treatment regimes for the area or the forest holding as will provide the best end result at the level of the whole area/holding, with respect to the objectives set for forest treatment and utilisation.

Prior to actual optimisation, objectives have to be set for forest treatment and utilisation, and each stand has to be provided with a host of alternative treatment regimes for the duration of the planning period (typically 10 to 20 years). The outcomes of the alternative regimes are computed through simulation of forest development.

When applying HERO, selecting the best alternative may be divided into two stages: estimation of the utility model and maximisation of this model. HERO uses an additive utility model, the variables of which are management objectives whose coefficients are the objectives' relative importance (weights), scaled to sum equal to one.

$$U=\sum_{i=1}^{m} a_i u_i(q_i) \qquad (1)$$

where U is the total utility; m is the number of objectives; a_i is the relative importance of objective i; u_i is the partial utility function; i.e. the sub-priority function of objective i; q_i is the quantity that the plan produces or consumes objective variable i; objectives are either forest products and values, such as timber, amenity and biodiversity, or resources, such as costs and labour requirement.

The weights are estimated applying pairwise comparisons carried out by the decision maker. The relative importance of the objectives can be computed using the eigenvalue method of ratio scale estimation. In that, the objectives are compared pairwise using a graphical interface, instead of the verbal scale as proposed by Saaty. The relative importance of two objectives at a time are defined by adjusting the lengths of the horisontal bars on the computer screen. Other modes of comparison can also be applied, if desired.

Also multiplicative parts can be added to the utility model. This is necessary, if the objective variables have too strong interrelationsships, which sometimes might be the case not only in theory but also in practical decision support. Multiplicative parts are of use also, if the objective variables are not interchangeable; i.e. if a good gain in one variable can not compensate for an inferior result in another. For example - considering biodiversity - if the population size of the population of some species or the population viability is considered as a part of a purely additive utility model, there is no guarantee that the viability, for example, is above zero in the plan chosen to be the best. (Kangas and Kangas 1998).

The relative worth of the planning alternatives with respect to the objectives are presented in the form of sub-priority functions. The sub-priority function depicts the change in the utility as a function of the objective variable. The utility model consisting of objective variables and sub-priority functions is a tool enabling different objective parameters to be made commensurable.

The method enables the presentation of objective variables in a hierarchical manner. When this is done, the objective is described using a hierarchically inferior model, whose variables are the components describing the objective in more detail and whose coefficients are the components' relative importance. For instance, the net income from wood production can be divided into net incomes of different periods, and biodiversity can be operationalised by means of the components explaining it. This being the case, the sub-priority functions are defined to depict the impact of the components on the utility felt through the objective variables they explain.

When estimating an ordinary HERO sub-priority function (i.e. not a multiplicative part of the model), first the maximum and minimum values that can be achieved for the objective parameter or component are computed. In addition, a few intermediate values are selected. The desirability of these values is then estimated, e.g., by means of pairwise comparisons and the values are then allocated relative priorities. These priority values define the sub-priority function. A sub-priority function can be non-linear, which is often the case, e.g., when describing the relationships between biodiversity and environmental variables. The sub-priority functions, with maximum value equalling one, is estimated separately for each component. Estimation can be based equally on expertise or subjective value information. Ratios of values of objectively measurable variables can also be used. Further advantages connected to this manner of estimation are that one operates in real case-wise production possibilities, and ease of operation.

At the 'maximisation of overall utility' stage, one uses the heuristic direct-search algorithm to search for the best treatment regime for the forest area. In the beginning of the optimisation process, one treatment schedule is selected randomly for each compartment. The values and the sub-priorities of the objectives are computed, as well as the total utility. After that, one compartment at a time is examined to see whether another treatment regime would increase the utility. If this is the case, the current treatment regime is replaced by the one that increases utility. Once all the treatment regimes of all the compartments have been studied in this way, the process is repeated again, starting from the first compartment, until no more schedules increasing the utility are found. The whole maximisation stage is repeated several times, and the solution with the greatest utility value is taken as the optimal solution.

When applying HERO, the user has to adhere to objective variables that can be described within the planning system applied. The MONSU forest planning software, programmed by Pukkala, makes use of HERO. It currently enables the examination of conventional forestry parameters depicting the growing stock (e.g. volumes by tree species) as well as parameters depicting the amount of decaying timber (standing and fallen), mushroom and berry crops, recreational and landscape values, and certain habitat indices for wild game as well as indices for biological diversity. The range of components is constantly growing as the planning software evolves.

For more details on techniques and qualities of HERO, readers are referred to Pukkala and Kangas (1993). So far, applications of HERO include interactive planning of private non-industrial forestry, public participation in forest management planning (Kangas et al. 1996b), taking risk and attitude towards risk into account in planning calculations (Pukkala and Kangas 1996), incorporating biodiversity into numerical forest planning (Kangas and Pukkala 1996), taking variation in forest characteristics at both stand and area level into account in calculations (Pukkala et al. 1997), integrating spatial analysis with optimisation (Pukkala et al. 1995), and modelling ecological expertise to be used in optimisation (Kangas et al. 1998).

Perhaps the greatest advantage of HERO, from the practical standpoint, is in its flexibility especially regarging setting of objectives and taking case-wise subjective preferences into account in planning. In that, HERO fits with the idea of value focused thinking. This is important both in customer-oriented planning of private forestry and in participatory planning. A futher advantage is that, due to the sub-priority estimation procedure, also case-wise expert knowledge is easy to utilise in HERO calculations. Because HERO does not put any demands for the form of sub-priority functions, it can cope with area-level spatial variables having non-linear utility effects more easily than mathematical programming, and with other non-linearities.

A drawback of any heuristic approach is that the solution may not always be optimal, but only an approximation to it. According to the tests carried out, and experiences gained in applications, this kind of inefficiency does not seem to be a serious problem: most often the global (technical) optimum is found, and the solutions are always at least close to optimum.

References

ANSELIN, A., MEIRE, P. AND ANSELIN, L. 1989. Multicriteria techniques in ecological evaluation: an example using the Analytical Hierarchy Process. Biological Conservation 49:215-229.

ALHO, J. AND KANGAS, J. 1997. Analysing uncertainties in experts' opinions of forest plan performance. Forest Science 43:521-528.

ALHO, J., KANGAS, J. AND KOLEHMAINEN, O. 1996. Uncertainty in the expert predictions of the ecological consequences of forest plans. Applied Statistics 45:1-14.

DYER, R.F. AND FORMAN, E.H. 1992. Group decision support with the analytic hierarchy process. Management Science 36:249-258.

HÄMÄLÄINEN, R. P. 1992. Decision analysis makes its way into environmental policy in Finland. Interfaces 18:12-23.

HEINONEN, P. 1997. Balancing forest uses at regional level: the case of State forests in western Finland. EFI Proceedings 14:203-210.

KAJALA, L. (ed.). 1996. Lapin metsästrategia. [Strategy for the forests of Finnish Lapland]. Maa-ja metsätalousministeriön julkaisuja 2/1996.

KANGAS, A. AND KANGAS, J. 1998. Ekologiset mallit ja ekologisten riskien hallinta metsäsuunnittelussa. [Ecological models and the management of ecological risks in forest planning]. Folia Forestalia 2:207-222.

KANGAS, J. 1991. A method for utility comparison of forest drainage alternatives. Suo 42:49-59.

KANGAS, J. 1992a. Choosing the regeneration chain in a forest stand: A decision model based on multi-attribute utility theory. Publications in Sciences 24. University of Joensuu.

KANGAS, J. 1992b. Multiple-use planning of forest resources by using the Analytic Hierarchy Process. Scandinavian Journal of Forest Research 7:259-268.

KANGAS, J. 1994. An approach to public participation in strategic forest management planning. Forest Ecology and Management 70:75-88.

KANGAS, J., ALHO, J.M., KOLEHMAINEN, O. AND MONONEN, A. 1998. Analyzing consistency of experts' judgments - case of assessing forest biodiversity. Forest Science 44:610-617.

KANGAS, J., KARSIKKO, J., LAASONEN, L. AND PUKKALA, T. 1993a. A method for estimating the suitability function of wildlife habitat for forest planning on the basis of expertise. Silva Fennica 27:259-268.

KANGAS, J. AND KUUSIPALO, J. 1993. Integrating biodiversity into forest management planning and decision-making. Forest Ecology and Management 61:1-15.

KANGAS, J., LAUHANEN, R. AND STORE, R. 1996a. Assessing the impacts of ditch network maintenance on water ecosystems on the basis of expert knowledge and integrating the assessments into decision analysis. Suo 47:47-57.

KANGAS, J., LOIKKANEN, T., PUKKALA, T. AND PYKÄLÄINEN, J. 1996b. A participatory approach to tactical forest planning. Acta Forestalia Fennica 251.

KANGAS, J. AND PUKKALA, T. 1992. A decision theoretic approach applied to goal programming of forest management. Silva Fennica 26:169-176.

KANGAS, J. AND PUKKALA, T. 1996. Operationalisation of biological diversity as a decision objective in tactical forest planning. Canadian Journal of Forest Research 26:103-111.

KURTTILA, M., PESONEN, M., KANGAS, J. AND KAJANUS, M. 1999. Utilizing AHP in SWOT analysis: A hybrid method and its application. Manuscript.

LESKINEN, P. AND KANGAS, J. 1998. Analysing uncertainties of interval judgment data in multiple-criteria evaluation of forest plans. Silva Fennica 32:363-372.

MENDOZA, G. A. AND SPROUSE, W. 1989. Forest planning and decision making under fuzzy environments: An overview and illustrations. Forest Science 35:481-502.

MILLER, G. A. 1956. The magical number seven plus or minus two: some limits on our capacity for processing information. The Psychological Review 63:81-97.

PESONEN, M. 1995. Non-industrial Private Forest Landowners' Choices of Timber Management Strategies and Potential Allowable cut: Case of Pohjois-Savo. Acta Forestalia Fennica 247.

PUKKALA, T. AND KANGAS, J. 1993. A heuristic optimisation method for forest planning and decision-making. Scandinavian Journal of Forest Research 8:560-570.

PUKKALA, T. AND KANGAS, J. 1996. A method for incorporating risk and risk attitude into forest planning. Forest Science 42:198-205.

PUKKALA, T., KANGAS, J., KNIIVILÄ, M. AND TIAINEN, A-M.1997. Integrating forest-level and compartment-level indices of species diversity with numerical forest planning. Silva Fennica 31:417-429.

PUKKALA, T., NUUTINEN, T. AND KANGAS, J. 1995. Integrating scenic and recreational amenities into numerical forest planning. Urban and Landscape Planning 32:185-195.

PUKKALA, T. 1998. MONSU-metsäsuunnitteluohjelma. Ohjelmiston toiminta ja käyttö. [User's guide of MONSU forest planning software]. Joensuu.

PYKÄLÄINEN, J. AND LOIKKANEN, T. 1997. An application of numeric decision analysis on participatory forest planning: the case of Kainuu. EFI Proceedings 14:125-132.

SAATY, R.W. 1987. The Analytic Hierarchy Process - what it is and how it is used. Mathematical Modelling 9:161-178.

SAATY, T. L. 1977. A scaling method for priorities in hierarchical structures. Journal of Mathematical Psychology 15:234-281.

SAATY, T. L. 1980. The Analytic Hierarchy Process. Planning, priority setting, resource allocation. McGraw-Hill. 283 p.

SAATY, T. L. AND KEARNS, K.P. 1985. Analytical Planning. The Organisation of Systems. Pergamon, Oxford.

SCHMOLDT, D. L., PETERSON, D. L. AND SILSBEE, D. G. 1994. Developing inventory and monitoring programs based on multiple objectives. Environmental Management 28:707-727.

TARP, P. AND HELLES, F. 1995. Multi-criteria decision making in forest management planning - an overview. Journal of Forest Economics 1:273-306.

VARIS, O. 1989. The analysis of preferences in complex environmental judgments - A focus on the Analytic Hierarchy Process. Journal of Environmental Management 28:283-294.

ZAHEDI, F. 1986. The Analytic Hierarchy Process - A survey of the method and its applications. Interfaces 16:96-108.

Chapter III

Stakeholder Involvement

Multiple Objective Optimisation in Agro-ecological Land Use Planning

Per J. Agrell

Abstract

A new interactive decision support system for agro-ecological land use planning in developing countries is presented. The optimisation system is based on the Agro-Ecological Zones (AEZ) model of FAO and operates with multiple objectives in an flexible and user friendly fashion. The decision variables are the agricultural cell activities and the intensity of livestock farming and forestation. An application to the Bungoma region in Kenya illustrates the approach, from the stage of defining decision criteria; related to food support, resiliency, economic growth and ecological sustainability; to the final decision making phase with local decision makers. Implemented on-site, the system is to facilitate analysis, communication and enforcement of local land development plans. Rather than forwarding the hierarchically defined values and trade-offs of a governmental agency or foreign-aid service, the model shifts the power to the regional authorities, which are more likely to commit to and follow up agreed-on land-use plans.

Keywords: Multicriteria Decision Making, integrated land use planning and management, decision support systems

1. Introduction

Nowhere are the conflicting objectives of man's endeavor so evident as in the usage of land. The land given to us is not only capable of producing our nutrition, it also serves as a source of energy, provides us with living quarters, infra-structure and recreation, it shelters us from the forces of nature, toxic waste and noise, it regenerates our drinking water and stores our disposals. Each and every use of land has consequences in the short and long-run, regardless of the formal ownership of the land and its resources. Thus, in countries with traditions of effective governmental intervention to regulate land-use, including agriculture, settlement and industrial zoning, the interests of the individual are put up against the common good of the society. However, in the developing world, with weak governmental authority and low confidence in the representativeness of common values, this process is far from perfected. Villagers in countries with poor infrastructure and low involvement in social processes have no reason to believe that their interests are taken into account in environmental legislation or that their support would be guaranteed if they obliged to a less intensive cultivation. The result is a widening gap between the need for sustainable practices and the evident overgrazing, deforestation and pollution of scarce land resources in Africa, South America and Asia. This discrepancy between the apparent necessity of land use management and the prevailing ad hoc practices is one of the motivations behind this paper. It reports on an initiative to bring state-of-the art land-use optimisation into regional use in order to inform, inspire and educate the population of sensitive areas of the inevitable trade-offs that are immanent in land-use management.

The outline of the paper is as follows. First, an introduction is given to the specific problems and challenges that the Kenya is facing. Next, the Kenya AEZ model (Agrell et al., 1997) is reviewed. The paper ends with a recount of a decision-making experiment.

2. Land-use Management in Kenya

Kenya is chosen as an example of this initiative, a developing country exhibiting rapid population growth, a relatively open economy and a traditional dependency on export of agricultural goods such as coffee, cotton and sugar cane. Currently, Kenya has about 2 million farm holders, with a typical lot size ranging from 2 ha to 15 ha. Given a steady population growth rate of 3% annually (UN 1997) and a weak economy suffering from high inflation that soared to 29% in 1994 (IMF 1995), the need for an accurate regional planning strategy is apparent. Agriculture has a high domestic importance, about 80% of the working population depend on farming for their survival and 50% on the export earnings stem from cash crops. The increasing pressure on the finite supply of land lead to over-exploitation and land degradation. Deforestation is accelerating, causing serious environmental problems, and the dependence on rainfall for natural irrigation renders the entire support system vulnerable. Irrigation in Kenya only accounts for less than 2% of the cultivated land. In the short-term, the challenge is to assure adequate support for the growing population and to maximise agricultural exports, while the long-term strategy is to ensure sustainability by preserving an ecological balance and attaining a satisfactory long-term growth in national income. A recurrent conflict in the policy making for land use concerns environmental versus economical objectives, e.g., whether it is preferable to use scarce resources to rehabilitate degraded land or to improve agricultural land, whether small-holder settlements or large-scale mechanised farming will better support the expanding population, the threat from encroachment of urban development onto high quality agricultural land, and the most appropriate use of scarce water resources.

The Bungoma district, a flourishing agricultural area in West Kenya, serves as a good example of the conflicting goals and objectives inherent to land use management. The district is quite suitable for agricultural activity, as about 222,000 of the total area of 319,300 ha are arable. The district's rich soil, plentiful and well-distributed rainfall and a fairly long growing season (270 days) allow for a wide variety of different agricultural crops. In addition to the current cereal production of maize, wheat, barley and finger millet, the region is well set for cassava, beans and potatoes, not to mention cash crops such as coffee, tea and pyrethrum (Fischer and Antoine 1994). The population is growing rapidly, from about 679,000 in the 1989 census to a projected 920,000 in 2000. The challenge for the local development committee is to improve the productivity level of the agriculture to cope with the boom, simultaneously considering the economical growth of the area and the need to stall deforestation and other adverse ecological effects of intensified farming. Although very fertile, the district has shallow soils in the southern part and on the slopes on Mt Elgon. These endangered areas are also important as water catchment zones and for forestry. Poor farming practices have caused an accelerated deforestation that may threaten the overall agro-ecological balance in the district.

The original model by Agrell et al. (1997), which draws upon a long-term collaboration between the Food and Agriculture Organisation (FAO) and the International Institute for Applied Systems Analysis (IIASA) (FAO 1981; 1984; FAO/IIASA 1991; 1993a), is intended as an agro-ecological decision support system for regional planning on site (Fischer and Antoine 1994). The underlying methodology, the Agro-Ecological Zones (AEZ) was developed in 1976 by a collaboration between FAO and IIASA (FAO 1981; FAO/IIASA 1991; 1993b). AEZ provides decision support for various problem related to land use appraisal for planning sustainable agricultural development. The AEZ methodology utilises a land resources inventory to assess, for a given level of input, all feasible agricultural land use options as well as expected production of relevant and agro-ecologically feasible cropping activities. On the basis of this agronomic evaluation, and using specific socio-economic data to specify constraints, targets and production options, the attainment of spatial resource allocation objectives can be optimised. The optimisation results provide perspectives on the capability of Kenya's land resources, technology, and policy, etc., to improve as well as sustain agricultural production. Before the application to Kenya

(FAO/IIASA, 1991), the methodology has been used in land use assessments in Bangladesh, China, Mozambique, Nepal, Nigeria, the Philippines, and Thailand.

3. Decision Maker Objectives

Sustaining food and cash-crop production is a major agricultural development policy of the Kenya government as set out in various government documents (e.g., Republic of Kenya 1986, 1994a; 1994b). The agricultural development policy of Kenya defines specific objectives and targets that include but are not limited to: growth of agricultural GDP, increasing foreign exchange earnings, enhancing food security and nutritional status, increasing employment and revenue generation, combating rural poverty and ensuring regional equity, increasing farm incomes, improving resource conservation (Onyeji and Fischer 1996). The Ministry of Land Reclamation, Regional and Water Development in Kenya has stated the following goals; generating employment with emphasis on increasing labour intensive production activities; obtaining maximum economic multiplier effects from other ongoing projects, especially through rural-small town linkages; increasing food production, generating foreign exchange earnings, slowing down rural-urban migration; developing the skills of technical and professional staff through training programmes and improvement and conservation of the environment. Locally, the Bungoma District Development Committee (DDC), in particular the subcommittees for agricultural and environmental management, has been very active in the collaboration to identify economic and agricultural opportunities, environmental threats and counter measures. In summary, there has been agreement upon a set of objectives that capture i) the food generating capability, to counter the population increase; ii) the food security and resilience, to create stability and independence, iii) the economic consequences of agriculture for the region, to promote economic growth and investments; iv) the land reserve, to protect and cherish the remaining game reserve and the unique ecosystem on the mountain slopes; v) the soil degradation as a result of land-use, to manage the serious problem of soil erosion.

It was decided that FAO and the International Institute for Applied Systems Analysis would develop, test and implement a decision support system to support the DDC in its regional planning. The system is to promote negotiation, learning and policy making at a regional level and hence it should be interactive and portable.

As a first step in the decision analysis, the five groups of criteria above have been operationalised as nine criteria: FOOD and FOODL for the total annual food production under normal and poor conditions, respectively; SSR for the self-sufficiency of selected crop groups; VALUE, COST and NETREV, for the total annual production value, production cost and net revenue, respectively; ARABLE and HARVEST for the total arable land used and harvested, respectively; and EROS and EROSMX for the total soil erosion and maximum cell soil erosion, respectively. Next, these criteria are defined in more detail.

4. Criterion Formulation

The original model (FAO/IIASA 1993b) optimises a single criterion, although in the documentation five different objectives are suggested for separate optimisation. In Agrell et al. (1997) this model is enhanced by multiple decision criteria and an interactive methodology to solve these in the AEZ framework. The land use allocation decision process essentially involves determining the appropriate share of each crop, the appropriate allocation of livestock and the best way to feed the livestock with available crops, crop residues and bye-products, and grass. Hence, the model includes four groups of decision variables, the price subsidies P_{ig} of crop i in market g, the land use shares x_{kj} of cropping activity k in each cell j, the number of livestock units L_{sz} of livestock system s kept in livestock zone z, and the feed ration f_{iltsz} in period (season) t of feed item l from crop i allocated to livestock system s in livestock zone z.

The AEZ database (FAO/IIASA 1993b) contains detailed information on 64 food and cash crop types, pastures, 31 fuelwood species and 10 livestock systems. The AEZ model includes 36 commodities, of which 26 are crop and 10 are livestock production commodities, and divides Kenya into a grid of 90,000 agro-ecological cells. Each cell is assumed reasonably homogeneous in terms of the relevant characteristics and is assigned to one or more cropping, livestock and market zones.

As an intermediate variable, introduce, a_{ikj}, as the area of cell j in which crop i is harvested as part of cropping activity k in cell j,

$$a_{ikj} = x_{kj} g_{ik} E_j cf_{kj}\text{, for all } i, k, j, \tag{1}$$

where E_j is the size (surface) of cell j, cf_{kj} is the cultivation factor for cropping activity k in cell j and g_{ik} represents the proportion of crop i in activity k. Note that crop activity k includes both simple and multiple sequential crop combinations, so that several crops i may belong to a given crop combination.

The first criterion (FOOD) simply maximises the total annual net food production (in billion kcal), whereas the second (FOODL, also in billion kcal) takes different moisture regimes into account, in order to assess the stability of the plan during years of draught and threatening famine. FOOD expresses the net annual calories made available through agricultural production. Specifically, FOOD in (2) measures food output as the net weighted sum of converted and processed food energy and protein available for human consumption. The first term in (2) applies to crop production, the second to livestock production,

$$\text{FOOD} = \sum_{j=1}^{J}\sum_{i=1}^{N}\sum_{k=1}^{K}\sum_{t=1}^{T} \kappa_i \varepsilon_i (1 - w_i) y_{ikj} a_{ikj} d_{ikt} + \sum_{h=1}^{H}\sum_{i \in N_h}\sum_{\varsigma=1}^{Z}\sum_{s=1}^{S} \kappa_i \varepsilon_i (1 - w_i) y_{is} L_{s\varsigma}, \tag{2}$$

where $y_{ikj}a_{ikj}$ represents the gross production of crop i, cropping activity k, cell j; y_{is} is the output of livestock product i per reference livestock unit (RLSU) of livestock system s, $L_{s\varsigma}$ is the number of RLSU of system s in livestock zone ς, so that $y_{is}\ L_{s\varsigma}$ is the gross production of livestock product i per RLSU of livestock system s, zone ς; w_i is the production waste proportion of crop i; κ_i is the calorie content per unit of food item i; ε_i is the extraction rate, i.e., the conversion rate of produce of crop/livestock product i to food product i; d_{ikt} is the share of production from crop i as part of activity k harvested in season t; N_h is the index set of crops that belong to crop group h.

FOODL is similar to FOOD, but the coefficients ys_{ikj} and ys_{ik} in (3) correspond to yields under severe climactic conditions, rather than average or good,

$$\text{FOODL} = \sum_{j=1}^{J}\sum_{i=1}^{N}\sum_{k=1}^{K}\sum_{t=1}^{T} \kappa_i \varepsilon_i (1 - w_i) ys_{ikj} a_{ikj} d_{ikt} + \sum_{h=1}^{H}\sum_{i \in N_h}\sum_{\varsigma=1}^{Z}\sum_{s=1}^{S} \kappa_i \varepsilon_i (1 - w_i) ys_{is} L_{s\varsigma}, \tag{3}$$

where $a_{ikj}ys_{ikj}$ is the gross production under severe climactic conditions of crop i, cropping activity k, cell j; ys_{is} is the output under severe climactic conditions of livestock product i per reference livestock unit (RSLU) of livestock system s, all other constants are as in (2).

One of Kenya's most important strategic objectives is to maintain a high level of food self-reliance. The self-sufficiency ratio SSR, stated mathematically in (4), measures the extent to which broad commodity groups (e.g., cereals, root crops, meats) are produced within the Bungoma region. SSR maximises the minimum self-sufficiency ratio across all crop commodity groups $h \in H_C$ and livestock commodities $h \in H_L$, so that the 'worst' value is maximised. H_C is the index set of crop commodities, and H_L is the index set of livestock commodities. The effect of this criterion is to balance the production mix to assure local resiliency and relieve the underdeveloped infrastructure.

$$\text{SSR} = \min_h \begin{Bmatrix} \frac{1}{D_h} \sum_{i \in N_h} \sum_{k=1}^{K} \sum_{j=1}^{J} \sum_{l=1}^{T} \varepsilon_i (1 - w_i) y_{ikj} a_{ikj} d_{ikl} : & \text{for all } h \in H_C, \\ \frac{1}{D_h} \sum_{i \in N_h} \sum_{s=1}^{S} \sum_{\varsigma=1}^{Z} \varepsilon_i (1 - w_i) y_{is} L_{s\varsigma} : & \text{for all } h \in H_L \end{Bmatrix}, \tag{4}$$

where D_h is the demand target (in 10^9 kcal) for commodity group h and the other parameters are as defined above.

VALUE, the expected gross value of the output (in 10^6 Kenyan Shillings, KSh), measures the extent to which the production provides the stake holders and municipalities with monetary funds. As higher levels of VALUE are preferred, *ceteris paribus*, this is a maximisation criterion. VALUE is defined as the average market valuation for the crops and livestock products produced,

$$\text{VALUE} = \sum_{i=1}^{N} \sum_{j=1}^{J} \sum_{k=1}^{K} (1 - w_i) p_{ikj} y_{ikj} a_{ikj} + \sum_{h=1}^{H} \sum_{i \in N_h} \sum_{s=1}^{S} \sum_{\varsigma=1}^{Z} (1 - w_i) p_{is} y_{is} L_{s\varsigma}, \tag{5}$$

where p_{ikj} and p_{is} are the per calorie price of crop produce of crop i, cropping activity k, cell j and livestock product i from livestock system s, respectively.

The expected production cost criterion COST in (5), defined as the expected average costs (in 10^6 KSh) for all production inputs of cropping and livestock systems, represents the monetary requirements needed to produce a given production mix. For a developing country like Kenya it is particularly important to keep production costs as low as possible. This is achieved by using domestic production inputs, such as labour, power and manure generated by local livestock as efficiently as possible, in order to reduce dependence on foreign money and production inputs.

$$\text{COST} = \sum_{m=1}^{M} c_m \left\{ \sum_{j=1}^{J} \sum_{k=1}^{K} \sum_{i=1}^{N} a_{ikj} v_{ikjm} + \sum_{s=1}^{S} \sum_{\varsigma=1}^{Z} L_{s\varsigma} v_{sm} \right\}, \tag{6}$$

where c_m is the cost/weight per unit of production input m, v_{ikjm} is the amount of input type m required per ha of crop i cultivated as part of activity k in cell j, and v_{sm} is the amount of input type m required per RLSU of livestock system s. The NETREV criterion in (7) expresses the expected net revenue (in 10^6 KSh) resulting from the planning decision, as the difference between the expected gross value of the production and the expected production cost,

$$\text{NETREV} = \text{VALUE} - \text{COST}. \tag{7}$$

Calculated as the sum across all cells and crop activities of the share of cell j devoted to cropping activity k, weighted by the size of the cell E_j, ARABLE represents the amount of arable land (in ha) used for agricultural production, including grazing and land that is fallow at a given point of time. Lower values of ARABLE are generally preferred from an ecological viewpoint, as a too extensive land use can harm the soil and environment,

$$\text{ARABLE} - \sum_{k=1}^{K} \sum_{j=1}^{J} x_{kj} E_j. \tag{8}$$

The HARVEST criterion in (9) represents the area (in ha) actively used for harvesting during the period, and is of particular interest in determining the environmental impact and the seasonal labour requirements resulting from the production. Like ARABLE, from a conservation viewpoint lower values of HARVEST are preferred.

$$\text{HARVEST} = \sum_{k=1}^{K} \sum_{j=1}^{J} \sum_{i=1}^{N} x_{kj} g_{ik} E_j cf_{kj}. \tag{9}$$

In (9), the cultivation factor cf_{kj} determines the fraction of time that cell j may in fact be used for cropping activity k. Western Kenya, where the Bungoma region is located, has two seasons, the long and the short rains, and almost yearlong production potential. Thus, except for cash crops, the land may be used for a second harvest of a different crop. In the current formulation, HARVEST includes the primary harvested area only. Thus, the difference ARABLE – HARVEST denotes the land area under fallow.

The two cell erosion criteria, EROS and EROSMX, are important means to assess the environmental impact of agricultural production. Whereas EROS minimises the total soil loss (in 10^3 tons of soil per year) for all cells in the region combined, EROSMX minimises the worst (i.e., maximum) soil loss (in tons per ha per year) in any cell used for agriculture. These erosion measures provide information about the long-term agro-ecological sustainability of the proposed production plan. EROS and EROSMX also complement the economic information by supplying the downside of extensive mono-cropping of cash crops, which is very rewarding economically. The expected soil loss per ha and year in cell j from activity k is estimated by e_{kj}, so that EROS and EROSMX can be expressed as (10) and (11), respectively,

$$\mathrm{EROS} = \sum_{k=1}^{K} \sum_{j=1}^{J} e_{kj} x_{kj} E_j , \tag{10}$$

$$\mathrm{EROSMX} = \max_j \left\{ \sum_{k=1}^{K} e_{kj} x_{kj} \right\} . \tag{11}$$

5. The Constraint Set

The pre-established upper (UA_h) and lower (LA_h) bounds in (12)-(13) restrict the harvested areas dedicated to each commodity group $h \in H$ to pre-determined levels,

$$\sum_{i \in N_h} \sum_{k=1}^{K} \sum_{j=1}^{J} a_{ikj} \le UA_h , \quad h = 1, \ldots, H , \tag{12}$$

$$\sum_{i \in N_h} \sum_{k=1}^{K} \sum_{j=1}^{J} a_{ikj} \ge LA_h , \quad h = 1, \ldots, H . \tag{13}$$

Similarly, in order to allow for reasonable non-agricultural land use, housing and infrastructure, game parks and forestry, the total arable land is limited by the upper bound TA, as in (14),

$$\sum_{k=1}^{K} \sum_{j=1}^{J} x_{kj} E_j \le TA . \tag{14}$$

The distribution of land across crop groups is controlled by imposing upper and lower bounds on the harvested area for each crop group N_h, determined as crop group-specific lower ($L\rho_h$) and upper ($U\rho_h$) bounds on the share of total arable land, as in (15)-(16).

$$\sum_{i \in N_h} \sum_{k=1}^{K} \sum_{j=1}^{J} a_{ikj} \le U\rho_h \sum_{i=1}^{N} \sum_{k=1}^{K} \sum_{j=1}^{J} a_{ikj} \tag{15}$$

$$\sum_{i \in N_h} \sum_{k=1}^{K} \sum_{j=1}^{J} a_{ikj} \ge L\rho_h \sum_{i=1}^{N} \sum_{k=1}^{K} \sum_{j=1}^{J} a_{ikj} \tag{16}$$

Specific group-specific food demand requirements for each livestock commodity are included in the model through the upper (UD_h) and lower (LD_h) bounds in (17)-(18), where h is a livestock commodity in the set H_L.

$$\sum_{i \in N_h} \sum_{s=1}^{S} \sum_{\zeta=1}^{Z} \varepsilon_i (1 - w_i) y_{is} L_{s\zeta} \le UD_h , \qquad \text{for each} \qquad h \in H_L . \tag{17}$$

$$\sum_{i\in N_h}\sum_{s=1}^{S}\sum_{z=1}^{Z}\varepsilon_i(1-w_i)y_{is}L_{sz}\geq LD_h, \qquad \text{for each} \qquad h\in H_L. \tag{18}$$

The ratio of calories to proteins from food products is regulated by constraints to assure an acceptable balanced diet regardless of policy in each climactic region *r*. In (19), *CAL* is the maximum calorie/protein ratio for human consumption, C_r is the set of agro-ecological cells in region *r*, Z_r is the set of livestock zones in region *r*, π_i is the protein content per unit of food item *i*, and the other parameters are as in (2).

$$\begin{aligned}&\sum_{i=1}^{N}\kappa_i\varepsilon_i(1-w_i)\left(\sum_{j\in C_r}\sum_{t=1}^{T}y_{ikj}a_{ikj}d_{ikt}+\sum_{z_r}\sum_{s=1}^{S}y_{is}L_{sz}\right)\leq\\&CAL\sum_{i=1}^{N}\pi_i\varepsilon_i(1-w_i)\left(\sum_{j\in C_r}\sum_{t=1}^{T}y_{ikj}a_{ikj}d_{ikt}+\sum_{z\in Z_r}\sum_{s=1}^{S}y_{is}L_{sz}\right), \qquad r=1,\ldots,R.\end{aligned} \tag{19}$$

To ensure that the animal feed consumption does not exceed the feed supply, aggregate feed availability constraints are imposed for every feed item in each livestock zone. These constraints dynamically connect the decision variables x_{kj}, the level of cropping activity *k* in cell *j*, and f_{iltsz}, the amount of feed item *l* from crop *i* allocated to livestock system *s* in season *t* and livestock zone *z*. To make the model structure more transparent, first two auxiliary variables are defined, the net production F_{iltz} of feed item *l* (in dry matter) from crop *i* in season *t* and livestock zone *z*, and, the supply FS_{iltsz} of feed item *l* (in dry matter) from crop *i* in season *t* and livestock zone *z*. These entities are defined in (20)-(22),

$$F_{iltz}=u_{ilt}r_{ilt}\sum_{j\in C_z}\sum_{k=1}^{K}y_{ikj}a_{ikj}d_{ikt}, \qquad \begin{cases}i=1,\ldots,N; l=1,\ldots,L,\\ t=1,\ldots,T; z=1,\ldots,Z.\end{cases} \tag{20}$$

where r_{ilt} is the production of feed item *l* from activity *k* relative to the production of primary produce of crop *i*, u_{ilt} is the feed utilisation factor of crop *i*, feed item *l*, in season *t*. The initial feed supply is given by (22),

$$FS_{iltz}=F_{iltz}, \qquad t=1, \tag{21}$$

For all succeeding periods *t*, the feed supply FS_{iltsz} is defined as in (23), where μ_{ilt} is the feed depreciation factor related to carry-over of feed supplies from crop *i*, item *l* from season *t*–1 into season *t*.

$$\begin{aligned}&FS_{iltz}=F_{iltz}+(1-\mu_{ilt})\left(FS_{il,t-1,z}-\sum_{s=1}^{S}f_{il,t-1,sz}\right),\\&i=1,\ldots,N; l=1,\ldots,L;\ t=2,\ldots,T; z=1,\ldots,Z,\end{aligned} \tag{22}$$

where f_{iltsz} represents the feed use from crop *i*, feed item *l*, allocated to livestock system *s* in season *t*, in livestock zone *z*. The set of aggregate feed consumption constraints on feed item *l*, for any crop *i*, by the livestock in zone *z* in period *t* is presented in (23),

$$\sum_{s=1}^{S}f_{iltsz}\leq FS_{iltsz}, \qquad \begin{cases}i=1,\ldots,N; l=1,\ldots,L,\\ t=1,\ldots,T; z=1,\ldots,Z.\end{cases} \tag{23}$$

The seasonal feed constraints assure that seasonal feed intake in all livestock systems and zones is within an acceptable range and that the total annual intake does not fall below average annual requirements. These constraints force a realistic distribution of the feed over time, rather than assuming that infinite quantities of feed can be stored for any length of time. The seasonal crude protein feed quality constraints are analogous to the seasonal feed constraints, and restrict the seasonal and annual intake of digestible crude protein. For further detail, cf. Agrell et al. (1997).

In order to prevent dramatic re-distributions of livestock types within each zone, to accommodate historical conditions and model adjustment speed, (24) and (25) constrain the number of livestock units to a pre-defined fraction of the total livestock in the zone. Let $U\beta_{sz}$ and $L\beta_{sz}$ be the maximum and minimum share of livestock system s supported in livestock zone z, respectively.

$$L_{sz} \le U\beta_{sz} \sum_{g=1}^{S} L_{gz}, \qquad s = 1,\ldots,S; z = 1,\ldots,Z \tag{24}$$

$$L_{sz} \ge L\beta_{sz} \sum_{g=1}^{S} L_{gz}, \qquad s = 1,\ldots,S; z = 1,\ldots,Z \tag{25}$$

A similar condition regulates the proportions between livestock systems in each broad climactic region. The proportion of animals (expressed as reference livestock system units, RLSU) in each region is restricted to a given fraction of the total livestock in the region. The climatic regions r are defined in terms of groups of length-of-growing period (LGP) zones, e.g., the sub-humid zone consists of all land with LGP values between 180 and 270 days.

$$\sum_{z \in Z_r} L_{sz} \le U\alpha_{sr} \sum_{g=1}^{S} \sum_{z \in Z_r} L_{gz}, \qquad s = 1,\ldots,S;\ r = 1,\ldots,R\,, \tag{26}$$

where $U\alpha_{sr}$ is the maximum share of livestock unit s in region r supported in relation to the total number of livestock units. In (27), $L\alpha_{sr}$ denotes the minimum share of livestock unit s in region r supported in relation to the total numbers of livestock units.

$$\sum_{z \in Z_r} L_{sz} \ge L\alpha_{sr} \sum_{g=1}^{S} \sum_{z \in Z_r} L_{gz}, \qquad s = 1,\ldots,S;\ r = 1,\ldots,R\,, \tag{27}$$

The resources needed for crop and livestock production are limited to the supply V_m of resource m (e.g., power, capital, fertiliser, labour) in the region. The constraints in (28) establish separate upper bounds for each resource m. On the left-hand side of (28), v_{ikjm} represents the input requirement of input m per ha of crop i cultivated through activity k in cell j, and v_{sm} is the input requirement of input m per RLSU of livestock system s. The input coefficients v_{sm} of those activities that generate resources, *e.g.*, draught power and manure from animals are negative.

$$\sum_{j=1}^{J} \sum_{k=1}^{K} \sum_{i=1}^{N} v_{ikjm} a_{ikj} + \sum_{s=1}^{S} \sum_{z=1}^{Z} v_{sm} L_{sz} \le V_m, \qquad m = 1,\ldots,M\,. \tag{28}$$

Each cropping activity x_{kj} in cell j has a nonnegative share, and the sum of all activity shares in cell j in (29) equals at most unity. As non-activity is an option, (29) is expressed as an inequality. Moreover, in order to prevent an overly skewed crop mix, most individual cropping activities in cell j cannot take more than a share ux_j, implying the constraint set in (30). In specific cases, the limitation in (30) is waived, e.g., for specific perennial crops, environmentally compatible cropping activities such as beans, forage legumes and grasses, and cells with only one suitable activity (Fischer and Antoine 1994).

$$\sum_{k=1}^{K} x_{kj} \le 1, \qquad j = 1,\ldots,J\,, \tag{29}$$

$$x_{kj} \le ux_j, \qquad k = 1,\ldots,K; j = 1,\ldots,J\,, \tag{30}$$

Finally, non-negativity applies to all decision variables, x_{kj}, $L_{s\zeta}$ and $f_{ilts\zeta}$:

$$x_{kj}, L_{s\zeta}, f_{ilts\zeta} \geq 0, \qquad \begin{cases} i = 1,...,N;\ j = 1,...,J; \\ k = 1,...,K;\ l = 1,...,L; \\ s = 1,...,S;\ t = 1,...,T; \\ \zeta = 1,...,Z. \end{cases} \tag{31}$$

To model the risk aversion at cell level in presence of weak sustainability conditions, an additional set of risk constraints are implemented as follows. First, disaggregate (3) to the cell level cropping output FOODL_j (in billion kcal) as in (32) below.

$$\text{FOODL}_j = \sum_{i=1}^{N} \sum_{k=1}^{K} \kappa_i \varepsilon_i (1 - w_i) ys_{ikj} a_{ikj} \tag{32}$$

For each cell j, let Fx_{ikj} be an optimal solution to the problem to maximise FOODL_j subject to constraints (12) through (31). From this information it is possible to determine the maximum possible food output in cell j under severe climactic conditions, $\text{MXF}_j = \text{FOODL}'_j$. Analogously for value maximisation, VALUEL_j in (33) expresses the expected value generated through cropping during severe conditions in cell j.

$$\text{VALUEL}_j = \sum_{i=1}^{N} \sum_{k=1}^{K} (1 - w_i) p_{ikj} ys_{ikj} a_{ikj} \tag{33}$$

Now, let Vx_{ikj} and MXV_j be an optimal solution and the associated cell level objective function value, respectively. The risk constraints in (34) and (35) require that the crop combination selected in any cell must produce at least α MXF_j or α MXV_j, where the parameter α is empirically estimated. The risk constraints apply at the cell level, rather than the district level, because in the case of subsistence farming a particular household cannot take too high a risk of crop failure, although this might be better on average or in aggregate. The constraints are necessary due to the non-enforcing structure of the model and a rational micro-economic behavioral assumption on the farmers.

$$\sum_{i=1}^{N} \sum_{k=1}^{K} \kappa_i \varepsilon_i (1 - w_i) ys_{ikj} a_{ikj} \geq \alpha \text{MXF}_j, \quad \text{for all } j = 1, ..., J \tag{34}$$

$$\sum_{i=1}^{N} \sum_{k=1}^{K} (1 - w_i) p_{ikj} ys_{ikj} a_{ikj} \geq \alpha \text{MXV}_j, \quad \text{for all } j = 1, ..., J \tag{35}$$

6. An Interactive MCDM Approach

Various multiobjective linear programming (MOLP) solution procedures have been developed (see, e.g., Shin and Ravindran 1991). We follow Agrell et al. (1997) and select the Interactive Weighted Tchebycheff Procedure (IWTP) of Steuer and Choo (1983), due to its desirable mathematical and decision-support related properties. The IWTP is attractive computationally, since it can be implemented using existing commercial optimisation software, since it is guaranteed to provide the decision maker with non-dominated solutions at all times, and since it is user-interactive and provides flexible decision support for multi-criteria programming problems. The mode of operation is straight-forward. First, the extreme points are assessed relative to the model to give attainable ranges of outcome, cf. Table 1. Second, an iteration commences where the algorithm samples and filters a sample of non-dominated solution and presents these to the decision maker. The decision maker chooses one or more as preferred and this information is used by the algorithm to contract the area of search. The iterations stop when the decision maker is completely satisfied and a most preferred solution has been obtained. The method does not require the decision maker to converge to a final solution at a pre-determined rate, and a previously discarded solution can be reconsidered at any time, thus facilitating learning about the problem. The decision maker can also guide the solution process at any time by inserting bounds on the criteria. A comparative experimental study of interactive MOLP methods by Buchanan and Daellenbach (1987) found the IWTP to be favored over several competing methods, from the user's viewpoint. Not to burden the paper with excessive detail regarding the functionality of the method, the description of the ITWP is omitted and the reader is referred to, e.g., Steuer and Choo (1983).

7. Decision-Making Experiment

The experiment was carried out using a single decision maker who is experienced with the regional planning issues in the Bungoma district. All calculations were performed on a 66 Mhz PC with Intel 486 processor, the AEZ optimisation problem was solved using MINOS (Murtaugh and Saunders 1977) and the entire multi-criteria decision model was encoded in Visual Basic for MS Excel 5.0. The computational times were low, less than 2 minutes per iteration, and the experiment was carried out in close cooperation with the decision maker, who upon request was also presented with complementary (more detailed) information regarding the general policies suggested.

8. The Payoff Table

Initially, the boundaries of the attainable criterion values of the MCDM formulation are assessed by establishing a payoff table as in Table 1. This table contains the selfish criterion values, obtained by successive single-criterion optimisations. Each solution in Table 1 is non-dominated, i.e., none of the criterion values in a given solution can be improved without sacrificing at least one of the other criteria. From Table 1, we can draw some important observations regarding the conflicts inherent in the model. The food output criteria, FOOD and FOODL, appear to be in direct conflict with the land-usage and erosion criteria (ARABLE, HARVEST and EROS, EROSMX). This is not surprising, because a maximum food output requires the full utilisation of all available land, even at the cost of increased erosion. Net revenue, NETREV, is maximised when primarily cash crops are grown, at the expense of food output, erosion and to some extent SSR. Production cost and gross value are related, and their optimal solutions have similar characteristics. Interestingly, the maximum possible gross value (VALUE = 1,378 MKSh) is only marginally higher than the corresponding level of VALUE (1,377 MKSh) when maximising NETREV, although the latter vector displays considerable improvements in all but one criterion value. If the amount of arable (ARABLE) or harvested (HARVEST) land are minimised, FOOD and SSR are reduced substantially, under moderate erosion levels (EROS). This makes sense intuitively, because if less arable and harvest lands are used, the food output will be limited although the erosion per ha arable land will stay constant due to the intensity. Finally, erosion criteria that are at their minimum levels reflect an agricultural system with low output, dismal economic value and a high reliance on import. In summary, the initial payoff table already illustrates that the model is capable of providing a wide range of vastly different decision scenarios and embodies interesting trade-offs that the decision maker must resolve in order to achieve a realistic solution.

Table 1a The Initial Model Payoff Table

Objective Criterion	*max* FOOD	*max* FOODL	*Max* SSR	*max* VALUE	*min* COST	*max* NETREV	
max FOOD	1 197	1 139	906	930	765	945	10^9 kcal
max FOODL	970	1 010	654	717	543	732	10^9 kcal
max SSR	96	86	107	80	32	80	%
max VALUE	1 129	1 109	1 087	1 378	757	1 377	10^6 KSh
min COST	46	38	42	103	15	60	10^6 KSh
max NETREV	1 083	1 071	1 045	1 275	742	1 317	10^6 KSh
min ARABLE	165 385	161 142	157 247	135 436	124 703	129 234	ha
min HARVEST	125 448	122 388	119 940	104 196	95 210	99 685	ha
min EROS	3 207	3 256	3 527	2 676	2 856	2 620	kton/ha
min EROSMX	113	148	228	85	119	85	ton/ha

Table 1b The Initial Model Payoff Table (cont'd)

Objective Criterion	*min* ARABLE	*min* HARVEST	*min* EROS	*Min* EROSMX	
max FOOD	745	728	773	747	10^9 kcal
max FOODL	551	535	599	575	10^9 kcal
max SSR	32	32	32	32	%
max VALUE	814	814	817	810	10^6 KSh
min COST	25	25	25	27	10^6 KSh
max NETREV	789	789	792	783	10^6 KSh
min ARABLE	96 249	96 311	105 804	120 960	ha
Min HARVEST	73 498	73 462	81 059	92 203	ha
Min EROS	1 866	1 690	1 165	1 838	kton/ha
Min EROSMX	85	79	29	12	ton/ha

9. Iterations

The experiment was conducted in five iterations, all of which are explained in Agrell et al. (1997). It is sufficient here to note that the decision maker through the iterations gathered enough insight about the problem to arrive at a most preferred decision. The modus of operation was focused and occasionally using additional constraints. The three primary criteria are SSR, erosion (EROS, EROSMX) and food output (FOOD, FOODL), in that order. The other criteria are judged in relation to the 'primary' criteria and serve a more guiding function. After the second iteration, when the decision maker has been exposed to a variety of dispersed solutions, he declares that only solutions with a self-sufficiency rate of 85% or higher are acceptable. This request is accommodated through an additional constraint, as well as later inserted requests to limit maximum cell erosion and total erosion to 50 ton/ha and 35 ton/ha, respectively. The line of reasoning is first to cater for immediate survival and stability, secondly to assure sustainability and thirdly to induce economic growth in agriculture and indirectly in tourism. In later iterations, trade-offs between ARABLE and EROS at high levels of SSR, food and economic indicators were made explicit. The values for the final iteration (5) is given in Table 2, where the decision maker after some reflection regarding the marginal values selected 1-5 as the most preferred. The decision maker was quite pleased with the interactive decision process in general. An interesting finding in all most all high-intensive land-use solutions is the increased importance of maize and rice, currently not grown very much in Bungoma, due to its high energy content and good suitability for the local soil conditions. The outcome of the process thus underscores the rice-planting projects that have been initiated by the DDC.

Table 2 Final solution in decision making experiment (Agrell et al. 1997)

Iteration 5	1-5	2-5	3-5	4-5	5-5	6-5	7-5	
Max FOOD	1 109	1 112	1 078	1 081	1 052	1 121	1 147	10^9 kcal
Max FOODL	880	894	858	851	824	889	924	10^9 kcal
Max SSR	99	99	100	100	100	98	97	%
max VALUE	1 088	1 080	1 076	1 083	1 085	1 121	1 123	10^6 KSh
min COST	40	39	39	40	40	43	44	10^6 KSh
max NETREV	1 048	1 041	1 038	1 043	1 045	1 078	1 078	10^6 KSh
min ARABLE	160 677	162 126	161 483	161 000	162 307	160 000	160 000	ha
min HARVEST	121 635	122 950	122 586	122 064	123 093	121 466	121 255	ha
min EROS	2 478	2 900	3 000	2 733	2 900	2 896	2 664	kton/ha
min EROSMX	35	35	35	35	30	40	39	ton/ha

10. Conclusion

Policy making in agri-economical land-use planning requires extensive analysis and communication with the stake holders. In countries where cultural and educational differences impede the interaction between the state and rural citizens, local voluntary committees may play an important role to discuss and anchor decisions of common land use. Computerised, user friendly tools may serve a useful role in this context. The ability to explicitly include analysis of several objectives, without incorporating the irrelevant preferences of the analyst, is fundamental requirement to be put on the model. Features that allow back-tracking and regret are also to be supported in a decision-support system in this application. The model presented in this paper, based on the Agro-Ecological Zones (AEZ) model (FAO 1981) and the Steuer and Choo (1983) method, applied to the Bungoma district in Kenya, fulfils these demands and demonstrates its benefits and limitations. The decision support is rich in detail for agro-economical land-use, including multi-cropping, livestock and crop sector integration, fuelwood production and soil erosion impact. Decision making experiments with the model show that the model refines recommendations that have been issued using other analyses.

References

AGRELL, P. J., STAM, A. AND FISCHER, G. W. 1997. Interactive Multi-Objective Agro-Ecological Land Use Planning: The Bungoma Region in Kenya. Working Paper WP-246, Department of Production Economics, Linköping Institute of Technology. Submitted to European Journal of Operational Research.

BUCHANAN, J. T. AND DAELLENBACH, H. G. 1987. A Comparative Evaluation of Interactive Solution Methods for Multiple Objective Decision Models. European Journal of Operational Research 29:353-359.

FAO 1976. A Framework for Land Evaluation. Soils Bulletin Vol. 32, Rome.

FAO 1981. Report on the Agro-Ecological Zones Project (1978-1981), Vol. I: Methodology and results for Africa World Soil Resources Reports 48/1, AGL-FAO, Rome.

FAO 1984. Population Supporting Capacity Assessment of Kenya. Mission Report, AGL-FAO, Rome.

FAO 1995. Planning for sustainable use of land resources. Towards a new approach. FAO Land and Water Bulletin 2, Land and Water Development Division, FAO, Rome.

FAO/IIASA 1991. Agro-Ecological Land Resources Assessment for Agricultural Development Planning; A Case Study of Kenya: Resource Database and Land Productivity. Main Report and 8 Technical Annexes, AGL-FAO, Rome.

FAO/IIASA 1993a. Guidelines for Land Use Planning. FAO Development Series 1, FAO, Rome.

FAO/IIASA 1993b. Agro-Ecological Assessments for National Planning: The Example of Kenya. FAO Soils Bulletin 67, Land and Water Development Division, FAO, Rome.

FISCHER, G. W. 1991. Agro-Ecological Land Resources Assessment for Agricultural Development Planning; A Case Study of Kenya. Technical Annex 7, World Soil Resources Reports 71/7, FAO/IIASA, Rome.

FISCHER, G. W. AND ANTOINE, J. 1994. Agro-Ecological Land Resources Assessment for Agricultural Development Planning; A Case Study of Kenya: Making Land Use Choices for District Planning. World Soil Resources Reports 71/9, FAO/IIASA, Rome.

INTERNATIONAL MONETARY FUND 1995. Kenya - Recent Economic Developments. IMF Staff Country Report No. 95/133, IMF, New York.

ONYEJI, S.C., G.W., FISCHER AND KAMAU, W. 1996. Agro-ecological assessment for national planning in Kenya: Database Structure for District Analysis. Working Paper WP 96-73. International Institute for Applied Systems Analysis, Laxenburg, Austria.

REPUBLIC OF KENYA 1986. Economic Survey 1986. Central Bureau of Statistics, Ministry of Planning and National Development, Kenya.

REPUBLIC OF KENYA 1994a. On Recovery and Sustainable Development to the Year 2010, Sessional Paper No. 1 of 1994.

REPUBLIC OF KENYA 1994b. On National Food Policy, Sessional Paper No 2 of 1994.

SHIN, W. S. AND RAVINDRAN, A. 1991. Interactive Multiple Objective Optimization: Survey I - Continu-ous Case. Computers & Operations Research 18:97-114.

STEUER, R. E. AND CHOO, E.U. 1983. An Interactive Weighted Tchebycheff Procedure for Multiple Objective Programming. Mathematical Programming 26:326-344.

UN 1997. World Population Prospects. The 1996 Revision.. UN, Population Division. New York.

Public Participation in Forest Management Planning Supported by Multi-Criteria Decision-Making Techniques

Vilis Brukas
Nico Hjortsø
Peter Tarp
Tove E. Boon

Abstract

Traditionally, commercial forests in Lithuania are managed to maximise timber output. In this paper we suggest a methodology for integrating economic efficiency and user group preferences in multiple-criteria forest management decision-making. The methodology includes i) a heuristic decision-making approach, ii) the use of an optimisation model – PEB - comprising political, economic, and biological aspects of forest management, and, iii) the use of a geographical information system (GIS) to facilitate communication between forest user groups and administration. The methodology is illustrated on multi-criteria short-term management problem in the central part of Lithuania.

Keywords: Public participation, forest management planning, Multi-Criteria Decision-Making

1. Introduction

State forests still comprise more than 80 per cent of the total forest area in Lithuania. They are managed according to laws and regulations, not leaving much space for economically substantiated decisions of forest managers. The aspiration to appropriately adopt market principles and to meet an increasing need for integration of material and immaterial values of forest creates a need to look for innovative methods in forest management planning (FMP) and decision-making. This study outlines a methodology for integrating public participation with multi-criteria decision-making techniques. An appropriate structure for interaction between forest planners and public user interests is recognised to be highly cultural dependent. In this study we present a heuristic approach for participation. This approach should not be seen as a final solution but as a point of departure from where local considerations can shape the process.

2. Features of the PEB Model

The optimisation model consists of two main sub-models. In the first sub-model, biological and economic technology parameters of the forest holding are expressed through linear and non-linear functions. The yield model of Kuliesis (1993) provides input data to express age-height, height-diameter, height-standing volume, and height-volume after thinning relationships, required for the PEB model (Tarp 1998). Assortment tables (Kuliesis et al. 1997) and log prices, announced by the Forest Market Economies Centre served to estimate the dependence of price (Lt/m^3) on the mean diameter of breast height. Growth, yield, and price curves have been estimated for the eight most common tree species in Lithuania. Based on the present information about the stand parameters, future parameters are estimated with the use of the following functional relationships:

$$S_{t+n} = S_t + dP_{t \to t+n}$$

where

S_t state now

S_{t+n} state after n years

$dP_{t \to t+n}$ increment of parameter P from time t to time t+n

Using actual characteristics of the stand as initial input data, contrary to general yield tables, decreases uncertainty with respect to the state of a parameter now and makes the estimation of the future state more reliable.

The first sub-model generates the input data for the second sub-model which is a linear programming (LP) solving algorithm. The LP software LINDO is used. The LINDO format comprises the elements shown below:

Max OF

Subject to

C1) ADV11+ADV12+...+ADV1j+...+ADV1n=area1
...
Ci) TPi1ADVi1+TPi2ADVi2+...+TPijADVij+...+TPinADV-GDVi=0
...
Cm) TPm1ADVm1+TPm2ADVm2+...+TPmjADVmj+...+TPmnADV-GDVm=0
End

where

OF objective function
Ci constraint i
ADV area decision variable
areaN area of stand N
TP technology parameter
GDV global decision variable

The technology parameter values are measures of a resource use or production output per unit/ha obtained by allocation of the resource denoted by the decision variable name. They may represent stand factors such as thinning volume, clear felling volume, regeneration establishment costs, expectation value, etc. The LP output consists of the optimal objective function value and values of decision variables.

Solution of the initial model, containing area constraints, shows the economic potential of the forest.During the planning process various management constraints are incorporated to narrow the gap between the model and reality. The duality properties of LP formulation allow for sensitivity and parametric analyses. Goal programming (GP) and multi-objective linear programming (MOLP) can also be included to consider forest management planning issues in a multi-criteria context.

3. PEB and the Problem Environment

3.1 Silvicultural regimes

The prescribed high rotation ages and large areas assigned to non-commercial forests reflect bio-ecological orientation in Lithuanian forestry. Severe conceptual issues arise during the transition stage in Lithuania when economic criteria, such as the soil expectation value, are incorporated in forest management planning. The PEB model facilitates the evaluation of objectives, at least in short-term optimisation. According to the Forest law (Ministry of For… 1994), the minimum rotation ages in commercial forests range from 35 years in white alder stands to 125 years in oak stands. In the PEB model, inclusion of constraints on minimal rotation ages permits to elicit their impact on profit maximisation.

Timing and intensity of thinnings heavily depend on tree species composition and site conditions. According to the usual forestry practice, rather intensive thinning regimes are applied in young (up to 20 years old) mixed stands, while older stands are managed less intensively. The PEB model allows to control the age of the first and the last thinning, and the frequency and intensity of thinnings. It is possible to transform the PEB from model I type to model II type so that the tree species composition is taken into consideration (Johnson and Scheuerman 1977). Final felling is assumed to be followed by regeneration of the same species in the present version of the PEB model and the integrity of the stand is preserved (model I).

3.2 Forest outputs

One of the most attractive features of the PEB model is its capability to account for both variety in forest functions and outputs, and diversity in preferences of forest managers and users. A certain level of quantitative and qualitative information about various forest outputs is needed to obtain applicable solutions.

Material forest outputs in Lithuania are dominated by timber production which accounts for about 87 per cent of the market value of the total material production (Rutkauskas 1997). Mushrooms, berries, medicinal herbs, and venison contribute with 8, 2, 0.1 and 1 per cent, respectively. Non-timber forest products do not normally add much to owner's revenue and thus do not have considerable impact on the strategy of the forest holdings. An exception could be game management. Nevertheless, the material value of venison and trophies accounts just for a small proportion of the game value associated with satisfaction and recreation, which is still not assessed in any units.

Among the immaterial forest outputs, recreational and protective functions of forests are considered to be the most important in Lithuania. Riepsas (1990) developed a method for recreational value assessment of stands. The stand suitability value for recreation is found on a scale between 0 and 100. The estimated recreational value varies according to tree species, site class, maturity group, and stocking index. The recreational value is also adjusted according to the presence of historical or cultural attractions, distance from water bodies, urban areas, and environmental pollution level. PEB enables an evaluation of trade-offs between the monetary value of timber and the recreational value, expressed in non-monetary units. A similar approach can be applied for evaluation of protective properties of forests, but in this case spatial parameters such as proximity to water bodies and cities are of greatest importance. Such elements may be analysed by use of a GIS linked to the stand list applied in the PEB model.

3.3. The case area

For illustration purposes the Raseliru Forest, covering forest blocks from 140 to 144 in Vaisvydava Forest District is chosen. Vaisvydava Forest District lies close to the second largest city Kaunas in the central part of Lithuania. The area of 306 ha is divided into 124 compartments, of which 109 are forest stands. The eight most common tree species in Lithuania are found in the Raseliu Forest.

During the forest inventory performed every 10 - 15 years in all forests, exhaustive data about various characteristics of stands are obtained. The PEB model uses the stand list data as input. Spruce covers more than half of the total area. A rather large share of forest land is occupied by ash and black alder, which are indicators of rich and wet soils. The high diversity in site fertility and humidity in the case area is representative for Lithuanian forests.

Middle-aged and immature stands prevail in Raseliu Forest which is typical for most forests in Lithuania. Mature stands cover 12 per cent of the area. Mature spruce, aspen, black alder, and white alder comprise 13.9, 11.0, 9.1, and 1.8 ha, respectively (Forest Man... 1997). Even flow of timber could be incorporated to improve the uneven age structure, depending on the scale of ownership.

4. Results

We will illustrate how the PEB model can be used in relation to short-term forest management planning emphasising conflict management and public participation. The planning task is concerned with a five years management plan for the Raseliu Forest. Three groups of stakeholders are included in the planning process: the local inhabitants (forest users), the local forest ranger, and the regional forest planning and policy agency. Local inhabitants use the forest for recreational purposes such as hiking and gathering of mushrooms and berries. The ranger is in charge of forest operations and emphasises a rational and efficient execution of planned activities. Regional planning and policy-making staffs want to secure a long-term profitable and sustainable forest as well as a forest serving local inhabitants demands. The profitability objective may be represented by implementation of economic optimal rotation ages. Economic optimal rotations are in general 10 to 30 years shorter than the rotation ages presently practised, using a discount rate at 2-3 per cent. Clear felling followed by natural re-growth or planting is the predominant management regime. Clear fellings should be of an efficient size due to effects of economy of scale. But large clear felled areas and clear fellings close to roads are, from an aesthetic point of view, not well received by local forest users. Non-timber outputs, such as berries and mushrooms are generally linked to mature and over-mature forest stands. Therefore, clear felling and silvicultural activities can reduce the potential output of such non-timber products. An overview of the stakeholders, their objectives, and management implications is given in Table 1.

Table 1 Stakeholders, their objectives, and implications on forest management

Stakeholder/ interest group	Objectives	Management consequences
Regional forest management authorities	Long-term profitability Minimising conflicts with other stakeholders Sustainability	Economic optimal rotations Public participation Harvest volume restrictions
Forest rangers	Stable workforce Motivated workforce	Even workload over time Efficient management and operations
Local forest users	Aesthetic experience of unspoiled nature Berries and mushrooms output	No clear felling along roads Mature and over-mature forests

We suggest a heuristic approach to the participatory planning task. This approach consists of six phases.

- Phase 1: objective analysis and agreement on a base problem formulation
- Phase 2: investigation of the problem to provide a common base of knowledge
- Phase 3: generation of a benchmark solution
- Phase 4: stakeholder dialogue, problem solving negotiation, and evaluation
- Phase 5: incorporation of stakeholder feedback into the management plan
- Phase 6: implementing the management plan

4.1 First phase

The first phase includes identification of stakeholders, their objectives, and suitable indicators for these objectives. A model type able to include the objectives is chosen. In this case a quantitative measure of the indicators exists and a quantitative technique, LP, is chosen as the base model. Different objectives may be recognised either through constraints on the problem formulation or through integration in an objective function. In this study, it is assumed that the stakeholders have agreed on incorporating the following objectives in the planning process:

- Forest profitability
- Sustainability of timber outputs
- Stable and motivating work conditions for forest workers
- Aesthetic value
- Local utilisation of berries and mushrooms

Management objectives concerning working conditions and sustainable timber output are addressed through constraints on the model. Sustainable timber output is regulated through a set of harvest volume restrictions. The harvest in any year is not allowed to fall below 90 per cent or exceed 110 per cent of the estimated increment that year. The working conditions are included through a set of restrictions, smoothing employment and harvest volume to maximum deviation of +/- 10 per cent from one year to another.

Forest profitability, recreational value, and non-timber outputs are considered explicit planning objectives. The NPV is used as an indicator of profitability, recreational value (the method by Riepsas (1990)) is used to indicate forest user utility, and the total annual yield (measured by weight) is used to indicate the utility associated with gathering mushrooms and berries.

4.2 Second phase

The second phase of the planning process includes an investigation of the decision problem and aims to provide a common base of knowledge about the decision problem. Multi-criteria decision-making (MCDM) methods can be applied in outlining the objective space of the management problem. So-called generating MCDM techniques can provide an overview of the range in which feasible solutions to the planning problem exists. The Constraint Method (Cohon 1978) based on parametric programming of the RHS is used to produce an approximation of the non-inferior (non-dominated or efficient) solution space. This approximation is shown in Figure 1.

A trade-off matrix is another method to indicate the range of the objective space. The trade-off matrix in Table 2 results from the maximisation of each of the three explicitly formulated objectives. The bold faced values are the maximum values of the objectives, respectively. Table 2 indicates to the decision-maker the range of performance of each of the three objectives, as well as the magnitude of the trade-offs involved.

Table 2 Trade-off matrix of the Raseliu Forest short-term management planning problem. Each row represents an optimal solution maximising the objectives listed in the first column

Maximising	NPV, Litas	Berries and mushrooms, kg per year	Recreational value, Points
NPV	**5,451,208**	2,966	6,940
Berries and mushrooms	5,132,136	**3,407**	7,646
Recreational value	5,011,522	3,310	**8,093**

The trade-off table and the objective space approximation provide a common reference and may facilitate mutual understanding among the different stakeholders in relation to evaluating the solutions to the planning problem.

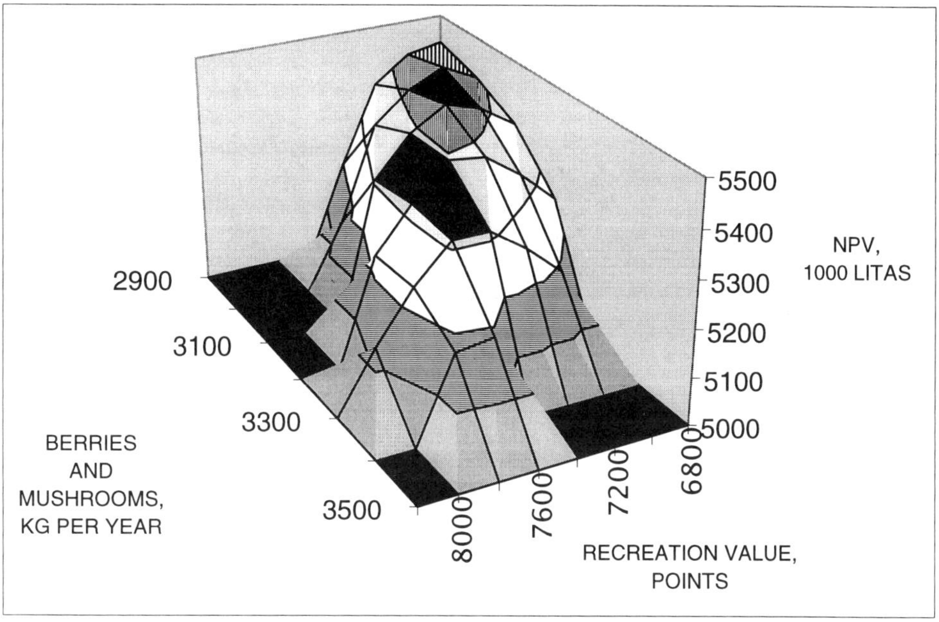

Figure 1 An approximation of the non-inferior solution set of the Raseliu Forest short-term management planning problem

In this study a direct link between objective and decision space exists since the Raseliu Forest is available on the GIS ArcView. Using the stand-based PEB model in combination with ArcView allows the planners to visualise to stakeholders the location and timing of forest management activities. This ability is used in Figure 2 to show the result of a weighted sum optimisation of the three objectives.

The solution in Figure 2 represents the situation where the value of the three objectives is made of the same magnitude (normalised) and objectives are then assigned equal weight in a mathematical sense. The resulting solution indicates the situation where the marginal numeric trade-off between the objectives is approximately of the same size. The objective space location of the solution is seen in Figure 5.

4.3 Third phase

When stakeholders have come to an agreement on a model type and initial problem formulation, a benchmark plan is generated. In the Raseliu Forest planning problem such a benchmark plan is achieved by incorporating the above mentioned sustainability restrictions concerning timber outputs and employment. The objective function values are: NPV: 5,451,280 litas; berries and mushrooms: 2,966 kilograms per year; recreational value: 6,940 points. The result in decision space is shown in Figure 3.

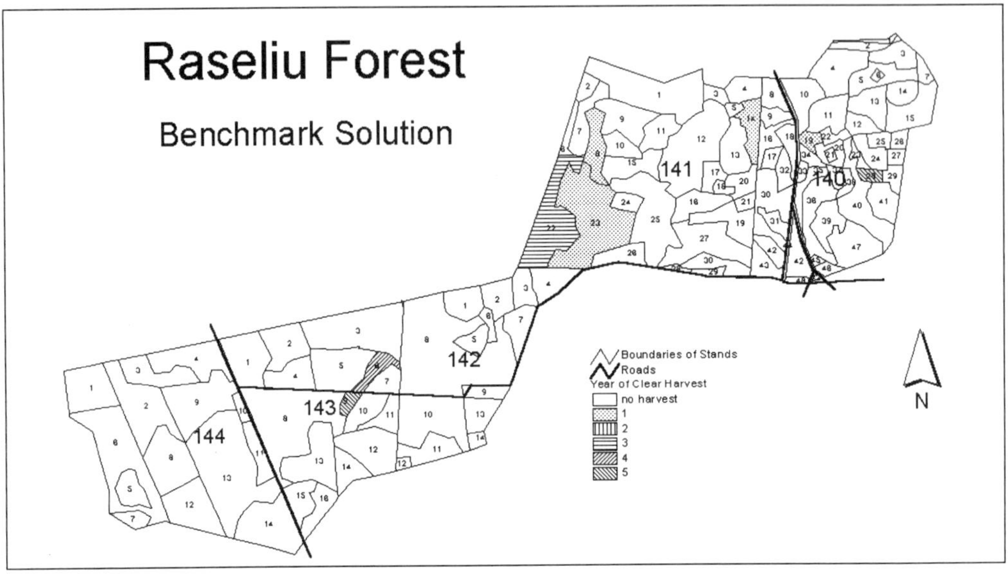

Figure 2 The map shows the clear felling areas during the five year planning period when using an equally weighting of the three objectives: NPV, berries and mushrooms yield, and recreational value

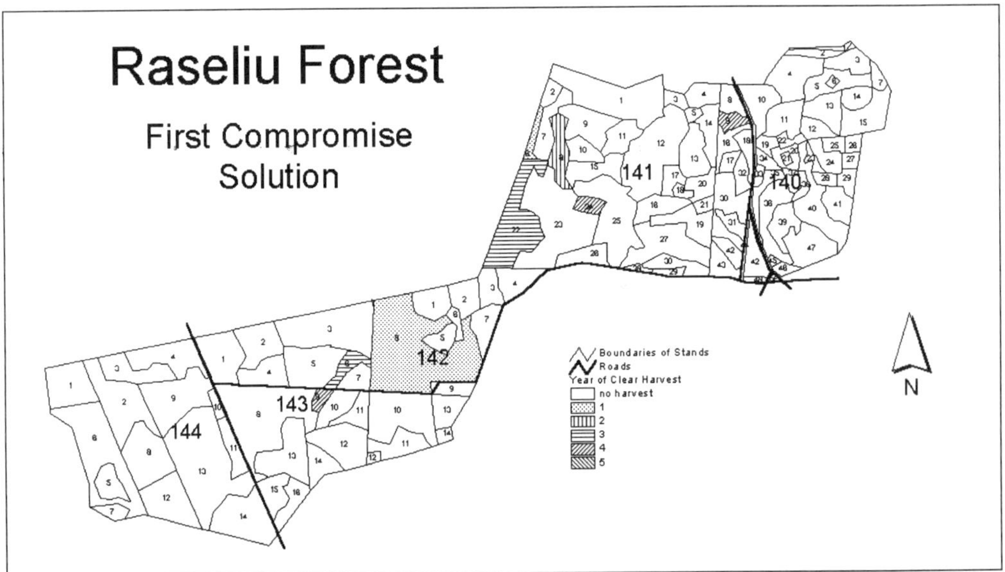

Figure 3 Spatial distribution of short-term (year 1-5) clear felling associated with the benchmark solution

The location of the benchmark solution in objective space is indicated in Figure 4. Sensitivity analysis on model parameters and changes in objective space may contribute to the general acceptance of using a certain model type as well as confidence in the results produced.

4.4 Fourth and fifth phase

The dialogue phase comprises presentation of the solution to stakeholders in the form of:

- forest maps representing decision space (such as Figures 2, 3, 5)
- graphical illustration of objective space (such as Figures 1, 4)

The solution space of the benchmark plan (Figure 3) and the related location in objective space (Figure 4) constitute an initial plan suggestion subject to further dialogue. The solution is discussed and evaluated and feedback from stakeholder groups is considered in phase 5, which forms the basis for generation of one or more revised management plans based on stakeholder input. An essential aspect of the proposed method is the ability of inter-stakeholder communication based on physical management alternatives (in decision space) as well as stakeholder overall preference level (in objective space) as input in the dialogue.

Figure 4 Approximation of the objective space of the benchmark plan and the location of the different management plan suggestions produced during the stakeholder dialogue, problem solving, and evaluation phase. Circles show the three solutions found in the trade-off Table 2

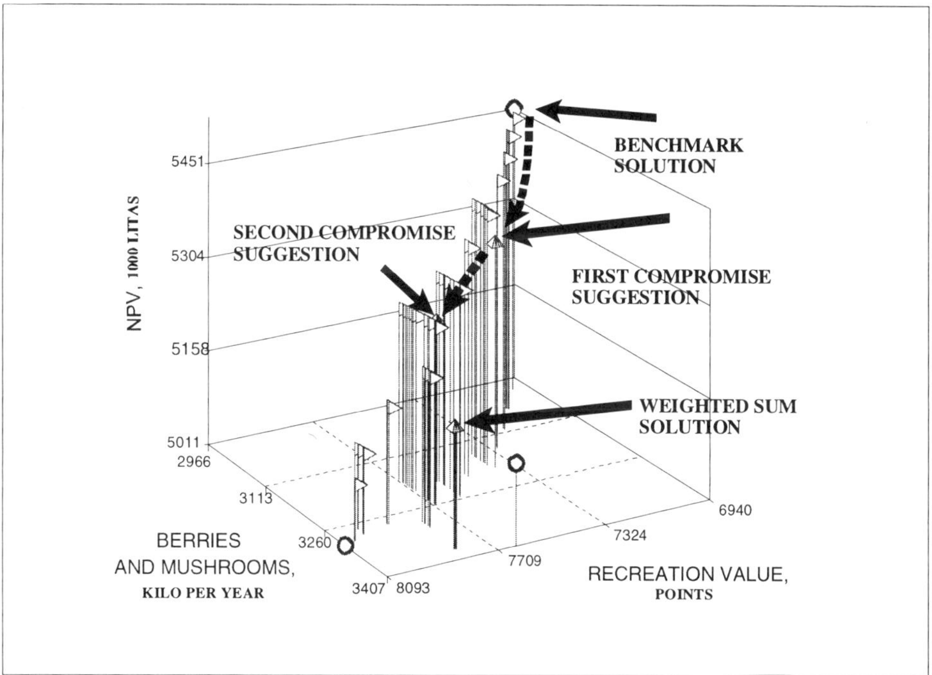

The main concern of the forest user group is related to the aesthetic values of the forest. The user group suggests a plan that avoids clear fellings adjacent to main roads. The suggestion includes a ban on clear felling in stands 140/09, 142/08, and 143/06,09. Moreover, an increase in berry and mushroom yield to 3,200 kilogram per year is found to be reasonable. A new solution incorporating these goals has the objective values: NPV: 5,361,779 litas; berries and mushrooms: 3,200 kilograms per year; recreational value 7,390 points. The new plan is then evaluated and the solution is not satisfactory from a forest management point of view, since the next years seem to offer good market opportunities for timber such as that from stand 142/08. A compromise suggestion could be to refrain from clear felling more than 50 per cent of the area in 142/08 and in the same time refrain from felling more than 50 per cent of area in 142/22. The last solution had two stands adjacent to the road. These stands are also excluded from the clear felling area. The new plan is shown in Figure 5. This solution may be accepted as the best compromise solution. If not, the process has to continue until a sufficient number of re-considerations of phases 4 and 5 results in an acceptable solution. The possibility that no agreement is reached and the process is blocked by one of the parties also exists. In this situation the process may have to be re-stated and another way may be found to reach a solution. The suggested procedure ends with the implementation of a best compromise solution approved by all stakeholders (phase 6).

An important feature of the proposed planning approach is that the stakeholders are explicitly contributing to a contentious process of designing an acceptable problem formulation that captures the multidimensional goals of the stakeholders. The learning aspect of this process should be emphasised as a source to greater mutual understanding. Phases 4 and 5 may be supported by the use of 'hard systems' models such as GP, MOLP, or LP. Each of the methods may provide a candidate solution or at least contribute to increase insight into the objective space combined with decision space information. Hereby, a compromise solution may be properly identified. At the same time it is recognised that these methods are not easy to understand for laymen, which puts a significant pedagogical load on the planning staff that wants to integrate such methods in participatory planning.

The use of indicators in multiple-use forest management planning is illustrated above and it is shown how they may function as important communication elements in MCDM forestry applications. The selection of indicators should be made to represent all important decision variables so that discussions about redundant decision variables/goals can be avoided and so that the degree of conflict in the final solution can be minimised through appropriate weighting of different criteria, including spatial ones.

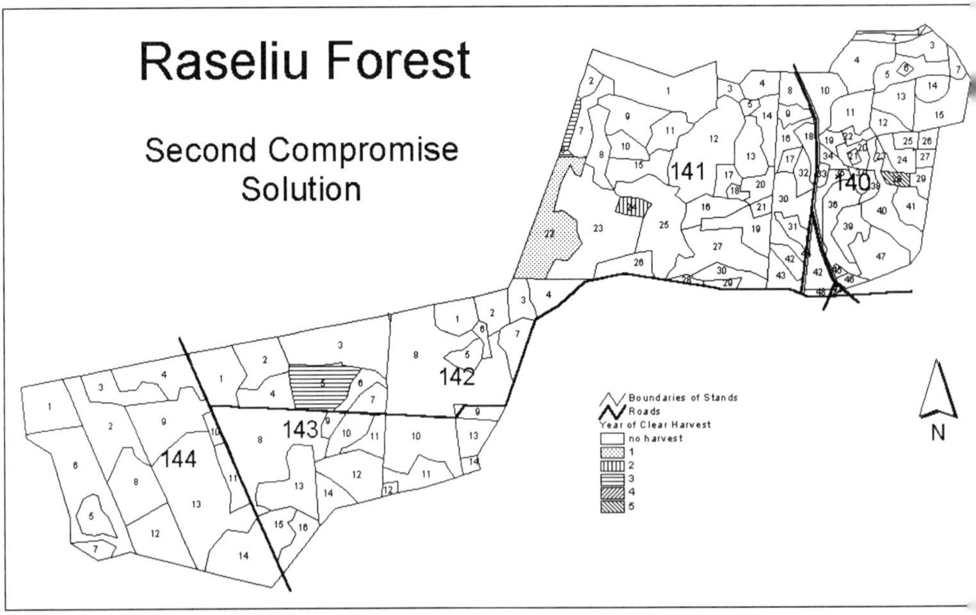

Figure 5 Second compromise solution to the short-term management plan for Raseliu Forest

5. Discussion and Conclusion

The present paper has focused on multi-criteria model development and stakeholder communication. The PEB model provides its strength in combination with participation of stakeholders. The advantage of the PEB model used as a MCDM model is that it reduces complex decision processes into sets of decision alternatives by means of transforming stakeholders' interests and values into sets of criteria and indicators, which are subsequently incorporated into a mathematical programming model. This may help to clarify the differing values and their interdependencies (trade-offs between them).

However, the decision process does not become more objective by transforming values into indicators/variable in an LP model. The output of the model depends on the choice of indicators and the weights given to them. This implies that formulating the model should take its point of departure in the actual planning case. Stakeholders should be identified and participate in the very design of the model formulation (including criteria and indicators), as well as the subsequent solution generation (weighting of indicators), and evaluation of decision alternatives.

5.1 Identifying interests and values

In identifying the indicators to be included in the model one encounters various challenges.

(1) Who are the stakeholders?

In the present paper, we have assumed only three groups of stakeholders, each with homogenous sets of interests. This most likely does not hold true in reality. Interview with key informants may reveal a range of stakeholders with differing and conflicting interests. For instance, the local users may also have an interest in maintaining job opportunities as well as recreational facilities, whereas the forest ranger may be a berry picker on Sunday's. However, once the model includes an interest in terms of an indicator it is irrelevant what stakeholder it represents, as long as the decision alternatives are considered by the stakeholders themselves.

(2) What interests and values do stakeholders bear?

In the present PEB model we have assumed the following interests to be present: profitability (NPV), recreation value (survey by Riepsas 1990), availability of berries and mushrooms (kg/ha), stable employment opportunities and aesthetics (ban towards clear felling along roads). However, more values are likely to be present.

Interview and questionnaires to the identified stakeholders can reveal what interests and values they bear. Besides the logic of asking stakeholders themselves what their stake in the case actually is, use of interviews would provide us with an opportunity to give voice to interests that in other situations (eg. debate among the various stakeholders) would not be acknowledged as a legitimate interest. This could be, e.g. forests having a symbolic value by its mere existence, or a site having a spiritual value for historical, traditional or religious reasons. Giving voice to formerly silent values may rise an ethical dilemma to the extent that the researcher, or planner, may alter power relationships by the inclusion of new, formerly unexpressed interests and values. That is what are the consequences of the choice of model and indicators? In the present example, it could be asked whether it is reasonable to design an optimisation model for multiple use in a so-called production forest, hereby legitimising claims for concerns to biodiversity, recreation etc., as

Lithuanian forestry has been based on a reserve thought, i.e. forests for soil and water protection, forests for maintaining biodiversity, forests for production etc.

5.2 Transforming values into indicators

Assuming that we know the interests and values held by the various stakeholders, we are still left with the challenge of transforming these values into measurable indicators.

A set of criteria for choosing indicators can be identified (Boon 1994):

- Relevance to policy and practice
- Precise description
- Precise to measure at
- Easy to interpret
- Economically, temporally, and technically feasible
- Comparable in time and space

In the present model we have considered recreation in terms of a countrywide survey (Riepsas 1990). Results from this survey have been transformed into a linear function, based on the mean recreational value of a given stand type. In reality, people's preferences differ, and only few may actually have preferences identical to the mean value. In fact, it seems plausible that what people would prefer is variation - and optimising recreational value according to a linear function based on mean values will have the opposite effect. The PEB model structure, on the other hand, allows for incorporation of structural restrictions suitable for modelling and managing species and age class variation as an indication of diversity.

5.3 Using the PEB model in the decision-making process

Using the PEB model in a decision process requires that people are equally able to understand the PEB model and negotiate on the basis of this. This is unlikely to hold true. On the other hand, the advantage of using the PEB model in a participatory decision process, as compared to a common debate simultaneously among all stakeholders, is, that it allows for a step-wise decision process within a common conceptual framework. One can make strategically composed smaller groups of stakeholders known to hold more similar views on interests related to forestry, allowing for a more open debate of the decision alternatives provided by the PEB model than in a group with more extreme viewpoints represented. Results from the different groups can subsequently be held up against each other in new groups of stakeholders.

6. Extensions of the PEB Model

The PEB model is designed to support foresters in the decision-making process. This is an ambitious task since forest management comprises an extremely great number of various factors, including complicated relationships between numerous components of forest ecosystems, and related economic factors as well as intra- and interpersonal conflicts of decision-makers. The result presented here could be improved by consideration of the following research areas with respect to forest management in Lithuanian forestry:

- Etimation of more precise relationships between non-timber outputs and stand and management characteristics
- Development and design of communication schemes suitable for participatory design and

implementation of the PEB model in forest management planning

- Timber yield dependence on sites conditions
- Modelling of different thinning regimes suitable for mixed stands
- Improving the input data for immaterial forest outputs
- Inclusion of spatial parameters by use of GIS
- Integration of short-term and long-term planning

Neither PEB, nor any other computer based optimisation programme can represent every detail of the reality. Experience, creativity, and intuition are important in forestry practice. Planning models can contribute to increased insight into the most sensitive and important goals of planning and hereby function as a valuable supporting basis for decision-making in forest management planning.

References

BOON, T. E. 1995. Indikatorer på bæredygtig skovdrift [Indicators of sustainable forest management]. The Royal Veterinary and Agricultural University, Copenhagen.

COHON, J. L. 1978. Multiobjective programming and planning. Academic Press Inc. New York.

FOREST MARKET ECONOMICS CENTRE 1997. MEC Naujienos. Vilnius. Centre of Forest Economics Information Bulletin 2(7):1-16.

FOREST MANAGEMENT INSTITUTE 1997. Dubravos misku uredijos taksaciniai duomenys. [Taxation data of Dubrava forest enterprise.] Forest Management Institute. Kaunas.

JOHNSON, K. N. AND SCHEUERMAN H. L. 1977. Techniques for prescribing optimal timber harvest and investment under different objectives - discussion and synthesis. Forest Science Monograph 18.

KULIESIS, A. 1993. Lietuvos medynu prieaugio ir jo panaudojimo normatyvai [Forest yield models and tables in Lithuania]. Kaunas.

KULIESIS, A., RUTKAUSKAS, A., PETRAUSKAS E., TEBERA A. AND VENCKUS A. 1997. Medienos turio lenteles. [Tables of timber volume]. Valstybinis miskotvarkos institutas. Kaunas.

MINISTRY OF FORESTRY OF REPUBLIC OF LITHUANIA. 1994. Forest law. Vilnius.

RIEPSAS, E. 1990. Rekreaciniu misku ivertinimas ir naudojimas. Metodines rekomendacijos. [Evaluation and use of recreational forests. Methodological recommendations]. Ministry of Agriculture of Lithuanian SSR. Vilnius.

RUTKAUSKAS, A. 1997. Non wood resources and utilisation in Lithuanian forests. Forest Management Institute. Kaunas. (manuscript)

TARP, P. 1998. Short-term optimisation of forest management planning based on stand data. Lecture notes. The Royal Veterinary and Agricultural University, Unit of Forestry, Copenhagen.

Chapter IV

Diversity Objectives and Option Value

Managing Forests for Tree and Landscape Diversity Part I: Predictions with Markov-chain Models

Ching-Rong Lin and Joseph Buongiorno

Abstract

A Markov-chain model was developed to predict the development of forest stands and landscapes. Applied to the Wisconsin maple-birch forests, it incorporated stochastic biological growth and prices. Transition probabilities between stand states were obtained with a non-linear matrix growth model, by Monte-Carlo simulation. Shannon's index measured tree and species diversity in each of 64 possible stand states. The frequency distribution of stand states throughout the maple-birch forest type measured the landscape diversity at different stages of forest development. Steady-state results showed that stand state 28 would dominate the landscape of an unmanaged forest, with or without catastrophic disturbances. Akin to old growth, state 28 had a high tree size diversity, high basal area in saw timber of shade tolerant and less tolerant species, and an expected residence time of 148 years. The landscape diversity of unmanaged forests increased as the period between natural catastrophes decreased from 1200 to 100 years, but then declined again with a period of 50 years. Continuing the current management regime, defined by the data of the last two forest inventories, would lead to a steady-state maple-birch forest with twice the landscape diversity of an unmanaged forest, but the proportion of old growth stands would be 4/5 lower, size diversity of trees 1/3 lower and species diversity slightly lower. The current management produced a low income, about $4/year on the average acre in the steady state, and an average current land expectation value of $120/acre. Part II deals with improving economic returns without ruining diversity.

Keywords: Forest management, ecology, Markov chain, simulation, risk

1. Introduction

Wisconsin has over fourteen million acres of commercial forest lands. About 27% of it (4 million acres) is the maple-birch forest type of northern hardwoods (Smith 1986). Traditionally, economic concerns have tended to dominate the management of industrial forests, because of the steady demand for wood and fiber products, but other private forests are owned for many non-financial reasons (Bliss and Martin 1988). Furthermore, as the ecological value of forest land is increasingly recognised and understood by ecologists and even by the public, biological diversity, generally defined as the variety of life forms, is of growing interest.

Since trees in a forest provide food, shelter, and substrate for other life forms (often a diverse and abundant entirety), maintaining a rich variety of both tree species and canopy structure helps to conserve biodiversity in a forest ecosystem. Stands with a high diversity of tree species have greater resistance to climatic threats or biological risk, while a high complexity in canopy structure can provide preferable habitat for wildlife (Pimental 1961, Hunter 1990, Norse 1990, Burton et al. 1992).

Some recent studies on conserving tree diversity have used deterministic models, at stand level (Buongiorno et al. 1994, Buongiorno et al. 1995). Yet, biological and economic processes are in large part stochastic. Solving forest management problems in an uncertain environment, with economic and ecological criteria, is the main thrust of this study.

Moreover, this paper expands the analysis to a landscape level, because the presence of stands in diverse structures and patterns across the landscape can enhance aesthetics and habitats for different plants and animals. After a major disturbance, a forest will regrow through a succession of stages, stand initiation, stem-exclusion, understory reinitiation, and old growth. As emphasised by Hansen et al. (1991) and Oliver (1992), each stage provides critical habitats for some plants and animals but not others, and paying attention only to old growth is misdirected for the sake of biodiversity. Oliver (1992) suggested that a target distribution of stands across the landscape can be achieved through silvicultural operations. This study will, in part, propose methods to predict the distribution of stand states throughout a large landscape, under different management, and in a stochastic environment.

A stationary Markov model (discrete time) is the main tool we use to study the stochastic process of development in forest stands. When decisions are introduced it becomes a Markov decision process model (MDP), in which a matrix describes the transition probabilities between stand states, conditional on a decision. This type of model has been used in the past to study the management of forest stands under uncertainty (Hool 1966, Lembersky and Johnson 1975, Lembersky 1976, Kaya and Buongiorno 1987, Lu 1992). It can handle simultaneously the uncertain nature of stand growth and markets. However, all past studies concentrated on the economic criteria. As biodiversity is an increasing concern of forest management, ecological criteria must and can be incorporated in the MDP model. Furthermore, it will be seen that the Markov-chain framework provides a natural way to describe the landscape diversity of disturbed or undisturbed forests, based on the proportional abundance of stands in different states.

Part I of this study describes a Markov-chain model and applies it to predict the development and economics of northern hardwood forests in Wisconsin. Part II moves to the stage of optimisation, with Markov decision process models. Part I starts by defining stand-market states and harvesting decisions for uneven-aged maple-birch forest stands in Wisconsin. The determinants of transitions between stand-market states are described, together with the technique to get the transition probabilities between stand states by stochastic simulation. The expected economic and ecological rewards are then defined and estimated. Next, the Markov-chain model is applied to explore the steady-state distributions of stand states, and landscape diversity, in forests with and without catastrophic natural disturbances. Last, the effects of the current management on economic returns, and diversity of trees and landscape are investigated.

2. Stand States and Transition Probabilities

In a Markov-chain model, a finite number of states defines all possible conditions of the system. Here, the condition of a stand of trees at a point in time is called a stand state. A well-defined stand state is a good compromise between operational feasibility and sufficiency of description. For our purpose, stand states were defined by the basal area of trees in each species-size class. Trees were classified into two species groups (shade tolerant and less tolerant: Preston, 1977, Lin 1995) and three size classes according to diameter at breast height (DBH): pole timber, 5-inch = DBH< 11-inch; small sawtimber, 11-inch = DBH< 15-inch; and large sawtimber, 15-inch = DBH.

This tree size classification is used commonly for inventory and management (Wisconsin Department of Natural Resources 1977). With only three size classes, a state is more easily recognised in the field, with a minimum of measurements. Yet, basal area levels for these size classes are practical measures of the stand diameter distribution (Leak and Gottsacker 1985). Here, two levels of basal area per acre were recognised, within each species-size class, leading to sixty-four possible stand states in total.

The data from 613 remeasured plots from the NC-FIA database (Hahn and Hansen 1985), representative of the maple-birch forest type in Wisconsin, were used to define stand states. A basal area per acre less than average was defined as 'low'; at least average was 'high'. For example, in Table 1, state 1 (LLL, LLL) had low basal area in all sizes and species, while state 57 (HHH, LLL) had high basal area in all sizes of the shade tolerant trees, but low basal area in all sizes of the less tolerant trees. The expected basal areas for each value of the state variables are in Table 2. For example, high pole basal area for shade tolerant trees means 48.6 ft^2/acre. For less-tolerant trees, it means 8.8 ft^2/acre in intermediate and 20.1 ft^2/acre in intolerant trees.

To employ the Markov-chain model, one must assume that the transition probabilities between stand states satisfy the stationary and Markovian properties: they depend only on the current state and on the decision, but are independent of time and of the path taken to get to the current state. Some past studies have argued that forest growth is not stationary and Markovian (Binkley 1980, Roberts and Hruska 1986, Johnson et al. 1991), but this depends largely on how detailed is the state definition. Here, the stationary and Markovian properties are verified by construction because the transition probabilities are obtained from a growth model in which the future stand state depends exclusively on the current state and is independent of time.

This approach was necessary because there were not enough plots in each state to compute directly the transition probabilities. Instead, they were derived by Monte-Carlo simulation. The core of the method is a stochastic stand growth model for maple-birch forests in Wisconsin (Lin et al. In press):

$$\mathbf{y}_{t+1} = \mathbf{G}_t(\mathbf{y}_t - \mathbf{h}_t) + \mathbf{c} + \mathbf{u}_t \qquad (1)$$

where t refers to a particular year, $\mathbf{y}_t$ is a column vector that denotes the number of live trees per unit area before harvest, by species and size, at time t. Similarly, the vector $\mathbf{h}_t$ denotes the harvest at time t. $\mathbf{G}_t$ is a matrix which contains the transition probabilities of trees between size classes, and the effect of trees of different species and size on ingrowth of different species. To express the effect of stand density on tree growth, the matrix $\mathbf{G}_t$ is a function of the stand residual basal area. The column vector $\mathbf{c}$ represents the stand-independent ingrowth. $\mathbf{u}_t$ is a vector of errors. One vector $\mathbf{u}_t$ was computed by applying the deterministic part of Equation (1) to each plot of the NC-FIA database. The prediction error on each plot gave an observation of $\mathbf{u}_t$. Analysis of these errors showed that they were unbiased (mean zero). Their variance-covariance matrix **?** showed that there was strong correlation in the errors of trees of adjacent size.

To obtain the transition probabilities between stand states over 5 years, five iterations of model (1) were applied to predict the sequent stand state $\mathbf{y}_{t+5}$. Each iteration extracted one random observation $\mathbf{u}_t$ from the multi-variate distribution of zero mean and variance-covariance **?** (Kaya and Buongiorno 1987), and added it to the deterministic part of Equation (1). Each plot in a stand state served as the initial condition ($\mathbf{y}_t$ - $\mathbf{h}_t$) of the simulation, with equal probability. Thus, choosing one plot and letting it grow for five years formed one replication. For each stand state, more than 20,000 replications were performed to get stationary transition probabilities. The relative frequencies of passages from the current state to another gave the desired transition probabilities.

Table 3 shows the 5-year transition probabilities between stand states. For example, if after harvest the stand was in state 1 (LLL, LLL), then the probability that it would still be in state 1 five years later was 0.80, while the probability that it would be in state 2 (LLL, LLH) was 0.02, and so on.

Table 1 Definition of possible states for maple-birch stands

State	Basal area					
	Shade tolerant			Less tolerant		
No.	Pole Timber	Small sawtimber	Large sawtimber	Pole timber	Small Sawtimber	Large Sawtimber
1	L	L	L	L	L	L
2	L	L	L	L	L	H
3	L	L	L	L	H	L
4	L	L	L	L	H	H
5	L	L	L	H	L	L
6	L	L	L	H	L	H
7	L	L	L	H	H	L
8	L	L	L	H	H	H
9	L	L	H	L	L	L
10	L	L	H	L	L	H
11	L	L	H	L	H	L
12	L	L	H	L	H	H
13	L	L	H	H	L	L
14	L	L	H	H	L	H
15	L	L	H	H	H	L
16	L	L	H	H	H	H
17	L	H	L	L	L	L
18	L	H	L	L	L	H
19	L	H	L	L	H	L
20	L	H	L	L	H	H
21	L	H	L	H	L	L
22	L	H	L	H	L	H
23	L	H	L	H	H	L
24	L	H	L	H	H	H
25	L	H	H	L	L	L
26	L	H	H	L	L	H
27	L	H	H	L	H	L
28	L	H	H	L	H	H
29	L	H	H	H	L	L
30	L	H	H	H	L	H
31	L	H	H	H	H	L
32	L	H	H	H	H	H
33	H	L	L	L	L	L
34	H	L	L	L	L	H
35	H	L	L	L	H	L
36	H	L	L	L	H	H
37	H	L	L	H	L	L
38	H	L	L	H	L	H
39	H	L	L	H	H	L
40	H	L	L	H	H	H
41	H	L	H	L	L	L
42	H	L	H	L	L	H
43	H	L	H	L	H	L
44	H	L	H	L	H	H
45	H	L	H	H	L	L
46	H	L	H	H	L	H
47	H	L	H	H	H	L
48	H	L	H	H	H	H
49	H	H	L	L	L	L
50	H	H	L	L	L	H
51	H	H	L	L	H	L
52	H	H	L	L	H	H
53	H	H	L	H	L	L
54	H	H	L	H	L	H
55	H	H	L	H	H	L
56	H	H	L	H	H	H
57	H	H	H	L	L	L
58	H	H	H	L	L	H
59	H	H	H	L	H	L
60	H	H	H	L	H	H
61	H	H	H	H	L	L
62	H	H	H	H	L	H
63	H	H	H	H	H	L
64	H	H	H	H	H	H

Note: Pole timber = 5 inches = DBH<11 inches, Small sawtimber = 11 inches = DBH<15 inches. Large sawtimber = 15 inches

Table 2 Expected basal area (ft²/acre) for each value of the state variable

		Species		
		Shade tolerant	Less tolerant	
Tree size	Basal area level		Intermediate	Shade intolerant
Pole timber	Low	9.6	1.8	1.8
	High	48.6	8.8	20.1
Small sawtimber	Low	2.0	0.6	0.6
	High	22.3	5.1	9.7
Large sawtimber	Low	1.3	0.4	0.3
	High	21.7	6.8	6.3

Note: Pole timber = 5 inches = DBH<11 inches
Small sawtimber = 11 inches = DBH<15 inches
Large sawtimber = 15 inches = DBH

3. Timber Price Changes

The stumpage prices of pulpwood and sawtimber in Wisconsin (Peterson, 1967-1990) served as the prices of pole and sawtimber trees. The real price of pulpwood changed little in the past, so that the price of pole trees was assumed constant, in real terms. However, the real prices of sawtimbers showed much fluctuation.

Still, the price of sugar maple (the most common species) was highly correlated with the price of the other species. Consequently, the price of sugar maple was used as the indicator of stumpage prices for sawtimber. The following equation described price changes in 5 years, from 1967 to 1990:

$$P_{t+5} - P_t = \underset{(25.19)}{104.49} - \underset{(0.27)}{1.16} P_t \qquad (2)$$
$$\bar{R}^2 = 0.48 \qquad DW = 1.92 \qquad n = 19$$

where P_t was the price in year t. The numbers in parentheses are the standard errors of the coefficients estimated with 19 observations. The coefficient of P_t was not significantly different from -1, which implied that the price in year t+5 was independent of the price in year t. The Durbin-Watson statistic (DW) showed no autocorrelation. Thus, the real price of stumpage sawtimber was essentially equal to the mean price \$91.61 per thousand board feet (Mbf, Scribner) plus a random term v_t. A chi-square test (χ^2 = 0.57, d.f. = 3) suggested that v_t could be assumed to be normally distributed with zero mean and constant variance.

The market state was then defined by the level (high or low) of the stumpage price of sugar maple sawtimber, relative to its mean. Since prices were not autocorrelated, the transition probabilities between market states were all 1/2. The expected unit prices by market state, for trees of different size and species, are shown in Table 4.

Table 3 Probabilities of transitions between states of maple-birch stands in Wisconsin in five years

Starting state	Ending state (probability)				
1	1(0.80)	2(0.02)	3(0.02)	5(0.07)	6(0.01)
	8(0.01)	9(0.01)	17(0.02)	33(0.04)	
2	2(0.67)	4(0.08)	6(0.08)	8(0.01)	10(0.07)
	18(0.01)	34(0.08)			
3	3(0.37)	4(0.29)	7(0.04)	8(0.05)	12(0.01)
	15(0.03)	19(0.06)	20(0.01)	27(0.01)	28(0.01)
	35(0.06)	36(0.03)	40(0.03)		
4	2(0.01)	4(0.67)	8(0.08)	12(0.16)	20(0.05)
	28(0.01)	40(0.01)	44(0.01)		
5	1(0.02)	4(0.01)	5(0.61)	6(0.03)	7(0.21)
	8(0.02)	13(0.01)	21(0.01)	23(0.01)	37(0.05)
	39(0.02)				
6	2(0.01)	6(0.70)	8(0.20)	14(0.01)	38(0.03)
	40(0.05)				
7	3(0.03)	7(0.59)	8(0.24)	23(0.02)	39(0.04)
	40(0.08)				
8	4(0.06)	8(0.79)	16(0.04)	24(0.03)	40(0.07)
	56(0.01)				
9	9(0.76)	10(0.12)	25(0.06)	26(0.01)	41(0.02)
	42(0.03)				
10	2(0.01)	10(0.66)	12(0.07)	26(0.19)	28(0.01)
	42(0.02)	46(0.04)			
11	9(0.02)	11(0.58)	15(0.08)	25(0.02)	27(0.24)
	31(0.06)				
12	12(0.77)	16(0.15)	28(0.01)	44(0.06)	60(0.01)
13	5(0.03)	7(0.01)	13(0.50)	14(0.01)	15(0.34)
	29(0.01)	30(0.10)			
14	6(0.03)	8(0.06)	14(0.55)	16(0.36)	
15	7(0.01)	12(0.10)	13(0.06)	15(0.69)	16(0.12)
	47(0.02)				
16	12(0.02)	16(0.97)	32(0.01)		
17	1(0.04)	9(0.06)	17(0.45)	18(0.09)	19(0.02)
	20(0.01)	21(0.01)	22(0.01)	25(0.25)	26(0.02)
	49(0.04)				
18	2(0.03)	10(0.01)	18(0.57)	20(0.01)	26(0.27)
	28(0.01)	34(0.02)	50(0.08)		
19	3(0.01)	11(0.02)	17(0.08)	19(0.15)	20(0.12)
	27(0.25)	28(0.25)	49(0.03)	59(0.09)	
20	18(0.02)	20(0.33)	26(0.01)	28(0.52)	32(0.02)
	36(0.01)	52(0.09)			
21	21(0.15)	23(0.05)	24(0.06)	27(0.08)	29(0.12)
	30(0.01)	31(0.35)	32(0.03)	56(0.05)	61(0.05)
	62(0.02)	64(0.03)			
22	6(0.10)	8(0.03)	14(0.01)	22(0.47)	24(0.10)
	30(0.03)	32(0.20)	54(0.06)		
23	7(0.08)	8(0.04)	23(0.41)	24(0.21)	31(0.11)
	32(0.15)				
24	8(0.01)	20(0.02)	24(0.60)	32(0.37)	
25	9(0.02)	25(0.76)	26(0.14)	27(0.03)	57(0.05)
26	10(0.01)	26(0.80)	28(0.15)	30(0.03)	58(0.01)
27	25(0.05)	27(0.58)	28(0.30)	31(0.07)	
28	26(0.03)	28(0.97)			
29	21(0.01)	29(0.73)	30(0.04)	31(0.04)	32(0.14)
	61(0.02)	62(0.01)	64(0.01)		
30	26(0.03)	28(0.06)	30(0.39)	32(0.52)	
31	27(0.09)	31(0.36)	32(0.53)	64(0.02)	
32	28(0.04)	32(0.96)			
33	1(0.03)	3(0.01)	33(0.61)	34(0.01)	35(0.04)
	37(0.04)	41(0.03)	49(0.19)	50(0.03)	57(0.01)
34	2(0.02)	4(0.03)	34(0.52)	36(0.05)	42(0.12)
	50(0.25)	58(0.01)			
35	35(0.51)	36(0.24)	39(0.02)	44(0.01)	49(0.02)
	51(0.11)	52(0.06)	60(0.03)		
36	8(0.01)	20(0.01)	34(0.01)	36(0.63)	40(0.08)
	44(0.05)	52(0.17)	56(0.04)		
37	7(0.01)	37(0.48)	38(0.02)	39(0.30)	40(0.04)
	49(0.01)	53(0.09)	54(0.03)	55(0.01)	56(0.01)
38	38(0.21)	40(0.16)	48(0.14)	52(0.14)	54(0.16)

	56(0.18)	64(0.01)			
39	7(0.02)	35(0.02)	36(0.01)	39(0.50)	40(0.13)
	43(0.01)	47(0.02)	51(0.03)	53(0.01)	55(0.12)
	56(0.10)	63(0.03)			
40	8(0.06)	20(0.01)	24(0.03)	40(0.75)	52(0.04)
	56(0.11)				
41	9(0.04)	25(0.06)	26(0.03)	28(0.01)	41(0.42)
	42(0.08)	45(0.04)	50(0.01)	57(0.23)	58(0.07)
	60(0.01)				
42	26(0.07)	42(0.57)	44(0.13)	58(0.09)	60(0.14)
43	43(0.05)	44(0.95)			
44	42(0.26)	44(0.38)	49(0.03)	58(0.14)	60(0.19)
45	39(0.03)	45(0.28)	47(0.36)	61(0.33)	
46	14(0.02)	38(0.09)	46(0.89)		
47	47(0.33)	48(0.34)	64(0.33)		
48	16(0.01)	32(0.01)	44(0.01)	48(0.70)	60(0.06)
	64(0.21)				
49	17(0.04)	25(0.01)	49(0.52)	50(0.06)	51(0.07)
	54(0.01)	57(0.24)	58(0.03)	59(0.02)	
50	18(0.02)	34(0.01)	50(0.62)	52(0.03)	58(0.28)
	60(0.04)				
51	49(0.01)	51(0.72)	59(0.27)		
52	20(0.01)	52(0.42)	60(0.57)		
53	21(0.09)	23(0.05)	31(0.01)	53(0.52)	54(0.03)
	55(0.17)	56(0.12)	63(0.01)		
54	50(0.02)	52(0.04)	54(0.36)	56(0.28)	62(0.20)
	64(0.10)				
55	24(0.03)	43(0.01)	51(0.01)	52(0.04)	55(0.32)
	56(0.35)	59(0.06)	63(0.15)	64(0.03)	
56	20(0.04)	24(0.09)	52(0.12)	56(0.66)	64(0.09)
57	25(0.04)	26(0.01)	57(0.79)	58(0.10)	59(0.05)
	60(0.01)				
58	26(0.04)	30(0.01)	58(0.80)	60(0.12)	62(0.03)
59	59(0.87)	60(0.09)	63(0.04)		
60	28(0.06)	58(0.06)	60(0.88)		
61	29(0.11)	30(0.03)	57(0.01)	59(0.06)	61(0.51)
	62(0.04)	63(0.24)			
62	30(0.03)	32(0.13)	61(0.01)	62(0.43)	64(0.40)
63	32(0.01)	47(0.02)	59(0.01)	60(0.05)	63(0.56)
	64(0.35)				
64	32(0.07)	60(0.05)	64(0.88)		

Note: State numbers are defined in Table 1.

Table 4 Expected price per unit by species, size, and market state, in 1990 dollars

			Market state:	
Species	Size	Unit	High	Low
Shade tolerant	Pole timber	Cord	5.4	5.4
	Small sawtimber	Mbf	42.9	33.5
	Large sawtimber	Mbf	85.8	66.9
Intermediate	Pole timber	Cord	5.2	5.2
	Small sawtimber	Mbf	48.1	38.3
	Large sawtimber	Mbf	96.2	76.6
Shade intolerant	Pole timber	Cord	5.7	5.7
	Small sawtimber	Mbf	45.8	29.7
	Large sawtimber	Mbf	91.6	59.4

Note: Mbf = thousand board feet, Scribner rule.

4. Decisions and Transitions Between Stand-market States

Decisions are the finite number of possible harvests that can be made for each stand state, including removing nothing. For example, the possible decisions for a stand with high level of basal area in each species-size class are to cut it into a state with low level of basal area in one or more classes, or leave it intact.

Transition probabilities are the probabilities that the system (stand and market) will move from one state to another following a decision. Let $p(s'|s,k)$ be the probability that the stand moves from state s to state s' when decision k is made. All these probabilities are in Table 3. Furthermore, let $p(m'|m)$ be the probability that the market moves from state m to state m'. Since the timber market is independent of the stand state, the probability that the stand-market system moves from state $i = (s,m)$ to $j = (s',m')$ is:

$$p(j|i,k) = p(s'|s,k)\,p(m'|m)$$

with:

$$\begin{aligned} &\sum_j p(j|i,k) = 1 && \forall i,k \\ &\sum_{s'} p(s'|s,k) = 1 && \forall s,k \\ &\sum_{m'} p(m'|m) = 1 && \forall m \end{aligned} \tag{3}$$

As shown by the price model (2), the market state in year t+5 is independent of the state in year t. Consequently, $p(m'|m) = p(m') = 1/2$, the unconditional probability of a market state.

5. Economic and Ecological Rewards

5.1 Returns from decisions

The economic reward $r(i,k)$ is the immediate return from making a harvest decision k in stand-market state $i = (s, m)$. It depends on the volume of timber removed and its price per unit. The reward $r(i,k)$ is then:

$$r(i,k) = r((s,m),k) = \sum_a \sum_b P_{ab}(m) Q_{ab}(s,k) - F \tag{4}$$

where the $P_{ab}(m)$ is the price per unit of volume of trees of species a, size b when the market state is m (Table 4). The $Q_{ab}(s, k)$ is the volume of trees harvested of species a, size b when the stand is in state s and decision k is made. F is the cost of harvesting, independent of the amount harvested (e.g., timber sale preparation and administration).

The volumes by species and size gotten from a particular state and decision are the standing volumes before the decision minus those left after it. Standing volumes were derived from the state basal areas (Table 2). A sample of the results is in Table 5 which shows that, given state 64 (HHH, HHH), the decision to cut the stand down to state 1 (LLL, LLL), produced 17 cords/acre of pulpwood, 3.7 Mbf/acre of small sawtimber, and 4.8 Mbf/acre of large sawtimber of various species. The immediate return in 1990 dollars was \$491/acre at low market and \$635/acre at high market state.

Table 5 Example of economic rewards

Stand State No.	Decision k	Pole timber (cord) S	M	I	Small sawtimber (bf) S	M	I	Large sawtimber (bf) S	M	I	Income ($/acre), by Market state Low	High
64	1	10	2	5	2176	470	1008	2948	917	888	491	635
64	2	10	2	5	2176	470	1008	2948	0	0	368	465
64	3	10	2	5	2176	0	0	2948	917	888	442	565

Note: S, M, I = shade tolerant, intermediate, and shade intolerant species.

5. 2 Tree diversity

The ecological reward $\delta(s,k)$ was defined as the species and size diversity of the trees remaining in a stand that starts in state s, when decision k is made. Shannon's index (Pielou 1975) was chosen to measure tree diversity in a forest stand. It meets the major criteria needed for a diversity measurement (Hunter 1990).

The tree species diversity is, with Shannon's definition:

$$H_{species} = -\sum_{a=1}^{m} \frac{y_a}{y} \log(\frac{y_a}{y}) \tag{5}$$

where y is the total number of trees, and y_a is the number of trees of species a per acre.

The tree size diversity is the weighted mean of size-diversity averaged over all species:

$$H_{size} = -\sum_{a=1}^{m} \frac{y_a}{y} \sum_{b=1}^{n} \frac{y_{ab}}{y_a} \log(\frac{y_{ab}}{y_a}) \tag{6}$$

where y_{ab} is the number of trees of species a and size b per acre. Defined in that way, diversity of size and species are additive (Pielou 1977: 293-304).

Here, $m = 3$ species groups and $n = 3$ size classes. Maximum diversity of tree species or size occurs when trees are evenly distributed, then: $\max H_{species} = \max H_{size} = \ln(3) = 1.10$. Table 6 shows the tree diversity for each stand state defined in Table 1. The highest tree size diversity (1.02) was found in state 28 (LHH, LHH), which had low basal area in pole timber and high basal area in small and large sawtimber of both the shade tolerant and the less tolerant species. The lowest tree size diversity (0.11) occurred in state 37 (HLL, HLL), with high basal area in pole timber and low basal area in small and large sawtimber of both the shade tolerant and the less tolerant species.

The highest tree species diversity (1.06) was in stand state 24 (LHL, HHH): high basal area for all sizes of the less tolerant trees, low basal area for pole timber, high basal area for small sawtimber, and low basal area for large sawtimber of the tolerant trees. However, there were several stand states with tree species diversity very close to the maximum. The lowest tree species diversity (0.28) was in stand state 57 (HHH, LLL) with high basal area in all sizes of the shade tolerant trees, and low basal area in all sizes of the less tolerant.

Given the diversity of tree size and species for each stand state, and the relative frequency of each state in a forest, it is a simple matter to compute the average tree diversity over the entire forest. For example, Table 6 shows that the average tree diversity of the entire maple-birch forest type in Wisconsin at the time of the last two inventories was 0.75 for species, and 0.46 for size.

Table 6 Tree diversity by stand state and for the whole current maple-birch forest in Wisconsin

State No.	Tree diversity: Size	Species	% of Plots	State No.	Tree diversity: Size	Species	% of plots
1	0.39	0.79	15	33	0.14	0.33	7
2	0.55	0.91	2	34	0.19	0.42	1
3	0.52	1.01	1	35	0.20	0.54	2
4	0.69	1.04	2	36	0.28	0.61	1
5	0.17	1.03	7	37	0.11	0.89	4
6	0.33	1.03	1	38	0.19	0.91	0
7	0.44	1.02	4	39	0.25	0.93	2
8	0.57	1.02	2	40	0.33	0.95	2
9	0.69	0.70	2	41	0.33	0.31	1
10	0.80	0.83	1	42	0.38	0.41	1
11	0.74	0.94	0	43	0.38	0.52	0
12	0.87	0.99	1	44	0.44	0.58	0
13	0.31	1.05	1	45	0.24	0.87	0
14	0.45	1.05	0	46	0.32	0.89	0
15	0.55	1.04	1	47	0.37	0.92	0
16	0.67	1.04	1	48	0.45	0.93	1
17	0.69	0.61	2	49	0.44	0.30	4
18	0.78	0.74	1	50	0.48	0.39	1
19	0.73	0.87	1	51	0.48	0.50	1
20	0.85	0.92	1	52	0.54	0.56	1
21	0.35	1.05	1	53	0.33	0.85	1
22	0.47	1.05	0	54	0.40	0.87	1
23	0.56	1.05	0	55	0.45	0.90	1
24	0.67	1.06	0	56	0.52	0.91	1
25	0.93	0.56	3	57	0.60	0.28	5
26	0.99	0.68	2	58	0.64	0.37	1
27	0.93	0.82	1	59	0.63	0.48	1
28	1.02	0.88	1	60	0.68	0.54	1
29	0.49	1.04	1	61	0.45	0.83	1
30	0.60	1.05	0	62	0.51	0.86	0
31	0.68	1.05	1	63	0.56	0.88	1
32	0.78	1.05	1	64	0.62	0.90	1
Landscape diversity							3.63
Average tree diversity over the whole forest:							
Species							0.75
Size							0.46

Note: Diversity measured with Shannon's index. % rounded to nearest integer.

6. Steady-state Distribution of Stand States and Landscape Diversity

The concept of a quasi-equilibrium forest landscape has been proposed by Shugart (1984), where the proportion of the landscape in each age class or successional stage remains fairly constant over time. If the Markov chain of the transition probabilities between stand states contains one and only one irreducible (i.e., all the states communicate with each other) ergodic (i.e., positive recurrent, aperiodic states) closed set, there exists a unique limiting (or steady-state) probability that the stand will be in state s after a large number of transitions, and independently of the initial state (Winston 1991: 976-983, 998).

Let p_s denote the steady-state probability that the stand will be in state s, then the vector $\mathbf{p} = [p_s]$ must satisfy and can be solved from:

$$\begin{aligned} \pi_{s'} &= \sum_s \pi_s p(s' \mid s) \qquad \forall s' \\ \sum_s \pi_s &= 1 \end{aligned} \tag{7}$$

where $p(s' \mid s)$ is the one-step transition probability of moving from stand state s to state s'. Since choosing a harvesting policy (i.e., a set of decisions) means essentially choosing a particular matrix of transition probabilities, Equation (7) can be used to compute the steady-state probability vector $\mathbf{p}$ for any policy.

If all stands in a forest are managed according to the same and permanent harvesting policy (including doing nothing), then the probability (p_s) of observing a stand in state s in the steady state is also the proportion of stands found in this state s, on a large forest. As a result, we can get a measure of the landscape diversity, H_L, of forests managed according to a particular policy, in the long run, by substituting the proportion of each stand state into the formula of Shannon's index:

$$H_L = -\sum_s \pi_s \log \pi_s \tag{8}$$

With this definition, maximum landscape diversity, equal to ln(s) would occur if all the forest stands were evenly distributed among the 64 possible states (i.e., maxH_L = ln(64) = 4.16).

Furthermore, knowing the tree species or size diversity of each stand state, one can compute the steady-state average tree diversity as:

$$H_{L,species} = -\sum_s \pi_s H_{species,s} \tag{9}$$

or

$$H_{L,size} = -\sum_s \pi_s H_{size,s} \tag{10}$$

These measures have two interpretations. They give the average diversity of the entire forest in the steady state, when all stands are managed by the same policy. They also give the expected diversity of a single stand, observed over a very long time period.

The transition probabilities between stand states (Table 3) do form an irreducible (all states communicate with each other), ergodic (each state can return to itself in one interval, thus the process is aperiodic) Markov chain. Therefore, there is a unique steady-state probability (7) of observing a particular state in an undisturbed stand (no management, and natural disturbances similar to those that occurred between the two inventories).

The results in the second column of Table 7 show that, without disturbance, there is a probability of 74%, 12%, and 9%, in the steady-state, to observe stands in state 28, 26, and 32, respectively. There is also a 1% chance to observe states 16, 30, 58, and 60. The probability of observing the other states is less than 1%. Thus, if there were no management and no major disturbance, stands in state 28, the state of highest tree size diversity among all possible stand states, would dominate the whole forest after a long time. The results also imply that stand state 28, with high basal area in the two sawtimber classes, but low in the pole, may represent the climax stage of an orderly succession path, as a stand gets very old. This result is consistent with Bormann and Likens' (1979) observation that in areas where natural disturbance is uncommon, old forests, which have the most complex canopy structure, probably once dominated the landscape.

Table 7 also shows that the steady-state landscape diversity H_L was 0.94 in an undisturbed forest, about one-fourth the landscape diversity of the current forest (Table 6). The steady-state average tree species diversity was 0.86, fifteen per cent higher than that of the current forest. The steady-state average tree size diversity was 0.98, more than twice that of the current forest. Therefore, natural growth without disturbance would lead to a forest of much lower landscape diversity (in terms of the frequency of the different stand states), higher diversity of tree size, and slightly higher species diversity.

7. Effect of Major Natural Disturbances

Incorporating major natural disturbances in the Markovian stand model changes the transition probabilities between stand states, thus changing the stand dynamics and the forest landscape. According to Canham and Loucks (1984), the estimated rotation period for catastrophic disturbance in mesic hardwood forests of northern Wisconsin was about 1200 years. This is equivalent to a 0.4% chance that any one-acre stand will be destroyed by a natural cataclysm in 5 years, and 99.6% chance that the stand will be relatively undisturbed. Let us measure that a catastrophic disturbance would always move the stand from its current state to state 1 (LLL, LLL). Then, the transition probability matrix with disturbances, **S'**, is obtained from the transition probability matrix without disturbances, **S** (in Table 3) as:

$$\mathbf{S}' = (1 - q)\mathbf{S} + \mathbf{K} \tag{11}$$

where q is the probability that any one-acre stand will be destroyed by a catastrophic disturbance, and **K** is a transition matrix with parameters q for the entire first column and 0 for the rest.

The second and the third columns of Table 7 show the differences between stand-state distributions of unmanaged forests, in steady-state, with and without heavy natural disturbances. By introducing the historical disturbance regime of 0.4%/acre/5 years, the steady-state proportion of stands in state 28 decreased to 63%, though the state still dominated the whole forest. There were more (about 5%) stands in state 1, 8, 12, or 64, which were extremely rare in the undisturbed forest. Major natural disturbances created a more diverse distribution of stand states in the steady-state, and thus increased the expected landscape diversity from 0.94 (in the undisturbed forest) to 1.52. This helps provide habitats for plants and animals that do not prefer old growth stands. Disturbances at historical frequencies had no influence on the steady-state average diversity of tree species, and only a small effect on tree size diversity (Table 7).

Human efforts in stabilizing ecosystems, such as fire suppression, have caused a loss of biodiversity. As stated by Reice (1994), 'disturbance is a reinitialising step in the successional development of a community.' Furthermore, 'disturbance and heterogeneity, not equilibrium, generate biodiversity.' The present results, which document that the disturbed forest has a higher landscape diversity than the undisturbed one, support this argument.

Table 7 Stand state distribution in steady state, for the whole Wisconsin maple-birch forest, without harvest and with natural catastrophes

State	Frequency (%) with catastrophe rate of:				State	Frequency (%) with catastrophe rate of:			
No.	0%	**0.4%**	5%	10%	No.	0%	**0.4%**	5%	10%
1	0	**2**	21	36	33	0	**0**	2	3
2	0	**0**	1	2	34	0	**0**	0	0
3	0	**0**	1	1	35	0	**0**	0	0
4	0	**0**	2	2	36	0	**0**	0	0
5	0	**0**	3	5	37	0	**0**	0	1
6	0	**0**	1	2	38	0	**0**	0	0
7	0	**0**	2	2	39	0	**0**	0	1
8	0	**1**	5	5	40	0	**0**	2	2
9	0	**0**	1	1	41	0	**0**	0	0
10	0	**0**	1	1	42	0	**0**	0	0
11	0	**0**	0	0	43	0	**0**	0	0
12	0	**1**	2	1	44	0	**0**	0	0
13	0	**0**	0	0	45	0	**0**	0	0
14	0	**0**	0	0	46	0	**0**	0	0
15	0	**0**	0	0	47	0	**0**	0	0
16	1	**4**	5	3	48	0	**0**	0	0
17	0	**0**	1	1	49	0	**0**	1	1
18	0	**0**	0	0	50	0	**0**	0	1
19	0	**0**	0	0	51	0	**0**	0	0
20	0	**0**	0	0	52	0	**0**	1	0
21	0	**0**	0	0	53	0	**0**	0	0
22	0	**0**	0	0	54	0	**0**	0	0
23	0	**0**	0	0	55	0	**0**	0	0
24	0	**0**	1	1	56	0	**0**	1	1
25	0	**0**	1	1	57	0	**0**	1	1
26	12	**10**	4	2	58	1	**2**	3	2
27	0	**0**	0	0	59	0	**0**	1	1
28	74	**63**	18	7	60	1	**3**	5	3
29	0	**0**	0	0	61	0	**0**	0	0
30	1	**1**	0	0	62	0	**0**	0	0
31	0	**0**	0	0	63	0	**0**	0	0
32	9	**10**	7	4	64	0	**1**	1	1
Landscape diversity						0.94	**1.52**	2.93	2.81
Average tree diversity over the whole forest:									
Species						0.86	**0.86**	0.82	0.81
Size						0.98	**0.92**	0.63	0.52

Note: Catastrophe rate = probability that any one-acre stand will be destroyed by a natural catastrophe and move from its current state to state 1 in five years. Bold numbers pertain to historical disturbance rate. Stand state frequencies rounded to the nearest integer.

Stand dynamics change with the frequency of natural disturbances, leading to different distributions of stands over the landscape (Frelich and Lorimer 1991). The Markov model provides a way to predict the effects of disturbance frequencies other than the historic ones for northern Wisconsin. Disturbance frequencies of 5%, and 10% were tested, implying periods between catastrophic events of 100, and 50 years.

Table 7 shows that the steady-state proportion of stands in state 28 decreased, while that of stands in state 1 increased, with more frequent disturbances. Stand state 1 became dominant when the disturbance frequency reached 10%. Landscape diversity increased as the frequency of catastrophic disturbance increased from 0 to 5%, but then declined with a frequency of 10%.

Frelich and Lorimer (1991) found, by simulation, that stand types across the landscape are more diverse at double the historical disturbance rate, and less diverse at half of it. They also found that much more frequent disturbances would reduce the landscape diversity again. The results shown in Table 7 are consistent with their results, though the stand states were defined in different ways.

Unless stated specifically, transition probabilities between stand states with historical disturbances (period between catastrophic events of 1200 years, or frequency of 0.4%/ 5 years), were used throughout the rest of this study, to mimic the natural growth process of northern hardwood stands in Wisconsin.

Table 8 Probability that a stand that starts in state 1 will be in one of the 64 possible states after 10, 50, or 100 years, and implications for forest diversity

State	Probability (%) at year:			State	Probability (%) at year:		
No.	10	50	100	No.	10	50	100
1	65	13	3	32	0	3	11
2	3	2	0	33	6	2	0
3	2	1	0	34	0	1	0
4	1	3	2	36	0	1	0
5	10	4	1	37	1	1	0
6	2	2	1	39	0	1	0
7	2	3	1	40	0	5	3
8	2	10	6	42	0	1	0
9	2	2	1	49	1	1	0
10	0	1	1	50	0	1	0
12	0	2	3	51	0	1	0
16	0	4	9	52	0	1	1
17	2	1	0	56	0	2	2
20	0	1	0	57	0	3	1
24	0	1	1	58	0	3	5
25	1	2	1	59	0	2	2
26	0	4	5	60	0	4	11
28	0	5	20	64	0	1	3
Landscape diversity:					1.54	3.43	2.91
Average tree diversity over the whole forest:							
Species					0.80	0.81	0.83
Size					0.38	0.56	0.72

Note: Omitted stand states had probabilities equal to or very close to zero.

8. Development of Unmanaged Stands and Forests

The matrix of transition probabilities between stand states, **S'**, provides a ready mean to predict the probability that a stand would be in a particular state after a certain amount of time. Let $\mathbf{P}_0$ be the row vector of initial probabilities of stand state (1 for the state the stand is in at time zero, 0 for others). Then, the probability distribution of stand states after T periods is (Hoel et al. 1987, p.13):

$$\mathbf{P}_T = \mathbf{P}_0 \mathbf{S}'^T \tag{12}$$

For example, Table 8 shows the development of an unmanaged stand over 100 years. The prediction started from a young and low-density stand, i.e., at time $t = 0$, the stand was in state 1 with probability one. Table 8 shows the probability that this stand ended in different states after 10, 50, and 100 years. As would be expected, at year 10, the stand had a good chance (65%) of being still in state 1. Else, it would most likely be in state 5 (10% chance), indicating more ingrowth of less tolerant species, or state 33 (6% chance) with more ingrowth of shade tolerant species. At year 50, the probability that the stand was still in state 1 decreased to 13%, and it was possible to find the stand in more different states. After 100 years of development, the stand was most likely to be in state 28 (20% chance), followed by state 32 (11% chance), state 60 (11% chance), and state 16 (9% chance). As time went by, the probability distribution of stand states would converge towards the steady-state distribution (the third column of Table 7).

The evolution of the diversity indices in Table 8 implies that a large forest, with all stands starting in state 1, would become more diverse at the landscape level, as time elapsed, up to a certain degree, and then become less diverse again as it approached the steady state. In addition, the average tree diversity over the whole forest, especially the diversity of tree size, would increase with time (Table 8) and would culminate in the steady state when stands in state 28 would dominate the landscape (Table 7).

The expected residence time of a particular stand state s, defined as the average time that a stand will stay in that state sequentially, can be obtained from the expected number of sequential periods (m_s) that a stand holds in state s:

$$m_s = 1/(1 - p(s|s)) \tag{13}$$

where $p(s|s)$ is the one-step transition probability that a stand stays in the same state s (Solomon 1987: 106-108).

The results for an unmanaged stand, are shown in Table 9. States 16 and 28 had the longest expected residence time (148 years). State 32 had the second longest (114 years). Since the stand state 28 tends to dominate the steady-state landscape (63% of the whole forest, see Table 7), it is natural to find that it has also a long expected residence time. However, although states 16 and 32 had long expected residence time as well, they occupied a smaller proportion of the forest than state 28 in the steady state, because once they changed to other states, it took longer to get back.

Table 9 Expected residence time in each state for an unmanaged stand

State No.	Residence time (years)	State No.	Residence time (years)
1	25	33	13
2	15	34	10
3	8	35	10
4	15	36	13
5	13	37	10
6	17	38	6
7	12	39	10
8	23	40	20
9	21	41	9
10	15	42	12
11	12	43	5
12	21	44	8
13	10	45	7
14	11	46	44
15	16	47	7
16	148	48	17
17	9	49	10
18	12	50	13
19	6	51	18
20	7	52	9
21	6	53	10
22	9	54	8
23	8	55	7
24	12	56	15
25	21	57	23
26	25	58	25
27	12	59	37
28	148	60	40
29	18	61	10
30	8	62	9
31	8	63	11
32	114	64	40

9. Effects of the Current Management Policies

The object, here, was to predict the long-term consequences of continuing the current management in the Wisconsin hardwood forests, in terms of economic returns and ecological diversity. The current harvest regime was defined from the harvest that occurred between the last two inventories, as revealed by the data in the FIA database. To get a harvest sufficient to change stand states, the interval between decision time points was set at 15 years. Since the year of harvest was unknown, the policy could be tied only to the stand state and was independent of the market. The revealed decision process was described by a probability matrix $\mathbf{D}=[d_{ss'}]$, where $d_{ss'}$ is the probability that a stand in state s was cut to state s'. For example, given stand state 3, the probability was 0.125 to cut the stand down to state 1, and 0.875 to leave the stand intact (Lin 1995). Assuming this decision matrix to be constant gave predictions consistent with the continuation of the current management. To illustrate, let $\mathbf{P}_t$ denote the row vector of the stand state distribution in year t, $\mathbf{D}$ the decision matrix, and $\mathbf{S}'$ the 5-year transition matrix between stand states, then:

$$\mathbf{P}_{t+15} = \mathbf{P}_t \times \mathbf{D} \times \mathbf{S}'^{3} \tag{14}$$

Successive iterations with Equation (14) make $\mathbf{P}_t$ converge to the steady-state probability distribution (Hoel et al. 1987: 47-49). The results in Table 10 show how stand states would be distributed after a long passage of time if the current management continued. The steady-state landscape diversity that would result was then computed with Equation (8), and the average tree diversity (species and size) of the whole forest was calculated by Equations (9) and (10).

The results in Table 10 show that, compared with the unmanaged forest (Table 7), the current management regime would lead to a much more diverse landscape, in the steady state, while it would decrease the average diversity of trees (especially the size diversity). The steady-state landscape diversity index was 1.52 for the unmanaged forest, and 3.33 for the forest obtained with the current management. Although harvest enhanced the landscape diversity by creating stands of different states, it cut down the proportion of old growth stands (state 28) from 63% to 12% of the whole forest area. This, hence, reduced the average diversity of tree size from 0.92 (Table 7) to 0.60 (Table 10).

The expected net present value, V_i (\$/acre), from a stand in a given initial state, i, managed according to the current policy, was calculated with the following recursive equations (Ross 1983):

$$V_i^{t+1} = \sum_k p(k|i)[r(i,k) + d\sum_{j=1}^{N} p(j|i,k)V_j^t] \qquad i = 1,2,\ldots,N \qquad t = 1,\ldots,\infty \tag{15}$$

where N is the total number of stand-market states, and d is the discount factor $d = 1/(1+g)^5$ where g is the annual interest rate (set at 3 per cent/year in real terms), $p(k \mid i)$ is the probability of making decision k, given that the stand-market is in state i.

The expected stand value V_i, was estimated by applying Equation (15) recursively, starting from $V_i^0 = 0$ for all i, and iterating on t until the change in stand value was less than an arbitrarily small amount. The method of error bounds (Bertsekas 1976) was used to speed up the convergence.

Table 10 shows the expected stand values, by state, of continuing the current management policy. The forest value was lowest for a stand in poor condition (State 1; LLL, LLL), \$43/acre regardless of the market state. It was highest for a stand in state 44 (HLH, LHH), \$515/acre at the high market and \$395/acre at the low market state.

A measure of the economic value of the current entire forest under the current policy is the expected average net present value of returns (\$/acre), obtained by adding the expected present value of returns from all possible stand states weighted by their relative frequency (Table 6). This value for timber production, including trees and land, was \$118/acre with a high initial market state, and \$109/acre with a low initial market state. In the steady state, such a forest would yield an average return of \$62/acre/15 years.

Table 10 Steady-state distribution of stand states, ecological diversity, and expected income of the Wisconsin maple-birch forest managed with the current harvesting regime

Stand State No.	Frequency (%)	NPV($/acre): High market	NPV($/acre): Low market	Stand State No.	Frequency (%)	NPV ($/acre): High market	NPV ($/acre): Low market
1	7	43	43	33	1	73	73
2	1	80	80	34	0	114	114
3	1	98	95	35	0	97	93
4	2	153	133	36	1	111	111
5	2	55	55	37	0	85	85
6	1	106	95	38	0	210	166
7	2	82	76	39	1	111	106
8	5	134	117	40	3	94	94
9	1	117	108	41	0	175	155
10	1	151	138	42	1	195	174
11	0	301	249	43	0	253	216
12	2	282	232	44	0	515	395
13	0	105	105	45	0	102	102
14	0	290	229	46	0	61	61
15	0	200	161	47	0	214	187
16	6	176	156	48	0	113	113
17	0	91	89	49	1	106	103
18	1	179	148	50	1	215	183
19	0	145	135	51	0	136	120
20	0	165	159	52	1	145	145
21	0	118	115	53	0	120	117
22	0	104	104	54	0	125	125
23	0	104	104	55	0	136	132
24	1	119	119	56	1	280	230
25	2	154	138	57	2	168	152
26	6	172	158	58	5	194	174
27	1	181	171	59	2	161	148
28	12	301	260	60	8	252	221
29	0	151	138	61	0	110	110
30	0	132	132	62	0	125	125
31	0	161	154	63	0	206	187
32	13	216	189	64	2	212	189
Landscape diversity in steady state							3.33
Average tree diversity in steady state:							
Species							0.81
Size							0.60
Average expected NPV from the current forest ($/acre):							
In high market state							118
In low market state							109
Average steady-state return every 15 years ($/acre)							62

Note: NPV = net present value.

10. Summary and Conclusions

This paper has presented a Markov-chain model to predict the development of individual forest stands and forest landscape, incorporating stochastic prices and biological growth. In the application to the Wisconsin maple-birch forests, 64 stand states were defined by the basal area of trees in three size classes of two species groups. Market states were either high or low, depending on the price of sugar maple sawtimber.

The core of the model is the matrix of probabilities of movement between stand-market states over five years. It is the product of transition probabilities between stand states, obtained by Monte-Carlo simulation with a stochastic stand growth model, and transition probabilities between market states defined by the distribution of real timber prices.

Shannon's index was the measure of tree diversity of species and size in each stand state. It also served to measure landscape diversity, based on the relative frequency of stand states in the whole forest. Stand state 28, with high basal area in sawtimber of the shade tolerant and the less tolerant species, had the highest tree size diversity. Stand state 37, with low basal area in sawtimber of the shade tolerant and the less tolerant species, had the lowest tree size diversity. High species diversity was found in several stand states with high basal area of poles in the less tolerant species and low in the shade tolerant trees. The lowest species diversity was in stand state 57 with high basal area in all sizes of the shade tolerant species, and low basal area in all sizes of the less tolerant.

The steady-state distribution of stand states obtained from the Markov-chain model showed that, without heavy natural disturbances, stands in state 28, with the highest diversity of tree size, would occupy 74% of the unmanaged forest area after a long time period. Major disturbances at historical rates (period between catastrophic events of 1200 years) reduced the proportion of stands in state 28 to 63%, and created a more diverse distribution of stand states. The landscape diversity of unmanaged forests increased as the period between catastrophic events decreased from 1200 years to 100 years, but then declined again with a period of 50 years.

The Markov-chain model was also applied to show how a forest would evolve, with only natural disturbances, starting from a depleted forest. After 100 years, stand state 28 would occupy 20 per cent of the forest area. During the transition, tree size diversity would almost double, species diversity would increase only slightly. Landscape diversity after a century would be double that in the steady-state. Once a stand entered state 28, it would stay there for 148 years on average, and thus dominate the forest landscape. These results suggest that stand state 28 characterises the 'old-growth' or the 'climax' stage in the succession of a maple-birch forest.

The current management regime was defined by the observed frequency of state changes due to harvests, revealed by the inventory data. This decision probability matrix, coupled with the Markov growth model, was applied to predict the ecological and economic effects of continuing the current management policies. The predicted steady-state forest had twice the landscape diversity of the unmanaged forest, because harvesting varied the stand states. But, the proportion of old-growth stands (state 28) was reduced to 12%, size diversity was one third lower, and species diversity only slightly lower. The income produced by the current management was very low, about $4/acre/year in the steady state, with a current forest value of less than $120/acre.

Part II of this paper will investigate harvesting policies that could increase the expected economic returns markedly without ruining diversity.

Acknowledgements

The research leading to this paper was funded in part by cooperative agreement 23-92-42 with the USDA Forest Service, North-Central Forest Experiment Station, East Lansing. We thank Mike Vasievich for his support and collaboration. Additional funding came from the University-Industry Relations program, Mc Intire-Stennis grant C855, and the School of Natural Resources, University of Wisconsin, Madison. Parts of this paper appeared originally in Management Science 44(10):1351-1366.

References

BERTSEKAS, D. P. 1976. Dynamic programming and stochastic control. Mathematics in Science and Engineering, Volume 125. Academic Press, Inc., New York.

BINKLEY, C.S. 1980. Is succession in hardwood forests a stationary Markov process? Foriest Science 26(4):566-570.

BLISS, J. C., AND MARTIN, A.J. 1988. Identity and private forest management. Society and Natural Resources 1:365-376.

BORMANN, F. H., AND LIKENS, G.E. 1979. Pattern and process in a forested ecosystem. Springer-Verlag, New York.

BUONGIORNO, DAHIR, J.S., LU, H.C. AND LIN, C.R. 1994. Tree size diversity and economic returns in uneven-aged forest stands. Forest Science 40(1):83-103.

BUONGIORNO, J., PEYRON, J., HOULLIER, F. AND BRUCIAMACCHIE, M. 1995. Growth and management of mixed-species, uneven-aged forests in the French Jura: Implications for economic returns and tree diversity. Forest Science 41(3):397-429.

BURTON, P.J., BALISKY, A.C., COWARD, L.P., CUMMING, S.G. AND KNEESHAW, D.D. 1992. The value of managing for biodiversity. The Forest Chronicle 68(2):225-237.

CANHAM, C.D., AND LOUCKS, O.L. 1984. Catastrophic windthrow in the presettlement forests of Wisconsin. Ecology 65:803-809.

FRELICH, L.L., AND LORIMER, C.G. 1991. A simulation of landscape-level stand dynamics in the northern hardwood region. Journal of Ecology 79:223-233.

HAHN, J.T., AND HANSEN, M.H. 1985. Data Bases for Forest Inventory in the North Central Region. Gen. Tech. Rep. NC-101, USDA For. Ser. North Central Forest Experiment Station, St. Paul, MN. 57 pp.

HANSEN, A.J., SPIES, T.A., SWANSON, F.J., AND OHMANN, J.L. 1991. Conserving biodiversity in managed forests: lessons from natural forests. BioScience 41(6):382-392.

HOEL, P.G., PORT, S.C. AND STONE, C.J. 1987. Introduction to stochastic process. Waveland Press, Inc., IL.

HOOL, J.N. 1966. A dynamic programming-Markov chain approach to forest production control. Forest Science Monograph 12.

HUNTER, M.L. JR. 1990. Wildlife, Forests, and Forestry: Principles of Managing Forests for Biodiversity. Prentice-Hall, Inc. NJ.

JOHNSON, S. E., FERGUSON, I.S., AND LI, R.W. 1991. Evaluation of a stochastic diameter growth model for mountain ash. Forest Science 37(6):1671-1681.

KAYA, I., AND BUONGIORNO, J. 1987. Economic management of uneven-aged northern hardwood stands under risk: a Markovian decision model. Forest Science 33(4):889-907.

LEAK, W. B., AND GOTTSACKER, J.H. 1985. New approaches to uneven-aged management in New England. North. Journal of Applied Forestry 2:28-31.

LEMBERSKY, M.R. 1976. Maximum average annual volume for managed stands. Forest Science 22(1):69-81.

LEMBERSKY, M.R., AND JOHNSON, K.N. 1975. Optimal policies for managed stands: an infinite horizon Markov decision process approach. Forest Science 21(2):109-122.

LIN, C.R. 1995. Economics and diversity of trees and landscape in northern hardwood forests: implications of deterministic and stochastic model. Ph.D dissertation. University of Wisconsin, Madison. 154pp.

LIN, C.R., BUONGIORNO, J. AND VASIEVICH, J.M. In press. A multi-species, density-dependent matrix growth model to predict tree diversity and income in northern hardwood stands. Ecological Modeling 91:193-211.

LU, H.S. 1992. Economic management of Wisconsin's northern hardwood forest stands: a mixed-species model. Ph.D. dissertation, University of Wisconsin, Madison. 145pp.

NORSE, E.A. 1990. Ancient forests of the Pacific Northwest. Island Press, Washington, D.C.

OLIVER, C.D. 1992. A landscape approach achieving and maintaining biodiversity and economic productivity. Journal of Forestry 90(9):20-25.

PETERSON, T.A. 1967-1990. Annual Wisconsin forest products price review. Cooperative Extension Service, University of Wisconsin-Extension, Madison, WI.

PIELOU, E.C. 1975. Ecological diversity. John Wiley & Sons. New York. 165pp.

PIELOU, E.C. 1977. Mathematical ecology. John Wiley & Sons. Toronto. 385pp.

PIMENTAL, D. 1961. Species diversity and insect population outbreaks. Ann. Entom. Soc. Amer. 54:76-86.

PRESTON, R.J. 1977. North American Trees. The Iowa State University Press. 399pp.

REICE, S.R. 1994. Nonequilibrium determinants of biological community structure. American Scientist 82:424-435.

ROBERTS, M. R. AND HRUSKA, A.J. 1986. Predicting diameter distributions: a test of the stationary Markov model. Canadian Journal of Forest Research 16:130-135.

ROSS, S. 1983. Introduction to stochastic dynamic programming. Academic Press, New York. 164pp.

SHUGART, H.H. 1984. A theory of forest dynamics. Springer-Verlag, New York, USA. 278pp.

SMITH, W.B. 1986. Wisconsin's forth forest inventory:area. Resource Bulletin NC-97, USDA For. Ser. North Central Forest Experiment Station, St. Paul, MN. 48pp.

SOLOMON, F. 1987. Probability and stochastic process. Prentice-Hall, Inc. N.J. 426pp.

WINSTON, W.L. 1991. Operations research: applications and algorithms. 2nd ed. PWS-Kent Publishing Co., Boston. 1262pp.

WISCONSIN DEPARTMENT OF NATURAL RESOURCES. 1977. Mapping terms and symbols. Rev. Jan. 26, 1977. Manual Code 8625.2. Wisconsin DNR, Madison.

Managing Forests for Tree and Landscape Diversity Part II: Optimisation with Markov Decision Process Model

Ching-Rong Lin and Joseph Buongiorno

Abstract

Part I of this paper presented a Markov-chain model of the growth of maple-birch forest stands, and applied it to predict the consequences of natural and human disturbances on diversity of tree size and species, diversity of landscape, and income. Part II extends the analysis with a Markov decision process model to seek optimum management policies. An unconstrained economic harvesting policy gave more than five times the returns of the current management regime, with the same long-run landscape diversity and diversities of tree size and species less than 10% lower. Achieving maximum species diversity demanded harvesting, it lowered landscape diversity and gave about the same size diversity, with a third of the income of the economic policy, though nearly double that provided by continuing the current policy. Maximum tree size diversity could be nearly achieved by letting nature take its course, but this would entail giving up all income, and having a landscape half as diverse as that obtained by continuing the current policy. Compromise policies were found by maximising expected discounted returns with harvests that kept stands above a specified lower bound of diversity. Raising the lower bound on either the diversity of tree species or size decreased the other economic and ecological criteria. From an economic viewpoint, it was less costly to increase species diversity than size diversity because cutting large trees tended to increase species diversity. Setting high lower bounds on tree size diversity led to high opportunity costs and low landscape diversity.

Keywords: Forest management, ecology, Markov chain, optimisation, risk

1. Introduction

The general purpose of this study was to explore methods of predicting economic and ecological consequences of different management regimes on forests. The maple-birch forest type of Wisconsin provided the context. In Part I, a Markov-chain model was developed and applied to predict the development of stands subject to natural or human disturbances. One of the results showed that continuing the current management regime over the entire maple-birch forest type would lead to a steady-state forest with twice the landscape diversity of an unmanaged forest, but the proportion of old-growth stands would be 4/5 lower, the diversity of tree size 1/3 lower and tree-species diversity slightly lower. Continuing the current management produced a low income, about \$4/acre/year and an average current land expectation value of \$120 per year. One of the objectives of Part II is to determine if there are harvesting policies that could increase income, without ruining diversity. More generally, the purpose is to obtain data on the trade-off between diversity of forest stands and landscape, and their financial returns.

The focus continues to be on tree-selection silviculture (uneven-aged management), for which there is currently a very strong interest, in temperate as well as in tropical regions (Vanclay 1995). Management affects both species composition and canopy structure, thus influencing the biological diversity of forests. In maple-birch forests, even-aged treatments such as clear-cut or shelterwood can promote regeneration of less tolerant trees and a higher diversity of tree species (Tubbs 1977, Godman 1983, Walters and Nyland 1989, Niese and Strong 1992). But, because it has more impact on the sites (both biotic and abiotic), clear cutting is less emphasised as a harvesting technique for the goal of conserving biodiversity (Evans and Szaro 1990). And, as pointed out by Smith (1986), uneven-aged management does not necessarily lead to shade-tolerant species only. Under suitable site conditions, the establishment of less tolerant species can be enhanced by heavy basal area cuts (Tubbs 1977). An economic advantage of uneven-aged management for maple-birch forests is that it relies on natural regeneration, avoiding costly site preparation and planting so that its overall financial performance can exceed that of even-aged systems (Niese and Strong 1992). Purely economic management usually ends up with some preferred species, at the sacrifice of biological diversity. Earlier results suggest that, as far as structural diversity is concerned, no management (or at most very light cutting) can improve upon the natural succession. However, management regimes are possible that maintain high diversity with adequate income (Buongiorno et al. 1994, Niese and Strong 1992).

About one fourth of the forests of Wisconsin belong to the public (Raile 1985). Since these lands are receiving most of the attention for sustaining biodiversity their managers must know what interventions, if any, can enhance it. For private forests, income is often one of the management objectives. However, even when the financial objective is dominant, as on industry lands, they cannot overlook the public concerns for other values. Thus, it is useful to know to what extent financial goals conflict with diversity. Ultimately, private landowners and public land managers have to understand this trade off to be able to strike a balance between the two.

To help public and private foresters make these choices, one specific purpose of Part II is to determine how to reach and maintain a particular level of diversity, and how much this would cost, in terms of timber revenues foregone. The results can help decision makers choose management guides that are good compromises between economic and ecological objectives in the Wisconsin maple-birch forest type.

The methods of Part II extend those of Part I to optimisation. The object which was then to predict the consequences of a management policy, is now to find the best policy, according to a given set of criteria. Markov decision process (MDP) models are used for that purpose. This approach was first applied by Hool (1966) to even-aged forests, and then by Lembersky and Johnson (1975) and Lembersky (1976) to Douglas-fir plantations. Kaya and Buongiorno (1987), and Lu (1992) adapted the methods to uneven-aged management. In Part I, ecological as well as economic returns were attached to stand states and decisions. In addition to diversity of tree species and size, the landscape diversity of the entire forest, measured from the frequency distribution of each stand state, was also considered.

Part II begins by investigating purely economic policies, giving decisions by stand and market states that maximise net present value, without any constraint. This solution, with its implications for diversity of trees, and landscape, serves as the basis to compare the economic performance of other policies. The next set of policies deals with purely ecological objectives: maximisation of expected average diversity of tree size or species. The last Part deals with policies with a constrained economic objective: net present value is maximised, but the only acceptable decisions are those that keep diversity of tree species or size above a specific threshold.

2. Economic Harvesting Policies

In Part I, a maple-birch forest stand was observed at 5 year intervals and found in one of 64 possible states. A stand state was defined by the basal area in three size classes and two species classes. Simultaneously, the market was observed as being low or high. Therefore, the stand-market system could be found in one of 128 possible states.

If the stand-market state was i at time t and decision k was made (cutting the stand to another state), then an economic reward was received, $r(i, k)$, the value of the harvest, together with an ecological reward, $\delta(i, k)$, the species and size diversity of the trees left after harvest. The stand-market system then moved to another state, j, with the probability $p(j|\ i, k)$.

This Markov decision process model was applied at the end of Part I to investigate the ecological and economic consequences of continuing the current harvest policy.

2.1 Method

Let N be the number of possible stand-market states and let X_t denote the stand-market state at time t, then the optimality criterion for an economic objective is the total expected discounted dollar returns:

$$V_A(i) = E_A[\sum_{t=0}^{\infty} r(X_t,k_t)d_t \mid X_0 = i] \quad i = 1,2,3,...,N \tag{1}$$

where A is the policy followed, i.e., the rule stating which decision should be made when the stand-market state is in one of the N possible states, $r(X_t,k_t)$ is the immediate return generated by making a harvest decision $k_t = k$ in stand-market state $X_t = i$, and d is the discount factor $d = 1/(1+g)^{\theta}$, where g is the annual interest rate, and θ is the time interval (in years) between two decision events.

$V_A(i)$ is the expected discounted return for a stand that starts in state i, when policy A is followed. For a system that evolves according to a stationary Markov model, the optimum policy A^* is stationary, i.e., the best decision depends only on the state, and not on the time of the decision. Furthermore, the decision is unique for each state (Ross 1983: 29-30).

The optimum decision policy, or decision rule can be solved by various methods. One of the most efficient, and simplest numerically because it does not involve solutions of simultaneous equations, is the method of successive approximation (Ross 1983: 35,36; Winston 1991, p. 1032). It is based on the observation that if V_i^t is the maximum expected present value of returns from initial stand-market state i over t periods when the optimal decisions are followed, then V_i^t satisfies the recursive equations:

$$V_i^{t+1} = \max_k [r(i,k) + d\sum_{j=1}^{N} p(j|i,k)V_j^t] \qquad i = 1,2,...,N \tag{2}$$

where N is the total number of stand-market states. $p(j|i,k)$ is the probability that the stand and market will be in state j given that it is in state i and that decision k is made. That is, the best decision, k, for a stand in a given initial state and growing for t+1 periods, maximises the sum of the immediate return plus the discounted value of the income from the stand growing for t more periods, when the optimal policy is followed.

As t increases toward infinity, V_i^t converges to V_i, which is the expected discounted value of starting in stand market state i and applying the optimal policy forever. Therefore, V_i is the expected stand value, including land and growing stock. It depends on the initial stand state and market state.

One can obtain the expected stand value V_i, as closely as desired, and the corresponding optimal harvesting policy, by applying equation (2) recursively, starting from $V_i^0 = 0$ for all i, and iterating on t until the change in expected stand value is less than a small chosen amount. The solution gives simultaneously the expected stand value and the best decision policy, i.e., the decision k attached to each stand-market state i.

In this application, the algorithm of successive approximation (2) was written as a Pascal program incorporating the error-bounds method (Bertsekas 1976, p. 237) to speed up the convergence.

2.2 Results

Table 1 shows the decisions that maximise the expected present value of returns (stand value) for all 128 different stand-market states, without diversity constraint. The results suggest that, for 16 of 64 possible stand states, the best decision was to choose the same state, i.e. to do nothing, regardless of the state of the market. The states that required no harvest had generally low or very low basal area in the two sawtimber classes.

If the price was low, 12 more stand states were best left intact. For 34 of 36 stand states that called for a harvest if the market state was low, the decision was the same if the market state was high. Accordingly, for 50 of the 64 stand states (16 not cut; 34 cut to another state), the best decision was not influenced by the state of the market, as defined here.

Table 2 shows the maximum net present value NPV = V_i that would be obtained from each initial stand-market state i by following the economic policy over an infinite length of time. Compared with the current management regime (Table 10 in Part I), the economic harvesting policy led to much higher economic returns. The average expected net present value of returns from the current Wisconsin maple-birch forest, given the distribution of stands revealed by the current inventory (Part I, Table 6), was \$650/acre with a high initial market state, and \$624/acre with a low market state, almost six times that of the current regime. The average steady-state return was \$99/acre/5 years, almost five times that of the current management (see Table 10 in Part I).

The current management regime turned out lower economic returns for its much less intensive harvesting. As noted in Part I, with the current management regime, the harvest was so light that the interval between two decision time points had to be set at 15 years to get noticeable changes in stand states. With the economic policy, instead, a harvest could occur every 5 years, though in general it did not. The expected cutting cycle, determined as in Kaya and Buongiorno (1987), was 10 years for the optimal policy, and 44 years for the current regime. Furthermore, in the current management, stands with high basal area in the two sawtimber classes had a good chance of being left intact. For example, 33% of the stands in state 28, and 50% in state 32 would not be cut at all under the current regime. In contrast, the economic harvesting policy called for all stands in state 28 to be cut to state 19, and all stands in state 32 to be cut to state 21.

Table 2 also shows the steady-state frequency distribution of stand states in forests managed according to the economic harvesting policy. This distribution was obtained by successive matrix multiplication (similar to Equation 14 in Part I). The expected steady-state landscape diversity based on this distribution was 3.32, about the same as for forests managed according to the current policy (Table 10 in Part I). However, the long-run average tree species diversity was 4% lower than under the current policy. Tree-size diversity was 8% lower.

These results suggest that the current forests have not been managed for profit maximisation. Indeed, for public lands, which represent a quarter of the maple-birch forest in Wisconsin, management goals have been much broader. Even non-industrial private owners do not manage their forests for profit only, but also for recreation, scenery, wildlife, or other reasons (Bliss and Martin 1988, 1989). Interestingly, however, the results suggest that pursuing maximum economic returns would reduce only moderately the diversity of trees while it would not affect landscape diversity at all, with the definitions used here, compared with the current management regime.

Table 1 Decisions that maximise net present value

Stand state No.	Decision when market is: High	Low	Stand state No.	Decision when market is: High	Low
1	1	1	33	33	33
2	2	2	34	33	34
3	3	3	35	35	35
4	3	3	36	3	36
5	5	5	37	37	37
6	5	6	38	38	38
7	3	3	39	39	39
8	3	3	40	38	38
9	1	9	41	33	41
10	1	10	42	33	42
11	3	3	43	3	43
12	3	3	44	3	3
13	13	13	45	45	45
14	5	13	46	38	38
15	3	3	47	47	47
16	3	3	48	38	47
17	17	17	49	49	49
18	17	18	50	49	50
19	19	19	51	19	19
20	19	20	52	19	52
21	21	21	53	21	21
22	21	21	54	21	21
23	21	21	55	21	21
24	21	21	56	21	21
25	17	17	57	49	49
26	17	17	58	49	49
27	19	19	59	19	19
28	19	19	60	19	19
29	21	21	61	21	21
30	21	21	62	21	21
31	21	21	63	21	21
32	21	21	64	21	21

Note: Decision numbers refer to stand states after harvest, defined in Part I, Table 1.

Table 2 Steady-state distribution of stand states, diversity, and income of the Wisconsin maple-birch forest managed to maximise net present value

Stand State No.	Frequency (%)	NPV($/a): High market	NPV($/a): Low market	Stand State No.	Frequency (%)	NPV($/a): High market	NPV($/a): Low Market
1	9	241	241	33	1	387	387
2	1	371	371	34	0	505	498
3	5	531	531	35	1	570	570
4	3	641	600	36	1	702	666
5	2	407	407	37	0	653	653
6	0	521	491	38	0	906	906
7	1	526	526	39	0	762	762
8	1	682	639	40	1	936	916
9	2	435	401	41	0	592	580
10	0	589	565	42	0	754	708
11	1	730	678	43	0	789	772
12	0	887	792	44	0	953	854
13	0	639	639	45	0	925	925
14	0	768	715	46	0	1117	1062
15	0	769	716	47	0	1054	1054
16	0	932	834	48	0	1188	1137
17	10	453	453	49	8	541	541
18	2	565	530	50	1	663	629
19	5	754	754	51	1	767	767
20	4	870	837	52	1	929	891
21	2	1020	1020	53	0	1033	1033
22	0	1139	1096	54	0	1198	1153
23	1	1049	1028	55	0	1102	1081
24	1	1209	1145	56	1	1269	1203
25	4	654	601	57	3	750	695
26	1	812	716	58	1	915	815
27	8	958	905	59	3	1016	961
28	8	1119	1022	60	0	1183	1082
29	1	1226	1172	61	1	1283	1228
30	0	1389	1291	62	0	1453	1351
31	4	1296	1221	63	0	1355	1278
32	0	1462	1342	64	0	1526	1403

Landscape diversity in steady state	3.32 (100%)*
Average tree diversity in steady state:	
Species	0.78 (96%)
Size	0.55 (92%)
Average expected NPV from current forest ($/acre):	
In high market state	650 (551%)
In low market state	624 (572%)
Average steady-state return ($/acre/5yr)	99 (479%)

Note: * relative to current management policy. % rounded to nearest integer.

3. Ecological Objectives

3.1 Method

Two ecological criteria were considered for optimisation: tree size diversity, or tree species diversity. In that context, the market state is irrelevant, and no discounting applies: diversity at time t has the same weight as the same weight as diversity at time $t+n$, for all t and n.

The optimality criterion is defined as the expected average level of diversity observed over an infinite length of time, when the initial stand state is i and policy A is employed:

$$U_A(i) = \lim_{n\to\infty} \frac{E_A\left[\sum_{t=0}^{n} \delta(Y_t, k_t) \mid Y_0 = i\right]}{n+1} \cdots\cdots i = 1, 2, 3, \cdots, S \tag{3}$$

where δ is the diversity of a stand in state Y_t after decision k_t , and S is the total number of possible stand states.

With this definition, in contrast to the expected discounted return criterion, long-term expected average diversity is independent of the initial state (i.e., $U_A(1) = U_A(2) = U_A(3) = \ldots = U_A(S)$), but again the optimal policy is stationary, and non randomised: there is still a unique best decision for each state (Ross 1983, p. 89).

The average-reward criterion can be reduced to a discounted-reward criterion, if there is a state which is accessible from every other state regardless of the decision. This is the case here: because of natural catastrophes, there is a positive probability that the stand will be in state 1 after 5 years, regardless of its current state. Then, the method of successive approximation can be employed to solve the dynamic problem to maximise the expected average tree diversity (Bertsekas 1976: 342-349; Ross 1983: 98-99).

There is also a more general linear programming solution to find the policy that maximises the expected average tree diversity. Let π_{ik} denote the steady-state probability that the stand will be in state i and decision k will be made, i.e. the stand will be cut into another state, if policy A is employed. Then, the vector $\pi = [\pi_{ik}]$ must satisfy (Ross 1983, p. 100):

$$\begin{aligned} &\text{(i)} && \pi_{ik} \geq 0 && \forall i,k \\ &\text{(ii)} && \sum_i \sum_k \pi_{ik} = 1. && \\ &\text{(iii)} && \sum_k \pi_{jk} = \sum_i \sum_k \pi_{ik} p(j|i,k) && \forall j \end{aligned} \tag{4}$$

where equation (iii) gives the steady-state probability of finding the stand in any state j, over a very long time period, given the probability of moving to that state, conditional on the current state and the decision.

The expected average tree diversity is then:

$$\sum_i \sum_k \pi_{ik} \delta(i,k) \tag{5}$$

The problem of determining the policy that maximises the expected average tree diversity can, hence, be solved from the following linear programme:

$$
\begin{aligned}
&\max \sum_{i=1}^{N} \sum_{k} \pi_{ik} \delta(i,k) \\
&s.t. \quad \pi_{ik} \geq 0 \qquad\qquad \forall i,k \\
&\qquad \sum_{i=1}^{N} \sum_{k} \pi_{ik} = 1 \\
&\qquad \sum_{k} \pi_{jk} = \sum_{i} \sum_{k} \pi_{ik} p(j|i,k) \qquad \forall j
\end{aligned} \tag{6}
$$

Let $A_i(k)$ denote the probability that the policy A chooses decision k when the stand is in state i. Then:

$$A_i(k) = \pi_{ik} \,/\, \sum_{l} \pi_{il} \,. \tag{7}$$

If $\pi^* = [\pi^*_{ik}]$ maximises the objective function of the linear programme (6), for each i, π^*_{ik} is zero for all but one value of k, i.e., the optimal policy is non-randomised and there is one best decision for each stand state. The stationary optimal policy A^* is then given by each $A^*_i(k)$, which equals zero or one (Ross 1983: 100-101).

3.2 Results

Given the ecological criteria of diversity of tree species and size, many objective functions are possible in problem (6). Here, we present the results for two policies: maximising expected average tree size diversity or maximising expected average tree species diversity. They represent the extremes of a continuum that could be described by an objective function equal to the weighted sum of the two diversity indices, with the weights indicating their relative importance.

Model (6) was solved by linear programming, and checked by the successive approximation algorithm for the equivalent discounted-reward criterion. Both methods gave identical solutions.

Since the objective was maximisation of diversity, irrespective of economic consequences, the ecologically optimal policies were independent of the market value of timber. Table 3 lists the best decisions that would maximise either the species or the size diversity of trees in the maple-birch forests of Wisconsin.

The policy that led to the highest expected average species diversity called for some harvest in 49 of 64 stand states. The stand state definitions (Table 1 in Part I) show that 34 of them had shade-tolerant trees of at least one size class cut, while leaving less-tolerant trees intact. Only eight of them called for cutting less-tolerant trees only. The other seven states had trees of both species groups harvested.

To maximise the expected average diversity of tree size, only 6 of the 64 stand states required some harvest for sawtimber trees, while 39 stand states had trees of pole size cut.

The steady-state stand distributions of forests managed according to these two policies are in Table 4. When maximum tree size diversity was sought, 77% of the stands were in state 28, assimilated to old growth. Under tree species diversity maximisation, the forest was more varied, with a relatively high frequency of states 24 and 32, but no stand in state 28.

The ecological and economic consequences of the two policies are summarised in Table 5. The maximum expected average diversity of species was 33% higher than the diversity obtained with the economic policy. Maximising species diversity would cause a 70% loss in the average expected NPV/acre from the current forest, relative to the NPV- maximising policy. But this would be still 75% above the NPV that would come from continuing the current policy (Table 10 in Part I). Though the tree-size diversity would be 9% higher by maximising species diversity than with the economic policy, the landscape diversity of the whole forest would be 29% lower.

Obtaining the maximum expected average diversity of tree size (0.98) had a very high opportunity cost: more than 96% loss in NPV, and an average yearly return near zero. Furthermore, the steady-state landscape diversity (0.91) decreased significantly relative to the other policies by maximising tree-size diversity, since most of the stands would be found in state 26 (15%) and state 28 (77%), while rarely in other states.

If maximising diversity of tree size were the major goal of management, forests owners might choose to do nothing, since the unmanaged forests would have much higher landscape diversity (1.52), slightly higher species diversity (0.86), and a diversity of tree size (0.92) near the achievable maximum value (Table 7 in Part I).

Table 3 Decisions that maximise expected average tree diversity

Stand state	Maximising diversity of tree:		Stand state	Maximising diversity of tree:	
No.	Species	Size	No.	Species	Size
1	1	1	33	1	33
2	2	2	34	2	34
3	3	3	35	3	3
4	4	4	36	4	4
5	5	5	37	5	37
6	6	2	38	38	38
7	7	3	39	39	39
8	8	4	40	38	38
9	1	9	41	1	41
10	2	10	42	2	10
11	3	11	43	3	11
12	12	11	44	12	11
13	13	9	45	13	41
14	13	10	46	13	10
15	13	11	47	13	11
16	13	11	48	13	11
17	1	17	49	1	49
18	2	18	50	2	18
19	3	19	51	3	19
20	4	20	52	4	20
21	21	21	53	21	21
22	21	21	54	21	21
23	21	19	55	21	19
24	24	20	56	24	20
25	1	25	57	1	25
26	2	26	58	2	26
27	3	27	59	3	27
28	12	28	60	12	28
29	29	25	61	29	25
30	29	26	62	29	26
31	29	27	63	29	27
32	24	28	64	24	28

Note: Decision numbers refer to stand states after harvest, defined in Part I, Table 1.

Table 4 Steady-state distribution of stand states, for the Wisconsin maple-birch forests managed to maximise expected average tree diversity

Stand state	Frequency (%) by Maximising diversity of tree:		Stand state	Frequency (%) by Maximising diversity of tree:	
No.	Species	Size	No.	Species	Size
1	3	2	33	0	0
2	1	0	34	0	0
3	0	0	35	0	0
4	5	1	36	0	0
5	1	0	37	0	0
6	0	0	38	0	0
7	1	0	39	0	0
8	7	0	40	1	0
9	0	0	41	0	0
10	0	1	42	0	0
11	0	0	43	0	0
12	7	0	44	1	0
13	5	0	45	0	0
14	0	0	46	0	0
15	3	0	47	0	0
16	1	0	48	0	0
17	0	0	49	0	0
18	0	0	50	0	0
19	0	0	51	0	0
20	1	0	52	0	0
21	0	0	53	0	0
22	0	0	54	0	0
23	0	0	55	0	0
24	32	0	56	0	0
25	0	1	57	0	0
26	0	15	58	0	0
27	0	0	59	0	0
28	0	77	60	0	0
29	6	0	61	0	0
30	1	0	62	0	0
31	0	0	63	0	0
32	20	0	64	0	0

Note: % rounded to nearest integer.

Table 5 Effects on the Wisconsin maple-birch forests of managements that maximise expected average tree diversity

	Maximising diversity of tree:	
	Species	Size
Landscape diversity in steady state:	2.35 (71%)*	0.91 (27%)
Average tree diversity in steady state:		
Species	1.04 (133%)	0.84 (128%)
Size	0.60 (109%)	0.98 (178%)
Average expected NPV from current forest ($/acre):		
In high market state	206 (32%)	26 (4%)
In low market state	187 (30%)	24 (4%)
Average steady-state return ($/acre/5year):	44 (45%)	0 (0%)

Note: * relative to the economic policy (Table 2).

4. Economic Objective with Ecological Constraints

4.1 Method

Maximising economic returns, or maximising diversity represent the extremes of a continuum. In practice, any objective, even if it were optimised, would be rarely unconstrained by others. Here, we chose to investigate models in which the objective function is the economic criterion (expected net present value), while the diversity objectives appear as constraints. In Part, this is a matter of convenience, because it is difficult to express expected net present value (NPV) as a constraint within the MDP framework. But, the approach is also appealing because it gives a well-defined measure of the opportunity cost of an ecological constraint.

Placing the economic criterion in the objective function does not mean that priority is given to financial considerations. On the contrary, priority is given to the ecological constraints, since the constraints must be met absolutely before the net present value is optimised. Optimisation occurs only within the feasible space defined by ecological considerations. But, within that space, resources are used efficiently, in the economic sense.

To insure that a minimum threshold of ecological diversity was maintained, the following constraints were introduced in the recursive equations (2), limiting the set of acceptable decisions, *k*:

$$\delta(s,k) \geq \alpha \times \delta^* \qquad s = 1,\ldots,S \qquad\qquad (7)$$
$$0 \leq \alpha \leq 1$$

here $\delta(s,k)$ is the tree diversity (species or size) in a stand that starts in state s, after decision k is made. δ^* is the highest value of tree diversity among all 64 possible stand states. As shown in Table 6 of Part I, the highest tree size diversity (1.02) occurred in stand state 28, akin to old growth, and the highest tree species diversity (1.06) in stand state 24.

Table 6 Decisions that maximise net present value, keeping tree diversity within 70% of maximum

Stand state No.	Maintaining diversity of tree: Species		Size		Stand state No.	Maintaining diversity of tree: Species		Size	
	High Market	Low market	High market	Low market		High market	Low market	High market	Low market
1	1	1	1	1	33	1	1	1	1
2	2	2	2	2	34	1	2	2	2
3	3	3	3	3	35	3	3	3	3
4	3	3	4	4	36	3	3	4	4
5	5	5	1	1	37	37	37	1	1
6	5	6	2	2	38	38	38	2	2
7	7	7	3	3	39	39	39	3	3
8	3	3	4	4	40	38	38	4	4
9	1	1	9	9	41	1	1	9	9
10	1	10	10	10	42	1	2	10	10
11	3	3	11	11	43	3	3	11	11
12	3	3	11	11	44	3	3	11	11
13	13	13	9	9	45	45	45	9	9
14	5	13	10	10	46	38	38	10	10
15	3	3	11	11	47	47	47	11	11
16	3	3	11	11	48	47	47	11	11
17	1	1	17	17	49	1	1	17	17
18	1	2	18	18	50	1	2	18	18
19	19	19	19	19	51	19	19	19	19
20	19	20	19	20	52	19	19	19	19
21	21	21	17	17	53	21	21	17	17
22	21	21	18	18	54	21	21	18	18
23	21	21	19	19	55	21	21	19	19
24	21	21	19	19	56	21	21	19	19
25	1	1	25	25	57	1	1	25	25
26	1	2	18	26	58	1	2	18	18
27	19	19	19	19	59	19	19	19	19
28	19	19	19	19	60	19	19	19	19
29	21	21	25	25	61	21	21	25	25
30	21	21	18	18	62	21	21	18	18
31	21	21	19	19	63	21	21	19	19
32	21	21	19	19	64	21	21	19	19

Note: Decision numbers refer to stand state after harvest, defined in Part I, Table 1.

Table 7 Steady-state distribution of stand states for the Wisconsin maple-birch forest managed to maximise net present value, keeping tree diversity within 70% of maximum

Stand state No.	Frequency (%) with bound on diversity of tree: Species	Size	Stand state No.	Frequency (%) with bound on diversity of tree: Species	Size
1	19	6	33	1	0
2	2	6	34	0	1
3	8	1	35	1	0
4	6	4	36	1	0
5	5	0	37	1	0
6	1	1	38	0	0
7	5	0	39	1	0
8	2	1	40	1	0
9	0	3	41	0	0
10	0	4	42	0	0
11	0	3	43	0	0
12	0	1	44	0	0
13	0	0	45	0	0
14	0	0	46	0	0
15	1	0	47	0	0
16	0	0	48	1	0
17	2	5	49	1	1
18	0	13	50	0	2
19	4	4	51	0	0
20	3	4	52	0	0
21	3	0	53	0	0
22	0	0	54	0	0
23	1	0	55	0	0
24	1	0	56	1	0
25	0	8	57	0	0
26	0	13	58	0	0
27	7	7	59	2	2
28	7	8	60	0	0
29	2	0	61	1	0
30	0	0	62	0	0
31	6	0	63	0	0
32	1	0	64	1	0

Note: % rounded to nearest integer.

The constraints (7) ensure that the only harvests acceptable at a decision time are those that maintain the stand at a level of diversity within at least $\alpha \times 100\%$ of the possible maximum values. However, if, for some stand state, no decision could obtain the desired level of diversity, the decision that led to the highest diversity was chosen. To illustrate, consider a constraint which requires that 60% of the maximum tree size diversity be maintained after harvest:

$$\delta_{size}(s,k) \geq 0.60 \times 1.02 \qquad \forall s,k \tag{8}$$

If a stand started in state 2, no possible decision (cut down to state 1, or leave intact) could reach this desired level. Then, decision 2 (leave intact) was chosen instead of decision 1, since it gave the highest tree size diversity.

Diversity of tree size and species could also be considered simultaneously by this method, but it would require giving priority to one of them. Alternatively, a constraint could be built with a diversity index made of a weighted average of species and size diversity. The weights would measure the relative importance of the two goals. Here, we dealt separately with a size diversity constraint or a species diversity constraint, in parallel with what was done with the objective function, to highlight the differences between the two criteria. Comparison of the unconstrained maximum expected returns with those obtainable with ecological constraints provide a measure of the opportunity cost of maintaining a desired level of tree diversity.

4.2 Results

Harvesting policies that maximised expected net present value were obtained with the lower bound on species or size diversity set at α = 50, 70 or 90% of their observed maximum. The effects of the different constraint levels are summarised in Tables 8 and 9, but, for brevity, the policy and the steady state are shown for α = 70% only in Tables 6 and 7.

Table 6 shows that the α = 70% constraint on species diversity alone called for a harvest in the majority of stand states, 48 or 51 out of 64, depending on market state. Most harvests took shade-tolerant trees in at least one size class, while leaving the less-tolerant trees. With the same constraint on tree-size diversity instead, 51 to 53 stand states had harvests, depending on market state. The harvest tended to concentrate on the smaller trees.

The data in Table 7 show that, with a 70% lower bound on species diversity, the most frequent stand state in the steady-state forest would be stand state 1, of low basal area in all sizes and species classes, which would cover about 19% of the entire Wisconsin maple-birch forest. Instead, with the same bound on size diversity, states 18, 26 and 28 (akin to old growth) would be the most frequent in the steady-state forest.

The economic and ecological consequences of maximising expected net present value with different bounds on species or size diversity are summarised in Tables 8 and 9. Table 8 shows that tightening the species diversity constraint from 50 to 90% of the maximum increased the average species diversity in the steady-state forest from 0.85 to 1.00. But, all other criteria would decrease. Financially, there would be a loss in the expected average return per period on the entire forest of 6% to 38% relative to the unconstrained economic harvesting policy. A similar trend was found for the average expected net present value from the current forest. Increasing diversity of species tended to decrease the diversity of tree size, but the effect was not large. Instead, high species diversity was achieved at a large loss in landscape diversity of the steady state forest, which decreased by 39% with a 90% constraint, relative to the purely economic policy, because most stands would cluster in states with higher species diversity.

Introducing constraints of the same magnitude on the diversity of tree size would lead to greater financial losses (Table 9). The least constraining alternative: harvesting so that size diversity was at least 50% of the maximum, improved expected average tree size diversity by only 11% relative to the economic policy, for a decrease of about 32% in economic returns. As the lower bound on size diversity increased to 90%, the losses in expected average return per period, and average expected NPV, increased to about 85%.

Within the range of the alternatives in Table 9, increasing size diversity also tended to decrease species diversity. Indeed, to improve size diversity, harvesting of sawtimber had to be lighter, which would reduce the regeneration of intolerant species. Still, the species diversity level, with a bound on size diversity near 70% of the maximum, was about the same as that obtained with the unconstrained economic policy. Instead, landscape diversity was much lower with the tree-size diversity constraints. At the extreme, decisions that kept stands within 90% of maximum size diversity decreased landscape diversity by 38%, relative to the unconstrained economic policy.

5. Summary and Conclusions

The objective of this research was to determine the economic and ecological consequences of alternative management regimes in Wisconsin maple-birch forests. To that end, Part I presented a Markov-chain model of stand growth with price changes and applied it to predict tree diversity, landscape diversity, and income in forests managed as they have been in the past, or left alone. Noting the low returns obtained by current management, Part II tried to obtain optimum management policies for different objectives. The core of Part II is a Markov decision process model with an economic or ecological objective function and with a decision space possibly constrained by ecological goals.

The unconstrained economic harvesting policy, solved by successive approximation, maximised the net present value of economic returns for each initial stand-market state. The results showed that this policy could yield more than five times the returns of the current management regime, with the same long-run landscape diversity and diversities of tree size and species less than 10% lower.

Table 8 Effects on Wisconsin maple-birch forests of management that maximise net present value, with constraints on tree species diversity

	Lower bound on tree species diversity (% of possible maximum)		
	50	70	90
Landscape diversity in steady state:	3.23 (97%)*	3.11 (94%)	2.01 (61%)
Average tree diversity in steady state:			
Species	0.85 (109%)	0.93 (119%)	1.00 (128%)
Size	0.55 (100%)	0.48 (87%)	0.51 (93%)
Average expected NPV from current forest ($/acre):			
In high market state	620 (95%)	570 (88%)	467 (72%)
In low market state	593 (95%)	540 (87%)	435 (70%)
Average steady-state return ($/acre/5yr):	93 (94%)	89 (90%)	61 (62%)

Note: * relative to the unconstrained economic policy (Table 2).

Table 9 Effects on Wisconsin maple-birch forests of management that maximise expected net present value, with constraints on tree size diversity

	Lower bound on tree size diversity (% of possible maximum)		
	50	70	90
Landscape diversity in steady state:	2.97 (89%)*	2.90 (87%)	2.06 (62%)
Average tree diversity in steady state:			
Species	0.82 (105%)	0.79 (101%)	0.69 (88%)
Size	0.61 (111%)	0.74 (135%)	0.88 (160%)
Average expected NPV from current forest ($/acre):			
In high market state	454 (70%)	334 (51%)	91 (14%)
In low market state	431 (69%)	324 (52%)	87 (14%)
Average steady-state return ($/acre/5yr):	67 (68%)	64 (65%)	15 (15%)

Note: *relative to the unconstrained economic policy (Table 2).

The policies that maximised the expected average diversity of tree size or species, in the long run, were found by linear programming and confirmed by successive approximation. To maximise either kind of tree diversity meant giving up much timber income, relative to the economic harvesting policy. Achieving maximum species diversity demanded some harvest, it lowered landscape diversity, gave about the same size diversity, but increased income substantially, compared to continuing the current policy. Instead, maximum tree size diversity could be approached by letting nature take its course, but that would entail giving up all income, and having a less diverse landscape, compared to continuing the current policy.

Compromise policies were found by maximising expected discounted returns with harvests that kept stands above a specified lower bound of diversity. All solutions illustrated the trade-off between criteria. Raising the lower bound on either the diversity of tree species or size decreased the other economic and ecological criteria. From an economic viewpoint, it was less costly to increase species than size diversity because cutting large trees tended to increase species diversity. Setting high lower bounds on tree size diversity led to high opportunity costs and low landscape diversity. An example of an attractive policy for the State of Wisconsin, which promotes 'big-tree silviculture', is the policy that maximised expected discounted returns while keeping size diversity within 70% of its highest possible value. It would lead to an average tree-size diversity in the long-run about 25% higher than if the current pattern of harvest continued. It would keep species diversity at nearly the same level. It would more than triple the expected present value of the forest. Its main defect would be to lower landscape diversity by 13% in the long run. Other solutions might avoid this, but not by trying to raise the size of trees. As we have seen, managing for more tree size diversity in stands (which requires keeping more of the large trees) would tend, instead, to bring more uniformity to the forest landscape.

References

BERTSEKAS, D. P. 1976. Dynamic programming and stochastic control. Mathematics in Science and Engineering, Volume 125. Academic Press, Inc., New York.

BLISS, J. C., AND MARTIN, A. J. 1988. Identity and private forest management. Society and Natural Resources 1:365-376.

BLISS, J. C., AND MARTIN, A. J. 1989. Identifying NIPF management motivations with qualitative methods. Forest Science 35(2):601-622.

BUONGIORNO, J, DAHIR, S., LU, H. C. AND LIN, C.R. 1994. Tree size diversity and economic returns in uneven-aged forest stands. Forest Science 40(1):83-103.

EVANS, K.E., AND SZARO, R. C. 1990. Biodiversity: challenges for hardwood utilisation. Diversity 6(2):27-29.

GODMAN, R.M. 1983. Applying the shelterwood system. Northern hardwood notes. Edited by J. Hutchinson. USDA For. Serv. North Cent. For. Exp. Stn. Res. Note 2.04. pp.1-3.

HOOL, J.N. 1966. A dynamic programming-Markov chain approach to forest production control. Forest Science Monograph 12.

KAYA, I., AND BUONGIORNO, J. 1987. Economic management of uneven-aged northern hardwood stands under risk: a Markovian decision model. Forest Science 33(4):889-907.

LEMBERSKY, M.R. 1976. Maximum average annual volume for managed stands. Forest Science 22(1):69-81.

LEMBERSKY, M.R., AND JOHNSON, K. N. 1975. Optimal policies for managed stands: an infinite horizon Markov decision process approach. Forest Science 21(2):109-122.

LU, H.S. 1992. Economic management of Wisconsin's northern hardwood forest stands: a mixed-species model. Ph.D. dissertation, University of Wisconsin, Madison.

NIESE, J.N., AND STRONG, T. F. 1992. Economic and tree diversity trade-offs in managed northern hardwoods. Canadian Journal of Forestry Science 22:1807-1813.

RAILE, G. K. 1985. Wisconsin forest statistics, 1983. Resource Bulletin NC-94. USDA Forest Service, North Central Forest Experiment Station, St. Paul, MN.

ROSS, S. 1983. Introduction to stochastic dynamic programming. Academic Press, New York.

SMITH, D.M. 1986. The practice of silviculture. John Wiley & Sons, 8th ed. New York.

TUBBS, C.H. 1977. Natural regeneration of northern hardwoods in the Great Lakes region. U.S. For. Serv. Res. Pap. NC-150.

VANCLAY, J.K. 1995. Growth models for tropical forests: a synthesis of models and methods. Forest Science 41(1):7-42.

WALTERS, R.S., AND NYLAND, R. D. 1989. Clearcutting central New York northern hardwood stands. Northern Journal Applied Forestry 6:75-78.

WINSTON, W.L. 1991. Operations research: applications and algorithms. 2nd ed. PWS-Kent Publishing Co., Boston.

Groundwater Protection and Forest Management - an Option Value Analysis

Jens Abildtrup and Niels Strange

Abstract

Non-contaminated drinking water may become a scarce resource in the future. In that case, forest watersheds may play a significant role in ensuring the supply of clean drinking water. However, forest management may affect this supply considerably. Christmas tree production is an important income generator in the Danish forestry sector but may contribute to non-point contamination of groundwater resources because Christmas tree production involves the application of significant amounts of fertiliser and pesticides. This paper addresses the decision to convert a natural forest into Christmas tree production when groundwater contamination is irreversible, and future values of non-contaminated groundwater and Christmas tree production are uncertain. It is concluded that solving the decision problem as an optimal stopping problem, which takes advantage of information obtained by delaying irreversible resource commitments, implies that a more conservative conversion strategy is optimal. It is also shown that the expected value of the natural forest may increase when the optimal stopping rule is applied instead of a conventional expected net present value (NPV) analysis. Traditionally, NPV analysis treats the decision as a 'now or never' decision and ignores therefore the option to wait and the value of new information. These results are pertinent and may contribute with substantial information for forest management planning and water resource policy.

Keywords: Christmas trees, multiple-use, option value, contamination

1. Introduction

In most parts of the world natural and semi-natural forests represent a land-use in which fertiliser and pesticides (pollutants) have never been applied, or only in small amounts, implying that groundwater quality beneath such forests has remained unaffected by management practice. Introduction of non-native tree species or intensively managed silvicultural systems, e.g. clear cutting, are often based on application of fertiliser and pesticides. In Denmark deciduous tree species, especially beech, are the natural vegetation on most sites. However, a large part of the deciduous forest has been converted into coniferous forest for economic reasons. Today, such conversion of deciduous forest is highly controversial. Especially, the conversion into Christmas tree production has attracted much political attention because the amounts of fertiliser and pesticides are comparable with intensively managed agricultural land.

The Danish Government has pure groundwater resources as one of its major environmental goals (Ministry of the Environment and Energy 1998a). Since 1993, when the waterworks started analysing for pesticides, an increasing number of borings have been closed because of contamination (Ministry of the Environment and Energy 1998b). A recent report by the Ministry of the Environment and Energy (1997) concluded that over the next ten years every fourth waterworks will contain pollutants above the threshold values. Pollutants may not reach the groundwater resources until 30 to 40 years after application; it is therefore expected that the number of contaminated borings may increase even more in the future.

It is also expected that the following measures have to be taken if the supply should be able to meet future demand for drinking water: *i*) existing waterworks are moved to non-contaminated groundwater resources, *ii*) increased purification of water, and *iii*) increased transportation of non-contaminated water. The total costs of these measures are estimated at DKK 2.51m^{-3}, 0.8 m^{-3}, and 5.18 m^{-3}, respectively, for waterworks with a capacity between 100,000 and 1 mill. m^3 year^{-1}. For waterworks with a capacity below 10,000 m^3 year^{-1} the costs are approximately four times higher (Ministry of the Environment and Energy 1997). However, the future costs of investment in groundwater services are indeed uncertain.

Generally, the economic return on natural forestry is negligible compared to the return on Christmas tree production. However, Christmas tree production induces a risk of causing irreversible contamination of the groundwater and this risk should be considered before deciding whether or not a natural forest stand should be converted. Furthermore, the future uncertainty of the economic return on Christmas tree production and costs of non-contaminated groundwater should also be taken into consideration.

The paper presents an option-value analysis of this decision problem. It is assumed that the forest holder is a municipality, responsible for the drinking water supply. The forest contains a watershed which supplies the municipality with a considerable amount of water. In case of groundwater contamination, the measures mentioned above must be taken for the supply to meet future demand for drinking water. It is assumed that the existing forest is a natural forest where fertiliser and pesticides have never been applied. An increase in the profit from the forest can be achieved by conversion into Christmas tree production. This involves, however, the application of significant amounts of fertiliser and pesticides which additionally implies an irreversible risk of contamination and therefore increased expected costs of obtaining non-contaminated drinking water. The decision problem of the municipality is therefore to balance the expected value of Christmas tree production against the expected value of avoiding the risk of contamination. The latter is defined as the expected net present value of avoided costs from not implementing the above measures.

Arrow and Fisher (1974) and Henry (1974) did the pioneering work on irreversibility and future uncertainty in the field of environmental and resource economics. They considered the problem of developing wilderness when the future benefits from conservation and development are uncertain. Since development is irreversible and conservation is not, there is a value, i.e. an option value, associated with conservation which may be interpreted as the value of maintaining the option to develop open. The value of postponing the decision of implementing the irreversible alternative is generated by new information obtained by waiting. Most of the literature uses two-period models. However, more recent literature presents continuous and infinite time problems. Clarke and Reed (1988) present a model which allows development at any time over an infinite time horizon and in Clarke and Reed (1990) the model allows for partial development. These results are applied by Reed (1993) who determines the optimal decision rule for the choice between conservation and harvest of an old-growth forest when timber and amenity values are uncertain. Using empirical data Conrad (1997) determines the option value of an old-growth forest when the value of timber is certain and non-timber amenity value is uncertain. To our knowledge there are no examples in the literature of including option values of groundwater in the economic analysis of silvicultural alternatives. Option values are estimated for groundwater protection in Sun et al. (1992) and Edwards (1988). However, in these two studies option value corresponds to the concept introduced by Weisbrod (1964) and represents the extra value of ensuring future availability of the resource when future supply and demand are uncertain. Basically, the stopping rule is derived from Reed (1993). However, his study includes risk of sudden destruction which affects both the timber and amenity production. The problem is different in this paper, where the opportunity cost of conversion is uncertain due to the risk of contamination and to the size of the cost of obtaining non-contaminated drinking water given that contamination occurs.

In the next section the model is presented. The optimal timing of conversion is solved as an optimal stopping problem in section three. In section four sensitivity analyses are performed for relevant parameters.

2. Model

The decision problem is either to preserve the natural forest or convert it into Christmas tree production. It is assumed that fertiliser and pesticides have never been applied in the natural forest and there is therefore no risk of contamination if this forest is preserved. On the other hand, if at time t the natural forest is converted into Christmas tree production, we obtain V(t) (DKKha^{-1}) which are the expected present value of the economic benefits when expected costs of the risk of contamination is not included. We assume that V(t) is a stochastic process following a geometric Brownian motion (GBM), governed by the stochastic differential equation

$$dV(t) = \alpha_V V(t)dt + \sigma_V V(t)dz_V \tag{1}$$

where α_V is the drift, σ_V is the standard deviation, and dz_V is the increment in a standard Wiener process. It is also assumed that the avoided cost of non-contaminated groundwater, P(t) (DKKm^{-3}), is a stochastic process following a GBM, governed by

$$dP(t) = \alpha_P P(t)dt + \sigma_P P(t)dz_P \tag{2}$$

where α_P is the drift, σ_P is the standard deviation, and dz_P is the increment in a standard Wiener process. We assume that $\alpha_V, \alpha_P < \delta$, where δ is the instantaneous discount rate, and that the relation between the two stochastic processes V(t) and P(t) is described by the coefficient of correlation, ρ, where $\rho dt = E[dz_V dz_P]$.

Assume that the annual recharge of groundwater is G (m^{3}ha^{-1}) and corresponds to the annual supply of groundwater from the watershed. With the natural forest the expected net present value of future avoided costs of non-contaminated water is calculated as

$$A_n(t) = E\left[\int_t^\infty GP(s)e^{-\delta(s-t)}ds\right] = \int_t^\infty GP(t)e^{-(\delta-\alpha_P)(s-t)}ds = \frac{P(t)G}{\delta-\alpha_P} \tag{3}$$

using

$$E[P(s)\,|\,P(t)] = P(t)e^{\alpha_P(t-s)},\ s > t$$

where s is time and t is the time of reference. If the natural forest is converted into Christmas tree production, the annual recharge of groundwater is also G, but now there is a risk of contamination by pollutants. Let S(s | t) denote the probability that contamination has not occurred until time s, where s > reference time t. We assume that S(s | t) is determined by a constant hazard rate λ, i.e. it is assumed that the hazard of contamination is time-independent. In this case S(s | t) can be expressed as

$$S(s\,|\,t) = e^{-\lambda(s-t)} \tag{4}$$

The hazard rate is independent of the applied amounts of fertiliser and pesticides, and even if the natural forest is reestablished we assume that this would not reduce the risk of contamination. In reality, the risk of contamination may decrease over time. However, the time lag from application of pollutants to contamination may be more than 30 years (Ministry of the Environment and Energy 1997), implying that it is reasonable to regard the risk of contamination as constant and irreversible.

If the natural forest is converted into Christmas tree production at time t, applying (4) means that the expected net present value of the future avoided costs is calculated as

$$A_c(t) = E\left[\int_t^{\infty} GP(s)S(s\,|\,t)e^{-\delta(s-t)}ds\right] = \int_t^{\infty} GP(t)e^{-(\delta-\alpha_p+\lambda)(s-t)}ds = \frac{GP(t)}{\delta-\alpha_p+\lambda} \tag{5}$$

given that contamination has not occurred until t. We see that $A_c(t) \leq A_n(t)$ since $\lambda \geq 0$, indicating that the expected present value of avoided costs is higher for the natural forest.

3. The Optimal Decision Rule

In this section we derive the optimal decision rule for conversion and estimate the option value of non-contaminated groundwater, applying standard methods for optimal-stopping problems (e.g. McDonald and Siegel 1986).

First, we define the intrinsic value function which represents the value of the option to convert at time t

$$R\big(V(t),P(t)\big) = V(t) + A_c(t) - A_n(t) = V(t) + P(t)G\left(\frac{1}{\delta-\alpha_p+\lambda} - \frac{1}{\delta-\alpha_p}\right) \tag{6}$$

If the decision was a 'now or never' decision, we would convert the natural forest if $R(V(t), P(t))$ is higher than zero. However, the real decision problem includes the option to postpone the decision. In the following a decision rule as a function of V(t) and P(t) is derived, i.e. a rule that describes for which values of V(t) and P(t) it is optimal to preserve the natural forest and for which values it is optimal to convert.

Let W(V, P) be the value function

$$W(V,P) = E\left[e^{-\delta\tau}R(V(\tau),P(\tau))\,|\,V(0) = V, P(0) = P\right] \tag{7}$$

where τ is a random variable denoting the time at which the process first leaves the continuation region, i.e. the natural forest is converted at time τ, and the expectation is taken with respect to τ, $V(\tau)$, and $P(\tau)$. The value function W(V, P) presents the expected benefit from conversion net of expected cost of contamination, given that initially V(0) = V and P(0) = P and the optimal stopping rule is applied. Applying Itô's lemma we obtain a partial differential equation which can be solved using appropriate boundary conditions and leading to the stopping rule

$$C = \frac{(1+\theta)}{\theta}G\left(\frac{1}{\delta-\alpha_p} - \frac{1}{\delta-\alpha_p+\lambda}\right), \tag{8}$$

i.e., we will convert the natural forest if $V/P > C$. θ is a parameter estimated using the boundary conditions and it can be shown that $\theta > 0$. Equation (8) can be rewritten as

$$V(t)^* = \frac{1+\theta}{\theta}G\left(\frac{1}{\delta-\alpha_p} - \frac{1}{\delta-\alpha_p+\lambda}\right)P(t) \tag{9}$$

where $V(t)^*$ is the stopping boundary with respect to V(t) as a function of P(t). Since θ is positive, it is clear that conversion is optimal if V(t) exceeds the expected cost of the risk of contamination by a factor $(1+\theta)/\theta$ which is greater than one. Thus, the optimal decision rule is more conservative than that arising from conventional net present value (NPV) analysis which treats conversion as a 'now or never' decision.

4. Results

Table 1 presents the parameter values for the base case scenario. Using these parameters and equation (9), the optimal stopping rule is calculated and presented in Figure 1 as the dotted line. The solid line presents the decision rule based on conventional expected NPV analysis and is calculated as

$$V(t)^{NPV} = G\left(\frac{1}{\delta - \alpha_p} - \frac{1}{\delta - \alpha_p + \lambda}\right)P(t) = C^{NPV}P(t) \tag{10}$$

It is optimal to convert the natural forest when $V(t) > V(t)^* =$ DKK 181,000 ha^{-1}, where $V(t)^*$ is the stopping boundary of V(t) given that P(t) equals DKK 1.0 ha^{-1}. Treating the decision problem as a 'now or never' decision, the stopping boundary of V(t) reduces to $V(t)^{NPV} =$ DKK 103,000 ha^{-1} when P(t) equals DKK 1.0 ha^{-1}. Between the two stopping boundaries conventional expected NPV analysis prescribes conversion. However, it is optimal to maintain the natural forest because of the option value of non-contaminated groundwater.

The optimal stopping boundary C is presented as a function of the risk of contamination, λ (Figure 2). C increases proportionally to the opportunity cost of conversion, $A_n(t)$- $A_c(t)$, which increases when λ increases. Using (8) it is estimated that C goes to DKK 235,000 ha^{-1} when λ goes to infinity. In the case of no risk of contamination, i.e. λ equals zero, C will also be zero since the net return on Christmas tree production, V(t), is always nonnegative (from (1)) and the opportunity

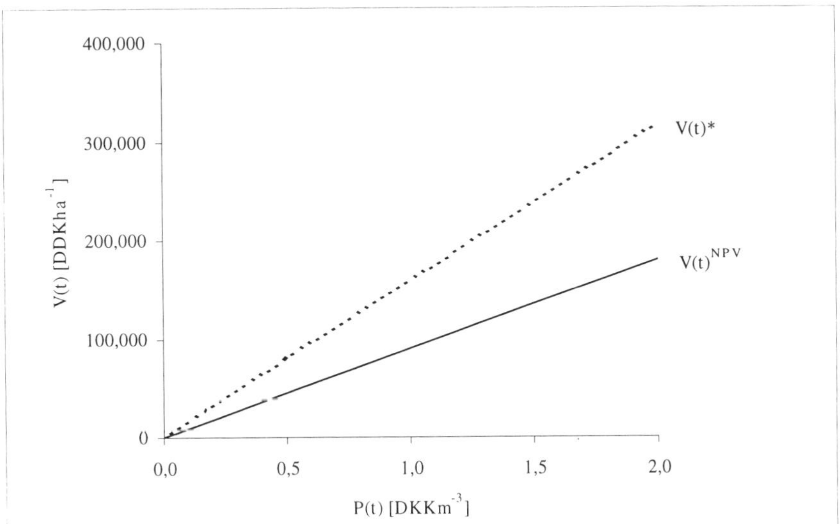

Figure 1 Optimal stopping boundaries using option value analysis (the dotted line) and using conventional NPV value analysis (the solid line)

cost is zero.

It can be shown that the variance parameters affect the stopping rule through the variable $\sigma^2 = \sigma_v^2 + \sigma_p^2 - 2\rho\sigma_v\sigma_p$ which is an estimate of the variance of the ratio V(t)/P(t). The value of waiting and thereby obtain new information is highest when the variance of the ratio V(t)/P(t) is high. Therefore, C increases when σ^2 increases, implying that C increases when σ^2_v and σ^2_p increase but decreases when the correlation between the two processes, ρ, increases (Figure 3).

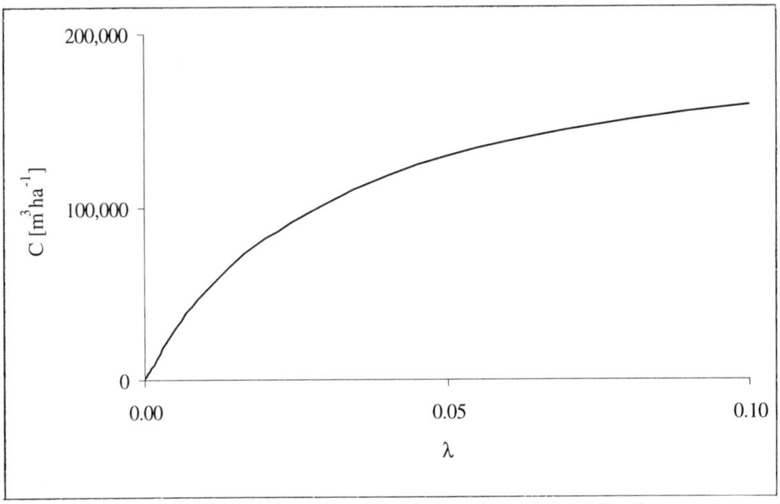

Figure 2 The optimal stopping rule, C, as a function of the risk of contamination

Table 1 Parameter values and stopping rules of the base case

Parameter	Value
σ_V	0.10
σ_P	0.10
ρ	0.0
α_V	0.0
α_P	0.0
δ	0.03
λ	0.10
G	4,000 m^3ha^{-1}
Result	
θ	1.303
C	181,000 m^3ha^{-1}
C^{NPV}	103,000 m^3ha^{-1}

In the base case it is assumed that the two drift terms, α_V and α_P, are zero. We find that C increases when the drift terms increase (Figure 4). When α_V is positive, the expected value of V(t) increases with time, implying that the cost of postponing the conversion decreases and thus the stopping rule C increases. An increasing α_P implies that the relation between the value of V(t) and

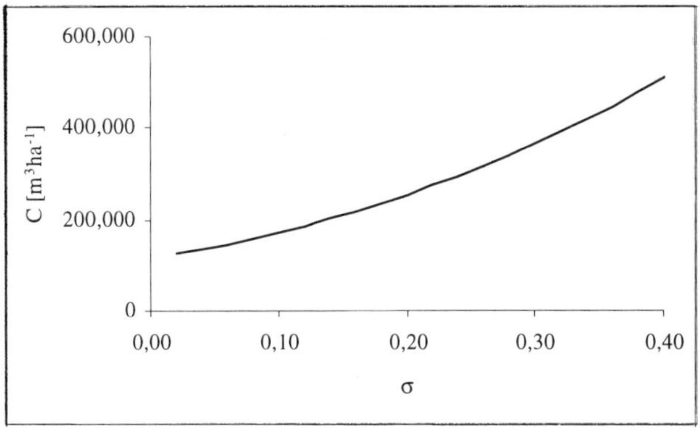

Figure 3 The optimal stopping rule, C, as a function of σ, where $\sigma^2 = \sigma^2_V + \sigma^2_P - 2\sigma_V\sigma_P\rho$

opportunity cost, $A_n(t) - A_c(t)$, decreases, implying that the cost of postponing conversion increases and therefore the option value decreases. However, the effect of a decreasing option value is more than compensated by the effect of increasing opportunity cost which implies that the stopping rule C as well as C^{NPV} increases.

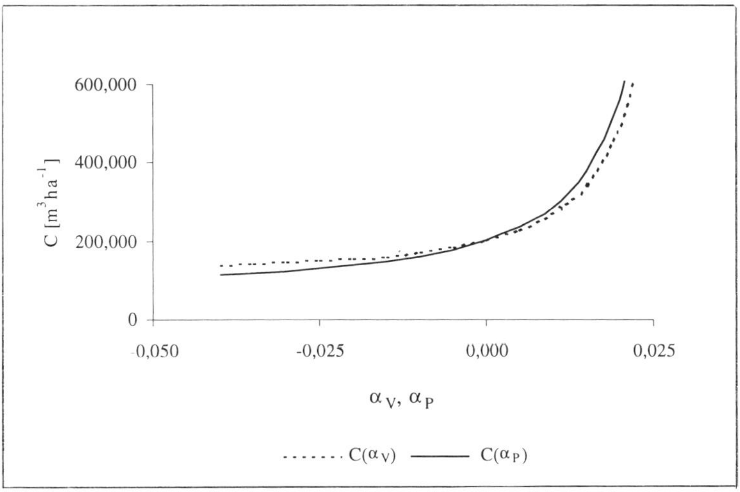

Figure 4 The optimal stopping rule, C, as a function of the drift terms α_V and α_P

5. Discussion and Conclusion

We claim that the option value of a non-contaminated forest watershed should be included in the analysis of decision alternatives which may cause irreversible contamination of groundwater resources. The decision to convert natural forest into Christmas tree production or other profitable and intensive production alternatives has been debated in Danish forestry for decades. The soil expectation value of Christmas tree plantation in Denmark is estimated at DDK 290,000 ha^{-1}, using a 3 per cent discount rate and 1995 prices (Danish Forest Society 1995). The corresponding soil expectation value of the timber production in a natural forest (beech) is close to zero. Therefore, ignoring the value of groundwater and other non-timber values of the beech forest, application of soil expectation value as profitability criterion provides an incentive for conversion.

We have made no attempt to estimate the parameters of the two stochastic processes (1) and (2) and the rate of occurrence, λ. Even if the parameters were estimated there would be no guarantee that the values would hold in the future. Instead our main goal approach has been to select plausible values and to perform a sensitivity analysis. The specification of the stochastic processes governing V(t) and P(t) is standard in the modelling of prices and asset values. One property of a GBM is that it can never be negative. It is reasonable to believe that the avoided cost of non-contaminated groundwater will not go below zero. It is more questionable that this also holds for the soil expectation value of Christmas tree production. It may be more realistic to assume a GBM. However, this would make an analytical solution intractable and numerical solution techniques are required.

The decision problem analysed in this paper can also be applied to other forest management planning decision problems where irreversible alternatives with uncertain return are considered. For example, the choice between a silvicultural system which allows environmental certification of the timber produced and one which excludes the timber from environmental certification. The choice of a system which excludes certification is irreversible in the sense it may take more than one rotation period before the silvicultural system can be converted into a certified system. As such there will be a value associated with postponing the decision and obtain new information about the return before an alternative is chosen that excludes certification.

In forest management the traditional concepts of silvicultural system mainly focus on the biological and economic decision parameters. Thus, we find that a decision basis is needed which considers the forest as part of a complex interacting resource system. Timber production cannot be considered in isolation and environmental values must be integrated in the decision-basis. In most studies, uncertainty about the future demand for timber and other services provided by forests is ignored. It is suggested that forest management must consider imposed externalities and flexibility of management alternatives. Where the conventional NPV approach fails to comprise uncertainty, the option value approach embraces this in a theoretically sound way. Application of the option value approach in forest management planning will typically imply more conservative choices of management strategies.

References

ARROW,K. J. AND FISHER, A. C. 1974. Environmental preservation, uncertainty and irreversibility. Quarterly Journal of Economics 88:312-319.

CLARKE, H. R. AND REED, W. J. 1988. A Stochastic Analysis of Land Development Timing and Property Valuation. Regional Science and Urban Economics 18:357-381.

CLARKE, H. R. AND REED, W. J. 1990. Land development and wilderness conservation policies under uncertainty: A synthesis. Natural Resource Modeling 4:11-37.

CONRAD, J. M. 1997. On the option value of old-growth forest. Ecological Economics 22: 97-102.

DANISH FOREST SOCIETY 1995. Skovøkonomiske tabeller. [Forest Economic Tables], (Unpublished).

EDWARDS, S. F. 1988. Option Prices for Groundwater Protection. Journal of Environmental Economics and Management 15:475-487.

HENRY, C. 1974. Investment Decisions Under Uncertainty: The 'Irreversibility Effect'. The American Economic Review 64(6):1006-1012.

MCDONALD, R. AND SIEGEL, D. 1986. The Value of Waiting to Invest. Quarterly Journal of Economics 101:707-727.

MINISTRY OF THE ENVIRONMENT AND ENERGY 1997. Udviklingen i den danske vandforsyningsstruktur. [Future developments in the Danish water supply]. Working Report No. 62. Danish Environmental Agency, Ministry of the Environment and Energy. (In Danish).

MINISTRY OF THE ENVIRONMENT AND ENERGY 1998a. Drikkevandsudvalgets betænkning. [Report from the Drinking Water Committee]. Report No. 1. Danish Environmental Agency, Ministry of the Environment and Energy. (In Danish).

MINISTRY OF THE ENVIRONMENT AND ENERGY 1998b. Status for lukkede boringer ved almene vandværker. [The number of closed borings at Danish waterworks]. Environmental Project No. 380. Danish Environmental Agency, Ministry of the Environment and Energy. (In Danish).

REED, W.J. 1993. The decision to conserve or harvest old-growth forest. Ecological Economics 8:45-69.

SUN, H., BERGSTROM, J. C. AND DORFMAN, J. H. 1992. Estimating the Benefits of Groundwater Contamination Control. Southern Journal of Agricultural Economics 24(2):63-71.

WEISBROD, B. A. 1964. Collective Consumption Services of Individual Consumption Goods. Quarterly Journal of Economics 77(3):70-77.

Chapter V

Spatial Objectives

How to Integrate Spatial Objectives into Private Forest Planning in Finland

Mikko Kurttila

Abstract

Until these days, only minor attention has been given to spatial layout of forestry operations and their effects on the spatial structure of Finnish private forests. Spatial analysis is needed in forest planning that is addressing questions of biodiversity management and landscape ecology. The needs concern the relations between landscape structure and viability of species populations. The populations are known or assumed to be members of the natural community in a certain spatio-temporal scale. Their viability is dependent, among others, on the amount of suitable habitats. If it is difficult to increase the amount of suitable habitats for economic reasons, an important question arises: How a certain amount of ecologically important habitats should be spatially arranged?

This study proposes a way to integrate spatial objectives into Finnish private forest planning. Some practical needs and objectives of monitoring the spatial structure of Finnish private forests are presented. In practical planning, also other than spatial objectives must be considered. This leads to multi-objective planning, where decision alternatives must be assessed numerically subject to all objectives and also to overall priority. One possibility of using numerical spatial objectives are indices that describe the structure of landscape. They are based on the sizes, shapes and proportions of and distances between habitat patches, and combinations of these features. The use of some of these indices as new objectives in Finnish private forest planning is suggested.

Keywords: Landscape ecological planning, private forest planning, spatial objectives, spatial structure of forests

1. Introduction

In Finnish private forest planning, the main focus during the past decades has been on controlling the development and exploitable quantities of the forest resources, mainly timber. No special attention has been given to spatial layout of forestry operations and their effects on spatial forest structure. The need to include spatial analysis in forest planning is mainly caused by biodiversity management reasons, the viewpoint being landscape ecological. However, there exist also economic and social reasons. Economically, wood supply is better from a forest, where desirable timber quantities are located in easily attainable areas (Baskent and Jordan 1991). Because the abundance, size and shape of habitats (e.g. key biotopes) vary, also their occurrence between private forest holdings varies (Carlsson et al. 1998). This variation should be taken into account when allocating treatment restrictions to certain habitats in private forests as it also contributes to the equity and social fairness of these actions.

Considering spatial objectives in forest planning comprises the fact that it is not self-evident where, e.g. clear-felled areas or old forests are located. Spatial structure is defined by spatial configuration (e.g., the size, shape and relative arrangement) and non-spatial characteristics (e.g., composition) of patches and interconnections among them (Baskent 1997). In Finnish forest planning, the goal of forest compartment division is to form homogenous forest patches with regard to wood production. The criteria for the division are site productivity, tree species composition, the development stage of the stands and also operational reasons. The result is not necessarily compatible with patch division based on landscape ecological principles (e.g. Uttera and Hyppänen 1997).

Biodiversity management aims at maintaining the populations of all those species that are known or assumed to be members of the natural community in a certain spatio-temporal scale (Mykrä and Kurki 1998). Landscape ecological theories and studies provide information of the relations between landscape structure and viability of different species and species groups. Viability is connected to the distribution, movement and persistence of species. The more specific are the habitat needs and the worse are the movement abilities of the species considered, the more important it is to take the spatial arrangement and its changes into account (e.g. Fahrig and Paloheimo 1988). When the distribution and pattern of the resources change, also resource utilisation or, more generally, the ecological processes change (O'Neill et al. 1988). In private forests, the main objective is often timber production and the acceptable costs of biodiversity management may not be very high (e.g. the area of protected habitats). From the landscape perspective, the best strategy may not be to try to spread small protected patches evenly over the planning area. It might be reasonable to aggregate the limited amounts of certain habitats to specified areas (Kurki et al. 1997, Mykrä and Kurki 1998) and try to synchronise the efforts of individual landowners of the area.

Spatial objectives can be included in forest planning by different ways. A standard LP model can be restricted by exogenously determined constraints that are of spatial nature (e.g. Nalli et al. 1996, Naesset 1997). There also exists a wide range of harvest scheduling approaches using adjacency constraint techniques (e.g. Roise 1990, Yoshimoto and Brodie 1994, Tarp and Helles 1997). It is also possible to formulate objective function and constraints for the land allocation problem that optimise spatial layout for wildlife and timber (e.g. Hof and Joyce 1992, 1993). The habitat needs of wildlife species are included in the constraints and the problem is solved by non-linear programming (e.g. Hof and Joyce 1992) or by mixed integer programming (e.g. Hof and Joyce 1993). In addition, landscape structure and critical components in it can be quantified by indices based on the sizes, shapes and proportions of and distances between habitat patches and combinations of these features (e.g. McGarigal and Marks 1994, Baskent and Jordan 1995). There exist metrics that measure the landscape at the patch level (e.g. fractal dimension, core area) and at the landscape level. The latter can be similar to patch level metrics but transformed for a single landscape level measure (e.g. total core area, mean nearest neighbour distance) or they can be calculated based on the whole landscape level data (e.g. spatial autocorrelation, aggregation). Indices can be useful in strategic planning when, for example, the outcomes of different (spatial) cutting strategies are compared (e.g. Li et al. 1993, Gustafson and Crow 1996, Baskent 1997) or if the existence of certain gaps between originally stated forest-level objectives and actual conditions is studied (Wardoyo and Jordan 1996). These spatial indices can be used also as objectives in forest planning.

The aim of this article is to outline means to integrate spatial objectives into Finnish private forest planning. The following sections illustrate the framework in which spatial objectives could be used, give landscape ecological reasons why to consider spatial analysis in Finnish private forestry, and discuss some spatial indices to be used as spatial objectives in forest planning.

2. Practical Framework to Include Spatial Objectives in Private Forest Planning

2.1 A need for regional planning

From the landscape ecological perspective, a typical Finnish private forest holding is not necessarily a relevant level for spatially explicit forest planning. The reason is that the emphasis of landscape ecological planning is on the maintenance of large, dynamic ecosystems, and the sizes of privately owned forests (on the average about 35 ha) and treatment units (on the average about 2 ha) are small for this purpose. Forest owners do not usually have control over the operations on the adjacent holdings and this can sometimes lessen the value of biodiversity management efforts in their own holdings. The distances between old forests, situated in the same forest holding are necessarily not very long and there can exist only a few old forest stands. Thus, alternative arrangements of old forests and clear-felled areas may be rather limited.

To increase the planning area, a new kind of planning approach to Finnish private forestry is proposed. So-called regional planning could be used to integrate the landscape ecological principles and needs into private forestry (Pukkala et al. 1997). In regional planning, a certain area, e.g. from 1,000 to 5,000 ha, covering several forest holdings, is taken under consideration as a whole. An aggregate solution is worked out for this area and allocated to the individual holdings, applying suitable techniques. One possible technique is to construct several feasible solutions for each planning unit, and among them, select the one that is optimal for the aggregate area (e.g. Davis and Liu 1991). Another is to produce a feasible solution for the aggregate planning area and try to attain the acceptance of individual forest owners by negotiations or by monetary compensations. For the Finnish circumstances, these techniques are under construction, and they are based on information sharing and group decision making (Pukkala et al. 1997). At regional planning level, analysis of the state and development of the spatial structure of forests are more appropriate. Therefore, methods allowing spatially explicit forest planning should be developed.

2.2 Operationalisation

In this article, spatial objectives will be included as additional objectives in the HERO's (Pukkala and Kangas 1993) heuristic optimisation algorithm. HERO consists of two main phases: estimation of the utility model and maximisation of the utility model. The objectives, i.e. the variables of the utility model can include a variety of parameters describing, e.g. the growing stock, timber removals, incomes, cutting areas and multiple-use values. The use of biodiversity indices in the utility model has been demonstrated (e.g. Kangas and Pukkala 1996, Pukkala et al. 1997), are based on combinations of non-spatial variables, e.g. volume of dead wood, volume of deciduous wood, stand age and within-stand variation, One variable indicating spatial diversity was also used (Pukkala et al. 1997). The difference between mean heights of neighbouring stands was used when calculating the length of clear (difference > 5 m) and distinct (difference > 10 m) boundary, which was intended to depict the variety of habitats and length of edge zones. The HERO method allows modification of sub-priority functions individually for all objectives, e.g. according to the principle of decreasing marginal utility. To further the regional planning, individual objectives can be targeted to cover the whole planning area or part of it (e.g. forest holding). In addition, several decision makers can formulate their own utility functions consisting of selected objectives and their importancies. Finally, the utility function for the whole planning area including the weights of all decision makers, is formulated and optimised. The graphical user-interface allows changing of the weights of objectives as well as the weights of decision makers in an interactive manner.

When spatial objectives are included in forest planning, spatial data are needed. Geographical information systems (GIS) can produce the needed spatial information. The problem is how to utilise this information jointly with optimisation algorithms. If the original compartment division remains unchanged during the planning period, spatial information needs to be incorporated only once. In this case, the needed spatial information comprises, e.g. the list of neighbouring compartments, the lengths of the borders between compartments and distances between stands. Other information needed on the initial state of forests is obtained from the inventory data of compartments and is processed by the forest planning software.

3. Examples of the Needs to Include Spatial Objectives in Forest Planning and Preliminary Solutions

3. 1 Case 1: Management of fragmented old forests

The Finnish private forests are rather fragmented. The amount of ecologically important habitats in a landscape is often small, and these habitats are often prone to disturbances from logging operations. For example, old forests are often distributed in small, isolated patches. In practice, the size of a certain habitat, e.g. old (broadleaved) forest, can be too small to be a functional habitat for some species. According to Andrén (1994), when the proportion of a suitable habitat varies between 10 - 30 %, the negative effects of patch size and isolation start to occur. When the proportion of habitat is further reduced from 20 %, an exponential increase in the distance between remaining habitat patches will result. Thus, after this point, the importance of taking spatial arrangement of patches into account increases also. Kurki et al. (1997) studied fragmentation of old forests in NE Finland. They found that the proportion of old forests (timber volume > 100 m^3/ha) was 15 % in the 250 km^2 study area and that continuous old forest areas were lacking already on scales that are known to be important to several bird species affected negatively by the fragmentation of old forests, e.g. capercaillie. Kurki et al. (1997) suggested to increase both the size and aggregation of cutting areas to decrease the fragmentation of old forest.

One index to measure the fragmentation is spatial autocorrelation (e.g. de Jong et al. 1984). It can be calculated for discrete variables (e.g. tree species or some binary classification of stands) and for continuous variables (e.g mean ages of stands or tree heights). Moran's I is a metric for continuous variables (for other equations to calculate spatial autocorrelation, see e.g. Reed and Burkhart 1985):

$$I = \frac{n}{2a} \frac{\sum_i \sum_j W_{ij}(x_i - \bar{x}) / (x_j - \bar{x})}{\sum_i (x_i - \bar{x})^2},$$

where n is the number of stands, a is the total number of neighbouring stands, x is the value of variable of interest, W_{ij} is the weight that indicates the spatial relation between stand i and j (1 adjacent, 0 otherwise). Usually, the criterion defining spatial relation of stands is adjacency condition. However, there are also different weighting functions that can be used instead, e.g. distance weight, area weight, boundary weight and combinations of these three (Chou et al. 1990).

In general, if autocorrelation is positive (on a scale between -1 and 1), similar stands occur in association, i.e. they are clustered, and when it is negative, similar patches are scattered across the area. Thus, the objective 'minimisation of spatial autocorrelation of clear-cutting areas' can be interpreted as a 'soft' adjacency constraint approach, because it does not unconditionally prevent adjacent clear-cuttings or set limit to maximum open area. Maximisation of spatial autocorrelation of cutting areas would produce clustered cutting areas. Maximisation of spatial autocorrelation of old forests would direct the locations of old forests towards more clustered structure. Wardoyo and Jordan (1996) suggested the use of spatial autocorrelation as an objective in forest planning that is based on forest landscape level objectives.

In regional planning, positive spatial autocorrelation of old forests could be pursued in the whole planning area while the other objectives could concern individual holdings. This would result in a plan where the forest owner's objectives, at the forest holding level, are satisfied as well as possible. In addition, the spatial distribution of old forests would be synchronised over forest holding borders and the connectivity between them would be better. The effect of the whole area level objective on forest holding level can be tested by altering its priority. Some biodiversity index (e.g. Pukkala et al. 1997) could be used and it could be connected to spatial objectives.

Some other indices illustrate the same phenomenon as spatial autocorrelation does, the relative locations of stands. For example, aggregation (the ratio of lengths of similar boundaries of certain stand types and total lengths of boundaries) or average proximity (quantifies the spacing of similar patches across the landscape)(Baskent and Jordan 1995) could be used instead.

3.2 Case 2: Preserving the spatial and temporal continuum of important habitats

When we examine a species with weak movement abilities and whose spatial scale covers only a single forest stand with specified characteristics, the spatial and temporal continuum of this stand type or its resources are important. These kinds of species are often dependent on deciduous and/or decaying trees of large diameter. If the stand is cut, the existence of this species in certain area can be lost, if there are no other adequate resources available within the species' dispersal ability.

One possibility to take resource needs of these species into account is to use distances between stands as an objective variable, e.g. nearest neighbour distance or mean nearest neighbour distance (McGarigal and Marks 1994). This objective would be targeted for the class of stands that can provide relevant features for the species in question. The sub-priority function of this objective could be such that when the distance is under a certain threshold value, the sub-priority of this variable is one or when the distance exceeds the threshold value, the sub-priority drops to zero. If only one stand within threshold distance provides enough of the resource, the value of the utility function will not increase although more suitable stands were saved inside this distance. It may be that the species in question can not use any foreign habitat when it migrates and only adjacent stands can be considered. Probably, certain transition period must be determined during which the stand can not be cut until some of the adjacent stands can provide needed resources.

Sometimes, the dispersal conditions of a species with moderate movement abilities (e.g. flying squirrel) need to be assisted so that it can migrate from one habitat to another. A creation of stepping stone network or corridors could be considered. In this situation, the maximum permitted distances between stepping stones or the width of the corridor must be determined. These measures depend on the species in question.

For a stepping stone network, only parts of the compartments may be needed and the rest could be harvested. One way to brake down the predetermined compartment division could be to simulate, by using GIS, a network of small compartments (with sizes and distances specified according to the needs of the species in question) covering the planning area and use them as potential stepping stones. The optimisation result would show which of them were chosen for stepping stones. In fact, something similar is already practiced in clear-cutting operations in Finland, since the present instructions of forest treatment suggest that some trees should be left in clear-felled areas, preferably in small groups. Other possibility could be partial treatments of compartments, where mixed strategies and combinations of methods, such as 70 % of the compartment area clear-felled and 30 % of the area thinned, are used (Pukkala et al. 1997). The maximum permitted distance between stepping stones or the width of the maximum open area may be strict, so that if it is exceeded, the usability of the stepping stone network diminishes sharply. When using heuristic optimisation, the strict objectives can be handled by using the method iteratively and interactively (Kangas and Pukkala 1996) or by using penalties that decrease the value of the objective function so that an unacceptable solution is never obtained.

4. Conclusions

In regional planning, several forest holdings are treated simultaneously, giving advantages through synchronising operations across forest holding borders. For example, a larger uniform old growth forest area could be created. According to Pesonen and Soimasuo (1998), the requirements for forest holding level sustainability and even or rising cutting and income flows do not decrease the allowable cut if the size of the planning unit exceeds 4000 ha, compared to smaller planning units. When the comparison was done between the actual ownership structure and treating the whole study area (15500 ha) as a single planning unit, the difference between allowable cuts was 13 %. These possible economic advantages could be considered as compensations for managing voluntarily for biodiversity. However, there also exist forest owners to whom biodiversity and other multiple-use objectives are more important than economic goals. By recognising these forest owners, the biodiversity management efforts could be concentrated particularly around their forest holdings.

In this article, the potential use of only some of the dozens existing spatial indices was briefly presented. When continuing the study, the usability of the other indices, e.g. the amount and distribution of core areas, will be examined. Spatial autocorrelation seems quite a versatile spatial index to be used as an objective because it connects locational and thematic information. A problem is that the indices are not necessarily very sensitive to changes in spatial structure. It may be hard to link the indices directly to the habitat needs of wildlife. Because one index measures only one aspect at a time, the simultaneous use of several indices may be needed. Therefore, including a set of habitat indices into optimisation, modified for different species, might be a suitable alternative. However, lacking species specific information may become a problem. Studies that provide this kind of information are currently in progress in Finland. In addition, assessments based on expert judgment can be utilised.

References

ANDRÉN, H. 1994. Effects of habitat fragmentation on birds and mammals in landscapes with different proportion of suitable habitat: a review. Oikos 71: 355-366.

BASKENT, E.Z. 1997. Assessment of structural dynamics in forest landscape management. Canadian Journal of Forest Research 27:1675-1684.

BASKENT,E.Z. AND JORDAN, G.A. 1991. Spatial wood supply simulation modeling. The Forestry Chronicle 67:610-621.

BASKENT, E.Z. AND JORDAN, G.A. 1995. Characterising spatial structure of forest landscapes. Canadian Journal of Forest Research 25:1830-1849.

CARLSSON, M., ANDERSSON, M., DAHLIN, B. AND SALLNÄS, O. 1998. Spatial patterns of habitat protection in areas with non-industrial private forestry - hypotheses and implications. Forest Ecology and Management 107:203-211.

CHOU, Y., MINNICH, R.A., SALAZAE, L.A., POWER, J.D. AND DEZZANI, R.J. 1990. Spatial autocorrelation of wildfire distribution in the Idyllwild quadrangle, San Jacinto Mountain, California. Photogrammetric Engineering and Remote Sensing 56(1):1507-1513.

DAVIS,L.S. AND LIU, G. 1991. Integrated forest planning across multiple ownerships and decision makers. Forest Science Vol.37(1):200-226.

FAHRIG, L. AND PALOHEIMO, J. 1988. Effect of spatial arrangement of habitat patches on local population size. Ecology 69(22):468-475.

GUSTAFSON, E.J. AND CROW, T.R. 1996. Simulating the effects of alternative forest management strategies on landscape structure. Journal of Environmental Management 46:77-94.

HOF, J.G AND JOYCE, L.A. 1992. Spatial optimisation for wildlife and timber in managed forest ecosystems. Forest Science 38(3):489-508.

HOF, J.G AND JOYCE, L.A. 1993. A mixed integer linear programming approach for spatially optimising wildlife and timber in managed forest ecosystems. Forest Science 39(4):816-834.

DE JONG, P., SPRENGER, J.C. AND VAN VEEN, F. 1984. On extreme values of Moran's I and Geary's C. Geographical Analysis 16(1):17-24.

KANGAS, J. AND PUKKALA, T. 1996. Operationalisation of biological diversity as a decision objective in tactical forest planning. Canadian Journal of Forest Research 26:103-111.

KURKI, S., NIKULA, A. AND MYKRÄ, S. 1997. Fragmentation of older forest in relation to scale in managed boreal forests in NE Finland. In: Kurki, S. 1997. Spatial variation in the breeding success of forest grouse: The role of predation in fragmented boreal forest landscapes. University of Turku, Academic Dissertation. Annales Universitatis Turkuensis, Ser. A II, 12.

LI,H., FRANKLIN, J.F., SWANSON, F.J. AND SPIES, T.A. 1993. Developing alternative forest cutting patterns: a simulation approach. Landscape Ecology 8(1):63-75.

MCGARIGAL, K. AND MARKS, B. J. 1994. FRAGSTATS: Spatial pattern analysis program for quantifying landscape structure. U.S. Forest Service General Technical Report, PNV 351.

MYKRÄ, S. AND KURKI, S. 1998. Landscape structure and local population viability in managed boreal forests of Fennoscandia: Ecological basis, management and planning. Manuscript.

NAESSET, E. 1997. A spatial decision support system for long-term forest planning by means of linear programming and geographical information system. Scandinavian Journal of Forest Research 12:77-88.

NALLI, A., NUUTINEN, T. AND PÄIVINEN, R. 1996. Site-specific constraints in integrated forest planning. Scandinavian Journal of Forest Research 11:85-96.

O'NEILL,R.V., MILNE, B.T., TURNER, M.G. AND GARDNER, R.H. 1988. Resource utilisation scales and landscape pattern. Landscape Ecology 2(1):63-69.

PESONEN, M. AND SOIMASUO, J. 1998. Tilakohtaisen kestävyyden vaikutus suuralueen kestäviin hakkuumahdollisuuksiin tapaustutkimus Satakunnan metsälautakunnan alueella. [The effect of sustainability requirements for forest holdings on the regionally sustainable allowable cut - A case study for the region of the Satakunta forest district.] Folia Forestalia 1:43-51.

PUKKALA, T. AND KANGAS, J. 1993. A heuristic optimisation method for forest planning and decision making. Scandinavian Journal of Forest Research 8:560-570.

PUKKALA, T., KANGAS, J., KNIIVILÄ, M. AND TIAINEN, A-M. 1997. Integrating forest-level and compartment-level indices of species diversity with numerical forest planning. Silva Fennica 31(4):417-429.

REED, D.D. AND BURKHART, H. E. 1985. Spatial autocorrelation of individual tree characteristics in loblolly pine stands. Forest Science 31(3):575-587.

ROISE, J.P. 1990. Multicriteria nonlinear programming for optimal spatial allocation of stands. Forest Science 36(3):487-501.

TARP, P. AND HELLES, F. 1997. Spatial optimisation by simulated annealing and linear programming. Scandinavian Journal of Forest Research 12:390-402.

UUTTERA, J. AND HYPPÄNEN, H. 1997. The relationship between forest management planning units and spatial distribution of forest habitat components. Silva Fennica 31(4):431-446.

WARDOYO, W. AND JORDAN, G.A. 1996. Measuring and assessing management of forested landscapes. The Forestry Chronicle 72(6):639-645.

YOSHIMOTO,A. AND BRODIE, J.D. 1994. Comparative analysis of algorithms to generate adjacency constraints. Canadian Journal of Forest Research 24:1277-1288.

A Spatial Optimisation Model for Wetlands Restoration Planning Integrating GIS and Linear Programming

Jan Juul Jensen

Abstract

Restoration of riparian wetlands, i.e. wet meadows along streams, is getting great political attention in Denmark. The attention is mainly due to the potential nitrate retention, but many different social and environmental interests are related to the use of the riparian areas, including both agricultural and landscape ecological aspects. Selection of restoration sites should be based on multiple criteria to reflect this. Geographic information systems (GIS) are well suited to analyse and visualise spatial relationships but are lacking options for solving spatial optimisation problems. To overcome this shortcoming a linear programming model for selection of wetlands restoration sites has been developed. The model is formulated as a Binary Integer Programming model and includes criteria regarding costs, nitrate retention and landscape ecology. Based on GIS analysis the spatial relationship between decision units, land use change and N-retention potential are derived from digital maps. The paper illustrates the problems related to spatial optimisation when decision units can not be defined by the individual cells on a regular grid and demonstrates the need for closer integration between GIS and optimisation tools.

Keywords: Wetlands restoration, optimisation, GIS, Linear Programming

1. Introduction

The human population growth together with the technological development have caused increasing demand for land for anthropogenic uses. Unfortunately, urban development, roads, agriculture, industry etc. cause unwanted environmental impacts. Degradation of water, air and soil quality, habitat fragmentation and species extinction are some of the results. In the past the different environmental consequences were neglected or treated independently of each other. However, recent developments within ecosystem and landscape ecology research have provided a better understanding of ecosystem functioning as a result of integrated processes at both ecosystem and landscape level. In land use planning space is recognised as a limited resource, subject to competing land use interests. Land use planning is the task of allocating one or more land uses to a set of geographic units, while trying to incorporate multiple and often conflicting environmental, social and economic objectives. Geographic information systems (GIS) and mathematical programming, e.g. linear programming (LP), are tools that can facilitate large-scale multi-objective land use planning.

Geographic information systems (GIS) are computer programs specialised for capture, storage, management, analysis and visualisation of spatial data. The main purpose of a geographic information system is to process spatial information. The data structure can be conceptualised as a set of 'floating maps' with common registration, allowing the user to 'look' down and across the stack of maps (Berry 1993). In land use planning one of the uses of GIS is for suitability analysis and mapping. GIS-based suitability analysis is a process in which factor maps, i.e. map layers expressing various siting criteria (e.g. slope, soil type, distance to river) are overlaid to generate composite suitability maps, depicting the relative suitability of a given area for achieving a particular objective, i.e. a specific land use. The variation within a factor map, such as soil classes or land use classes, is referred to as type. In most cases a rating of types is performed, in which a numerical value representing criterion score is applied to each type. Numerous multi-criteria-decision-making (MCDM) approaches exist for determination of scores and weights and for combination of criteria (see e.g. Eastman et al. 1993, Pereira and Duckstein 1993,

Jankowski 1995). One of the most commonly used techniques for suitability analysis is weighted linear combination, performed on grids. Each factor map is given a weight with which the criterion score (rating) of individual locations is multiplied before summation to create the suitability map:

$$S = Sw_ix_i$$

where

S = suitability
w_i = weight of factor i
x_i = criterion score of factor i

GIS-based suitability analysis is an effective way of ranking individual decision units according to the stated objectives. However, suitability analysis cannot address combinatorial problems, i.e. spatial relationships among decision units, e.g. selection of a set of contiguous units giving the highest score, or situations where the score of a given unit is dependent on whether or not another unit has been allocated, e.g. adjacency. It has been suggested to use heuristics to handle these holistic allocation problems (Tomlin 1990). Heuristics can be considered as a set of guidelines by which a problem can be explored, rather than a set of guideposts by which a problem can be solved. Although heuristics strive for an optimal solution, and probably will find good ones, they cannot guarantee an optimal solution.

During the last decades several examples of the use of mathematical programming (MP) in combination with GIS for optimal solution of spatial land use problems have been published (Cleveland et al. 1979, Diamond and Wright 1988, Diamond and Wright 1989, Janssen and Rietveld 1990, Carver 1991, Campbell et al. 1992, Kao and Lin 1996). MP has been applied in various fields like site selection for waste fills and power plants, wild life habitat protection and agriculture. Many of these address the problem of aggregating basic decision units to form sites that maximises the utility while satisfying specified spatial constraints, e.g. concerning contiguity, area and shape. The problem of spatial relationship between decision units is also addressed within the field of forest planning. The main focus has been on formulation of constraints to enforce adjacency delay of clearcutting. Recently, wildlife considerations have gained much research attention (Hof and Joyce 1993, Bettinger et al. 1996). Spatial measures of wildlife habitat protection, e.g. edge effects, fragmentation and compactness of uncut areas, have been implemented by various mathematical formulations. In most cases basic decision units are defined as the cells of a regular grid. Only few examples of irregular basic units exist (e.g. Diamond and Wright 1989).

Spatial optimisation problems have been solved by both linear and non-linear programming and by different heuristics, e.g. TABU search (Bettinger et al. 1997) and simulated annealing (Lockwood and Moore 1992, Bos 1993).

In this paper a binary integer programming model for optimised selection of wetlands restoration areas is presented. Binary variables ($X \in \{0,1\}$) are well suited for spatial planning as they can be used to explicitly represent whether a decision unit i has been selected ($X_i=1$) or not ($X_i=0$). Generation of input data is facilitated by a geographic information system, ArcView (ESRI, 1997) and the build in object-oriented programming facility, AVENUE. The river Vejle Å watershed (342 km^2) in eastern Jutland, Denmark, is to be used as case area for test of the optimisation model and the GIS-based procedures for data generation.

2. Wetlands Restoration

The loss of wetlands is a world-wide environmental problem. In Denmark the loss of wetlands is mainly the result of a drastic intensification of the agricultural production. By canalisation of streams, tile drainage, ditching and pumping wetlands, shallow lakes and coastal marine areas have been reclaimed for agricultural purposes. This is the result of the prevailing utilitarian view of the environment and lack of knowledge of the real values, i.e. fully assessed benefits, of wetlands. This fact can be attributed to general unawareness or ignorance of the environmental work done by wetlands, but also to the lack of systematic efforts to measure the non-market values of wetlands (Maltby et al. 1994). The costs of replacing benefits like natural nutrient retention have rarely been taken into consideration when wetlands have been reclaimed. However, a growing awareness of the environmental benefits of wetlands have placed wetlands restoration on the political agenda. According to the guidelines from the Danish Forest and Nature Agency (DFNA 1992) for restoration and management of nature areas, objectives like ecological functioning, geographic localisation, achievement of multiple goals and cost-benefit considerations should be emphasised when prioritising restoration projects. In that respect wetlands are recognised as very important ecosystems to protect and restore, as they provide a number of environmental benefits, e.g. nutrient retention, life support, flood control etc. (Hoffman 1998, Jansson et al. 1994). In Denmark restoration of wet riparian meadows is given special attention in the newly adopted second Danish Action Plan of the Aquatic Environment (Iversen et al. 1998) as one of the measures to combat nitrogen pollution of surface waters.

3. The Optimisation Model

3.1 Criteria

Each of the sites are characterised by a set of economic, environmental and spatial attributes. The ideal model would incorporate all environmental and social interests in the siting decision. However, in this study focus is limited to nitrogen retention and life support provided by riparian wetlands. The environmental objectives are weighed against the restoration costs. Hence, the objective is to maximise nitrogen retention and the life support function while minimising the costs.

3.2 Costs

The costs can be divided into the costs related to land acquisition, lost revenue due to changed agricultural production, direct restoration costs and management costs. Precise estimation of restoration costs require detailed site level assessments. For large-scale analysis this is a very comprehensive task. Given the availability of data and the limitations on resources (time and money) for land use planning, the loss in revenues caused by land use change is used as a reasonable cost indicator.

3.3 N-retention

The N-retention potential of riparian wetlands is a function of both catchment and wetlands characteristics. A GIS-based methodology developed by Jensen (1998) is used to determine N-load to and N-retention potential in restoration areas, based on catchment characteristics like size, soil type, land use and topography, and the width of restoration area. Calculations are performed on grids and a digital elevation model is used to determine the accumulated load to the potential wetlands. The use of grid based calculation method implies that N-retention can be summarised on decision units of any preferred size.

3.4 Life support

While costs and N-retention potential are well-defined objectives that can be evaluated in real quantities, the ecological objectives are of more subjective character. One of the problems of assessing land use allocation ecologically is related to the enormous diversity of species, and hence, the multiple scales on which the environment can be perceived. In general, an organism-centred approach is unsuitable, unless the allocation is targeted at the protection of a limited number of species. Ecological objectives in terms of wild-life habitat conservation with respect to a single or a few species have been used (Pereira and Duckstein 1993, Nevo and Garcia 1996, Bettinger et al. 1996), but also the application of more general terms for ecological objectives have been demonstrated, e.g. connectivity and edge effects (Hof and Joyce 1993) and proximity to desirable or undesirable land features (Gilbert et al. 1985,

Bos 1993).

In landscape ecological theory it is recognised that ecological processes and species richness and abundance are functions of both site specific conditions and landscape structure (Forman and Godron 1986), i.e. patterning of landscape elements (patches). A large number of assessment metrics at both patch and landscape level have emerged from landscape ecology theory (Turner and Gardner 1991). A comprehensive collection of landscape ecological metrics are implemented in the GIS-based FRAGSTAT software (McGarigal and Marks 1994). The metrics are related to either the composition or the configuration of the landscape. While composition refers to patch statistics like presence and number of patch types without being spatially explicit, configuration refers to the physical distribution or spatial character of patches within the landscape. Since the wetlands restoration problem is considering only one land use type and the objective is to compare the score related to alternative sites, the emphasis is on landscape configuration rather than composition. Some measures of configuration are related to the spatial character of a single patch, e.g. shape and core area, others are related to the distribution of patches within the landscape (Table 1).

Two problems are related to the use of landscape ecological metrics in relation to spatial optimisation model - a general and a practical problem. The general problem is to decide the relevance of each metric for wetlands evaluation. The practical problem is how the relevant metrics can be formulated within the formal syntax of Linear Programming. The FRAGSTAT metrics are developed for evaluation of landscapes with a fixed patch configuration. However, in the present case the metrics are to be used in a prescriptive way, i.e. dynamically to guide the selection of a set of potential restoration sites. Since the LP model does not operate on spatial variables, i.e. maps, all relationships between site alternatives must be pre-specified.

The choice and formulation of metrics are partly dependent on the representation of decision units. Decision units can either be represented as the cells of a regular grid or as predefined irregular sub-regions. The cellular model has a number of advantages regarding formulation of spatial constraints, e.g. adjacency, calculation of edge length, core area etc. However, the use of regular cells is unsuited for riparian wetlands restoration planning due to the nature of the problem. The main reason is the hydrology, implying that units cannot be chosen randomly. Since riparian wetlands are defined as being in contact with the water-course, the use of a cellular model would necessitate constraints enforcing this spatial restriction. Further, N-retention is a function of N-load and the flow length through the wetland area to the water-course and cannot be calculated for each cell independently. Therefore, decision units are represented as predefined sub-areas in contact with the water-course.

The life support criteria considered are:

1. Land use change - reflects the greater ecological benefits of converting, e.g. intensive crop land into wetlands compared to extensive pasture.
2. Width of area - measured as maximum Euclidean distance from stream, represents a measure for habitat diversity potential within a unit.
3. Distance to existing wetlands - areas close to existing wetlands are of greater value because close proximity increases wetlands habitat diversity.
4. Distance to roads, urban areas and agriculture - close proximity to these land use types increases risk of disturbance (human impacts, pesticides etc.).
5. Adjacency - selection of sites next to each other increases protectedness and diversity.
6. Proximity - selection of sites close to each other increases habitat diversity.

Life support criteria 1 - 4 are combined in a suitability index derived by GIS-based weighted summation suitability analysis, as described above, while criteria 5 and 6 are formulated as individual objectives.

Table 1 Examples of landscape configuration metrics computed in FRAGSTATS (after McGarigal et al. 1994)

Scale	Acronym	Metric (units)
Area metrics		
Patch	AREA	Area (ha)
Class	%LAND	Per cent of landscape (%)
Class/landscape	LPI	Largest patch index(%)
Patch density		
Class/landscape	NP	Number of patches (#)
Class/landscape	PD	Patch density (#/100 ha)
Edge metrics		
Patch	PERIM	Perimeter (m)
Patch	EDCON	Edge contrast index (%)
Class/landscape	CWED	Contrast-weighted edge density (m/ha)
Shape metrics		
Patch	SHAPE	Shape index
Patch	FRACT	Fractal dimension
Class/landscape	MSI	Mean shape index
Core area metrics		
Patch	CORE	Core area (ha)
Patch	NCORE	Number of core areas (#)
Class/landscape	MCA1	Mean core area per patch (ha)
Nearest-neighbour metrics		
Patch	NEAR	Nearest-neighbour distance (m)
Patch	PROXIM	Proximity index
Class/landscape	MNN	Mean nearest-neighbour distance (m)
Diversity metrics		
Landscape	SHDI	Shannon's diversity index
Landscape	PR	Patch richness (#)
Contagion metrics		
Class/landscape	IJI	Interspersion and juxtaposition index (%)

4. Model Formulation

The optimisation model for wetlands restoration areas selection is formulated as a Binary Integer Programming (BIP) problem. BIP is a special case of Linear Programming (LP), in which the decision variables only can attain one of two values, most commonly the values 0 (zero) or 1 (one). In the optimisation model the set of decision variables $X = \{x_1, x_2, \ldots\ldots, x_n\}$ represents the set of restoration unit. A decision variable is given the value 1 or 0 according to whether the unit has been selected or not. Optimisation of the multiple (and conflicting) objectives is based on the use of a weighted objective function. Use of a weighted objective function may not lead to a single optimal solution but to a set of efficient solutions giving the weighted objective function the same maximum value, but based on different scores of the mutually inclusive objectives (Pareto-optimal solutions).

The problem can be specified as:

> Given a set of decision units representing potential wetlands restoration sites, select the set of units which maximises suitability, nitrogen retention, adjacency and proximity

indices at the minimum possible cost, subject to upper and lower bounds on total area to allocate.

The cost objective is transformed into a maximisation objective by adding a negative sign.

$$Max\sum_i -C_i \cdot X_i \cdot W_c \qquad \text{('minimise' costs)}$$

$$Max\sum_i N_i \cdot X_i \cdot W_n \qquad \text{(maximise nitrogen retention)}$$

$$Max\sum_i S_i \cdot X_i \cdot W_s \qquad \text{(maximise suitability for life support)}$$

where:

i = set of restoration units
C_i = costs related to restoration of unit i
N_i = Nitrogen retention by restoration of unit i
S_i = suitability of restoration of unit i
W = weight of objective

$$X_i = \begin{cases} 1 \text{ if unit i is selected} \\ 0 \text{ if not} \end{cases}$$

The adjacency and proximity objectives are dependent on the concurrent selection of two units. For the adjacency objective the length of common edge between units is used as measure. The proximity objective is used to enforce selection of so-called functional connected sites. An upper bound on proximity reflects a maximum distance for functional connectivity. Proximity is expressed as the inverse distance between two units within functional distance and calculated as maximum functional distance subtracted minimum distance between two units, i.e. the shorter distance the higher value:

$$Max \sum_{(i,m)\in K} A_{lm} \cdot Z_{lm} \cdot W_a \qquad \text{(maximise adacency)}$$

$$Max \sum_{(f,g)\in L} P_{fg} \cdot R_{fg} \cdot W_p \qquad \text{(maximise proximity)}$$

where :

A_{lm} = length of common edge of unit l and m

P_{fg} = proximity of unit f to g

$$Z_{lm} = \begin{cases} 1 \text{ if two adjacent units (l, m) are selected} \\ 0 \text{ if not} \end{cases}$$

$$K = \{(l, m) \mid \text{restoration unit l \& m are adjacent}\}$$

$$R_{fg} = \begin{cases} 1 \text{ if a unit f within functional distance of g and not adjacent} \\ 0 \text{ if not} \end{cases}$$

$$L = \{(f, g) \mid \text{unit f is within functional distance of g and not adjacent}$$

W_k = weight of objective k (k = a or p)

Subject to :
1) adjacency selection (Z_{lm} =1 only if both X_l and X_m are selected).

$$\left.\begin{array}{l} Z_{lm} \le X_l \\ Z_{lm} \le X_m \end{array}\right\} = \text{for each } (l,m) \in K$$

2) functional connectedness selection (R_{fg} =1 only if both X_f and X_g are selected).

$$\left.\begin{array}{l} R_{fg} \le X_f \\ Z_{fg} \le X_g \end{array}\right\} = \text{for each } (f,g) \in L$$

3) upper and lower bound on allocation

$$\sum_i a_i X_i \le A_{up}$$

$$\sum_i a_i X_i \ge A_{low}$$

where:
a_i = area of unit i
A_{up} = maximum area to allocate
A_{low} = minimum area to allocate

5. Case Study

A case study is to be undertaken using the 342 km^2 drainage basin of river Vejle Å located in the eastern part of Jutland, Denmark. The landuse is predominantly agriculturel. Elevation is varying between 0 and 126 m above sea level. The terrain topography is characterised by the west-east tunnel valleys of river Vejle Å and river Grejs Å. Soil types vary from coarse sands in the western parts to loamy soils in the eastern parts. Organogenic soils are the dominating soil types in the tunnel valleys.

The potential restoration areas are identified by combining areas with organogenic soils and lowland areas. Organogenic soils are derived from digital maps of the soil in the upper (Madsen et al. 1992, MAF 1996) and lower soil layers, i.e 1 m below terrain (DGS, 1978). Lowlands are defined as areas identified as waterlogged or reclaimed on maps from the beginning of the century (Madsen et al. 1992; MAF 1996). Only areas adjacent to streams, i.e. riparian areas, are considered, thereby excluding hydrological isolated depressions (sinks) in recharge areas. Further, developed areas have been excluded. The total size of the potential wetland area is 2445 ha, equal to about 7 per cent of the catchment. The deliniated area is a mixture of small isolated areas and large contiguous areas in the river valley (Figure 1). While some of the smaller areas are of a size which can be considered appropriate for a decision unit, the large contiguous areas are of a size which necessitates subdivision. This subdivision can be carried out according to hydrological relationships, e.g. stream segments, or manually based on topographic maps, cadastral maps, orthophoto interpretation etc. No general guidelines can be given about the appropriate number and size range of decision units. However, the smaller the units, the more spatial variation can be captured and the more accurate will be the attribute description. On the other hand, the more the units, the more complex the problem becomes. Preliminary work indicates that the number of decision units in the study areas is about 1000.

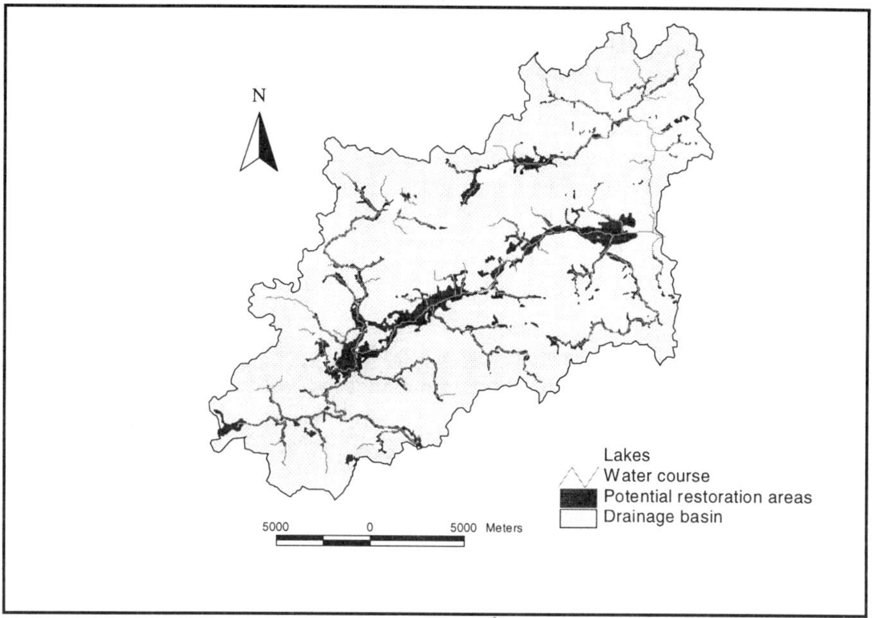

Figure 1 Potential riparian wetlands in river Vejle Å watershed

6. Discussion

The model approach is based on a loose coupling between LP and GIS. A loose-coupling is characterised by data exchange through external files. The idea is to take advantage of the specific strengths of each of the two tools - spatial analysis and optimisation, respectively.

LP is a mathematical programming technique that uses linear representation of objectives and constraint to search optimal solutions. LP does not directly use spatial information. Both this study and many of the referreed studies address the difficulties related to handling of spatial phenomenon in traditional LP. Binary integer programming can alleviate some of the spatial concerns. The main disadvantages of the use of BIP for spatial optimisation problems are that the number of decision variables becomes very large and that spatial constraints easily result in unwieldy model formulations.

Diamond and Wright (1988) state that there are basically two types of spatial optimisation models: those that assign single or multiple land uses 'optimally' among geographically predefined areas, and those that actually define the geographical region best suited for a single facility or land use. In this study the use of predefined irregular geographic units was found most appropriate, due to the nature of the wetlands allocation problem. Using predefined units also has the advantage that the total number of decision units is reduced compared to using a cellular model.

Two categories of criteria are included i) site specific and ii) spatial. The site specific criteria, i.e. costs, suitability and N-retention, are characterised by having a fixed value for each site alternative that can be determined prior to the optimisation. Site specific criteria are not necessarily defined solely by in-site features. Determination of the N-retention is an example of a site attribute that relates to conditions external to the site, e.g. land use and soil types in upslope catchment. The spatial criteria are characterised by site mutuality, i.e. that the 'value' added to the spatial objective of selecting one site is determined by the concurrent selection of one or more other sites. In the model the principle of quadratic assignment was used to express the mutual dependency of spatial criteria. The quadratic expression ($S_{lm}X_lX_m$, where S_{lm} = Score related to X_l and X_m selection) was reformulated as a linear expression by introduction of two auxiliary variables (Z_{lm} and R_{fg}) and two extra constraints per spatial relationship. The total number of extra variables and constraints depends on the spatial configuration of sites and the choice of maximum functional distance (the greater the distance more functional relationships to incorporate).

The two spatial objectives included in the model capture some essential landscape features for support of species protection and diversity: connectivity and functional connectedness. The model could be expanded to include additional or alternative objectives, but the number of possible solutions to a multiobjective programming model tends to grow exponentially with an increase in the number of objectives (Cohon, 1978). This could lead to computational problems with solving the LP. The combination of multiple spatial objectives should be considered carefully as spatial objectives may be partially redundant, i.e. objectives quantify identical or similar aspects of landscape structure. It could be argued that the two spatial objectives, adjacency and proximity, are promoting almost identical landscape features. However, the model formulation is made with the deliberate purpose of analysing how various weights on the two objectives would influence the result. The formulation of the two spatial objectives included is relatively simple. Many others would be much more difficult to implement, partly because geographic units of irregular size and form prevent the application of efficient grid-based procedures (see e.g. Kao and Lin 1996). It is likely that only a limited number of the landscape ecological metrics found in FRAGSTATS can be formulated in a suitable way to be used in LP.

The spatial siting criteria are defined as objectives, not as constraints. The model is therefore a multiple site selection model. Units might be adjacent but do not have to be so, i.e. the model do not explicitly search for a single contiguous site. Whether or not the solution will be a single contiguous site depends on a combination of the actual features of the area under study, i.e. site attributes and spatial configuration, and the weights on objectives. The larger the weight on adjacency, the more compact the set of selected sites will be.

Definition of optimality is based on a weighted summation of the multiple objectives. This technique belongs to the class of compensatory approaches (Jankowski 1995). It is based on the assumption that a high score on one criterion can compensate for a low score on an another criterion. This means that a high score of an alternative achieved on one criterion is traded off, according to the decision makers preferences, for the low score received on another. Hence, alternative sites with different properties and therefore functional very different but with same total score are considered of equal 'value'. The realism of this assumption can be questioned. One way to alleviate the negative effects of the additive approach could be to constrain the selection with minimum scores for the individual site specific objectives. Diamond and Wright (1989) used a 'weakest link' principle to express overall sub-region suitability as a function of individual geographical unit suitability. This principle is represented by the 'maxi-min' objective of maximising the minimum of suitability of any given sub-region for the proposed land use. They find that this is a more realistic function than weighted summation. However, the principle was applied in a model that allocated a single land use to one single contiguous sub-region. In the model presented here contiguity is not a constraint. Hence, the weakest link principle cannot be applied globally, but should be evaluated for individual sub-regions of contiguous units. Methods for solving this problem have not been found.

Ultimately, the choice of objectives and the related weights should reflect expert judgement and decision makers' preferences. Many different multi-criteria-decision-making methods exist for determination of preferences. Pairwise comparison (Saaty 1977) in the context of the analytical hierarchy process (AHP) is an example of a weighting technique, which is used in connection with GIS-based MCDM (Eastman et al. 1993). The usefulness of rigid mathematical methods for spatial planning can be discussed. It is difficult to express all aspects in quantifiable criteria. However, the use of MCDM methods forces decision makers to clearly define objectives and preferences. Thereby making the decision process more transparent. No formal methods for determination of weights are used for the model presented here. So far it will only be used analytically, ie. to explore how the spatial layout of the optimisation solutions changes when different weights are applied.

The model has been implemented in the SAS operation research software (SAS 1989), which uses the Branch and Bound technique for solving of integer LP problems. A small test data set (ten units) has verified model implementation. Next step is analysis on the full case area.

The technological development of GIS has greatly expanded the possibility of performing large-scale spatial analysis. The advantages of GIS are to be found in the capability of exploring the implicit spatial information in digital maps. The identification and delineation of geographic units representing choice alternatives were facilitated by the analytical capabilities in GIS. GIS also serves an important role for extraction of site attributes, i.e. criteria values. GIS analysis and modelling make use of an algebra which is similar to traditional algebra, except that the variables are entire maps and that the algebra has been augmented with a set of functions that are specific for GIS operations, e.g. distance functions. What is needed in land use planning is that the set of GIS functions is further augmented with fully integrated optimisation tools, operating on entire maps as variables. This tight coupling would take GIS one step further in the role as spatial decision support system. The inclusion of object-oriented programming languages in GIS, like Avenue in Arcview, is promising for development of a tight coupling. Eventually, the optimisation should be based on effective heuristic. Simulated annealing has shown promising for spatial phenomenon (Bos 1993). More research in this direction is needed.

References

BERRY, J.K. 1993. Cartographic Modeling - The Analytical Capabilities of GIS. In: Goodchild, M.F., Parks, B.O. and Steyaert, L.T. (eds.). Environmental modeling with GIS. Oxford University Press, New York: 58-74.

BETTINGER, P., JOHNSON, K.N. AND SESSIONS, J. 1996. Forest Planning in a Oregon Case Study: Defining the Problem and Attempting to Meet Goals with a Spatial-Analysis Technique. Environmental Management 20 (4), pp 565-577

BETTINGER, P., SESSIONS, J. AND BOSTON, K. 1997. Using Tabu search to schedule timber harvests subject to spatial wildlife goals for big game. Ecological Modelling 94, pp 111-123.

BOS, J. 1993. Zoning in Forest Management: a Quadratic Assignment Problem Solved by Simulated Annealing. Journal of Environmental Management 37: 127-145.

CAMPBELL, J.C., RADKE, J., GLESS, J.T. AND WIRTSHAFTER, R.M. 1992. An application of linear programming and geographic information systems: croplands in Antigua. Environment and Planning A, 24: 535-549.

CARVER, S.J. 1991. Integrating multi-criteria evaluation with geographical information systems. International Journal of Geographical Information Systems 5(3):321-339.

CLEVELAND, J.A., GROVER, R.B., PETRILLO, J.L. AND LADD, E. 1979. Using computers for site selection. Environmental Science and Technology 13(7):792-797.

COHON, J. 1978. Multiobjective programming and planning. Academic press, New York.

DFNA. 1992. Prioritering og Behandling af Naturforvaltningsprojekter [Prioritising and Management of Nature Management Projects]. Forest and Nature Agency, Ministry of the Environment.

DGS. 1978. Foreløbige geologiske kort (1:25.000) over Danmark. [Preliminary Geological Maps (1:25.000) of Denmark]. Geological Survey of Denmark, Series A, No.3: 19.

DIAMOND, J.T. AND WRIGHT, J.R. 1988. Design of an integrated spatial information system for multiobjective land-use planning. Environment and Planning B: Planning and Design 15:205-214.

DIAMOND, J.T. AND WRIGHT, J.R. 1989. Efficient Land Allocation. Journal of Urban Planning and Development 115(2):81-96.

EASTMAN, J.R., KYEM, P.A.K., TOLEDANO, J. AND JIN, W. 1993. GIS and Decision Making. Explorations in Geographic Information Systems Technology, 4, United Nations Institute for

Training and Research.
ESRI 1997. Arcview GIS version 3a. Environmental Systems Research Institute, Redlands, California, USA.
FORMAN, R.T.T AND GODRON, M. 1986. Landscape Ecology. John Wiley and Sons, New York.
GILBERT, K.C., HOLMES, D.D. AND ROSENTHAL, R.E. 1985. A Multiobjective Discrete Optimisation Model for Land Allocation. Management Science 31(12):1509-1522.
HOF, J.G. AND JOYCE, L.A. 1993. A Mixed Integer Linear Programming Approach for Spatially Optimizing Wildlife and Timber in Managed Forest Ecosystems. Forest Science 39(4): 816-834.
HOFFMAN, C.C. 1998. Nutrient retention in wet meadows and fens. PhD thesis. National Environmental Research Institute, Silkeborg, Denmark.
IVERSEN, T.M., GRANT, R. AND NIELSEN, K, 1998. Nitrogen enrichment of European inland and coastal waters with special attention to Denmark and Danish policy measures. Environmental Pollution 102 (S1): 771-791.
JANKOWSKI, P. 1995. Integrating geographical information systems and multiple criteria decision-making methods. International Journal of Geographical Information Systems 9(3):251-273.
JANSSEN, R. AND RIETVELD, P. 1990. Multicriteria Analysis and Geographical Information Systems: An Application to Agricultural Land Use in the Netherlands. In: Scholten, H.J. and Stillwell, J.C.H. (eds.). Geographical Information Systems for Urban and Regional Planning, Kluwer Academic Publishers, Dordrecht: pp. 129-139.
JANSSON, M., ANDERSSON, R., BERGGREN, H. AND LEONARDSSON, L. 1994. Wetlands and Lakes as Nitrogen Traps. Ambio 23(6): 320-325.
KAO, J-J. AND LIN, H-U. 1996. Multifactor Spatial Analysis for Landfill Siting. Journal of Environmental Engeneering 122(10):902-908
LOCKWOOD, C. AND MOORE, T. 1993. Harvest scheduling with spatial constraints: a simulated annealing approach. Canadian Journal of Forest Research 23:468-478.
MADSEN, H.B., NØRR, A.H. AND HOLST, K.AA. 1992. The Danish Soil Classification. Atlas of Denmark I,3. Reitzels Copenhagen.
MALTBY, E., HOGAN, D.V., IMMIRZI, C.P., TELLMAN, J.H. AND VAN DER PEIJL, M.J. 1994. Building a new approach to the investigation and assessment of wetland ecosystem functioning. In: Mitsch, W.J. (ed.). Global Wetlands: Old World and New. Elsevier Amsterdam. pp.637-658
MAF 1996. Geografiske data hos Afd. for Arealanvendelse 1996 [Geographic data at Department of Land use 1996]. SP rapport 6, Ministry of Agriculture and Fisheries, Inst. Plant and Soil Science, Foulum.
MCGARIGAL, K. AND MARKS, B.J. 1994. FRAGSTATS Spatial Pattern Analysis Program for Quantifying Landscape Structure - version 2.0. Oregon State University, Forest Science Dept.
NEVO, A. AND GARCIA, L. 1996. Spatial Optimisation of Wildlife Habitat - Model Formulation. Transactions of the ASAE 39(3):1189-1196.
PEREIRA, J.M.C. AND DUCKSTEIN, L. 1993. A multiple criteria decision-making approach to GIS-based land suitability evaluation. International Journal of Geographical Information Systems 7(5):407-424.
SAATY, T.L. 1977. A Scaling Method for Priorities in Hierarchical Structures. Journal of Mathematical Psychology 15:234-281.
SAS. 1989. SAS/OR User Guide, Version 6, First Edition, SAS Institute Inc., Cary, NC.
TOMLIN, C.D. 1990: Geographic Information Systems and Cartographic Modelling, Prentice-Hall, Englewood Cliffs, New Jersey.
TURNER, M.G. AND GARDNER, R.H. (eds.). 1991. Quantitative Methods in Landscape Ecology: the Analysis and Interpretation of Landscape Heterogeneity. Springer, New York.
WRIGHT, J.F. AND BUEHLER, K.A. 1993. Probalistic Inferencing and Spatial Decision Support Systems. In: Wright, J.F., Wiggins, L.L., Ravinder, K.J. and Kim, T.J., (eds.). Expert Systems in Environmental Planning, Springer Verlaag, Berlin, pp. 119-144.

A Cellular Automaton Approach to Optimising the Choice of Tree Species

Niels Strange and Henrik Meilby

Abstract

This paper presents a cellular automaton approach to optimising the choice of tree species in afforestation. The case study deals with 180 ha of recently afforested land which is located at the northwest coast of Zealand, Denmark. A two-dimensional automaton containing approximately 100,000 active cells is used for optimising the land use setup in this area. Four land use alternatives are considered: pasture, beech (Fagus sylvatica (L.)), Norway spruce (*Picea abies* (L.) Karst.), and oak (*Quercus robur* L.). Accordingly, the task of finding the optimal solution constitutes a tremendous combinatorial problem comprising a total number of approximately $4^{100,000}$ combinations. Obviously, no heuristic is able to solve such a problem. However, in this paper it is shown that appropriately formulated cellular automata may converge towards near-optimum solutions within a comparatively low number of iterations. It is concluded that a major advantage of self-organising algorithms is that the overall maximisation problem is solved as a decentralised optimisation problem where the use of the individual forest components is optimised in parallel. In that respect, self-organising algorithms contrast with the centralistic top-down approach known from most traditional optimisation methods. Geographical data on water retention capacity is used to estimate the soil expectation value of the suggested land use alternatives. The objectives are to maximise the total soil expectation value, minimise the deviation from a target diversity index, minimise a scale-dependent fixed cost function, and maximise the aesthetic value.

Keywords: Cellular automaton, heuristics, multiple-use, spatial optimisation, forest management planning, policy analysis, spatial economics

1. Introduction

Afforestation has become an important element in the Danish environmental policy. It is believed that afforestation is an important means of increasing the amount of renewable natural resources and environmental as well as social goods, e.g. recreational services, carbon sequestration, protection of groundwater resources, and timber production.

In the mid 1980s surplus agricultural production and the impact of the EU Common Agricultural Policy made afforestation a viable alternative land use and increased the incentive for a Danish policy within the field. The comments to the 1989 Act on Natural Resource Management included the target of doubling the Danish forest area from the present 12 per cent to about 25 per cent of the land area within one forest rotation (70-150 years). To reach this target an annual afforestation of about 5,000 ha is required. Therefore, since 1989 funding for state afforestation has been granted. Moreover, in 1991 subsidies for private afforestation were implemented.

Evaluation of multiple objectives is most often an integrated task in the planning of afforestation. The overall objective, from society's point of view, may be to maximise the total utility of afforestation. However, when maximising utility, information on costs and benefits of timber as well as non-market environmental values is required. Estimates of costs and benefits of timber production are usually based on current market prices, but in most cases involving environmental values there will be no evident source of estimates of associated costs and benefits. Some frequently applied methods for estimating non-market benefits of forests are the travel-cost method, contingent valuation and the hedonic pricing method. Usually, limited information on non-market values implies that the decision-maker must implicitly define the optimal or most satisfactory solution. A majority of the literature on maximisation of timber values and spatially related non-market values can be put into two categories: i) studies where the maximisation of timber values is subject to constraints which implicitly represent non-market production functions, typically minimum exclusion period between harvest of adjacent stands and maximum size of clear felled area (e.g., Torres-Rojo 1989; Roise 1990; Nelson and Finn 1991; Weintraub et al. 1994); ii) studies where non-market production functions are integrated into the objective function and weighted according to the choice of the decision-maker (e.g., Bare and Mendoza 1988; Hof and Joyce 1992).

As regards spatial optimisation, with or without conflicting multiple objectives, most studies conclude that this sort of problems quickly grow into such large dimensions that they become computationally intractable. Therefore, the number of suggested heuristic algorithms has increased during the years. Some examples are, lagrangian relaxation (Torres-Rojo 1989), column generation (Weintraub et al. 1994), map-colouring theory (Roise 1990), random search (Clements et al. 1990; Nelson et al. 1991), biased sampling search (O'Hara et al. 1989), random ordering and one-stage dynamic programming (Yoshimoto et al. 1994), simulated annealing (Lockwood and Moore 1993; Tarp and Helles 1997), and tabu search (Thompson et al. 1994; Bettinger et al. 1997). However, all these methods are centralistic approaches which decide the steps to be taken by evaluating the objective function of the entire system repeatedly. A classic class of centralistic approaches is the multi-level or hierarchial planning optimisation methods (Dirickx and Jennergren 1979; Weintraub et al. 1994) which originate from centralistic planning economies. Centralistic approaches possess the general advantage that they are easy to plan and manage. On the other hand, when the resolution of the system is high and the number of potential states of each element is considerable, then the number of potential solutions to the problem is tremendous. No centralistic procedure can avoid the 'curse of dimensionality', but an approach where decisions are taken in parallel may be able to solve huge problems within reasonable time. Therefore, we suggest a self-organising system (algorithm) as a promising tool for solving complex spatial problems.

In recent years the problem of handling complex dynamic systems has been a quickly growing research field in natural as well as social sciences. One branch within the field of Artificial Intelligence includes the application of self-organisation and evolutionary simulation. This paper presents an application of a classical, albeit slightly generalised, type of self-organising system, a two-dimensional cellular automaton approach designed for land use optimisation. The algorithm follows a decentralised procedure and optimises each compartment (cell) of the area in parallel. Unfortunately, due to the inseparability of the decisions such a procedure is not guaranteed to lead to a globally optimal solution. However, when supplied with appropriate guidance the cellular automaton proves to be efficient for solving complex nonlinear forest management optimisation problems of considerable size.

Cellular automata were originally introduced by Von Neumann (1966) to demonstrate self-reproduction. Generally, they consist of regular arrays where each cell is associated with a discrete variable. A cellular automaton evolves in discrete time steps, the value of the variable in each cell being affected by the values of variables in all the cells of a specified neighbourhood. The cell variables are updated simultaneously. Cellular automata may be suitable for economic modelling because they can be considered natural models of local interaction between economic agents within a specified neighbourhood.

An interesting model of a spatial market is presented by Prasad (1988) who describes an economy as a cellular automaton where the adaptation or self-organisation of the model system corresponds to an evolution towards a more efficient disposition of social resources. In the economic model each cell acts as an individual economic agent. A neighbour is an individual with whom the cell is connected, i.e. a potential buyer-seller relationship. It is shown to be possible to find a set of prices such that if each individual maximises his own valuation relative to the price system a stable or dynamic equilibrium is settled. There need not be any central authority that decides who to assign to which 'job' - once prices are discovered, the agents will do simple maximisation over the choices available to them.

Another economic application of cellular automata is presented by Keenan and O'Brien (1993) who demonstrate that measures of complex behaviour developed for cellular automata are meaningful measures of the self-organising tendencies of Edgeworth dynamics in an economy. Both Prasad (1988) and Keenan and O'Brien (1993) illustrate that cellular automata may be capable of solving large nonlinear optimisation problems with high speed. Moreover, it seems evident that cellular automata may form the basis of efficient solutions to problems which include spatial interrelationships.

In Balmann (1997) structural change in the agricultural landscape is analysed using a two-dimensional cellular automaton. Individual farms are located at different points in a region and compete on various markets. The overall maximisation problem is solved as a decentralised optimisation problem where each individual farm maximises profit in parallel subject to their expectations, financial state and assets. The model is mainly used to identify path dependencies, i.e. to show whether different initial configurations converge toward similar equilibria or not.

An evolutionary model of urban systems using the concepts of the theory of self-organisation has been developed by Engelen (1988). The model was used to explore the many possible futures and alternative structures of complex urban systems in Belgium, e.g. for calculating the magnitude and spatial distribution of future energy consumption. Moreover, it was used for testing the impacts of small as well as larger interventions, changing policies and technological evolution.

Spatial systems, whatever their dimension, are characterised by a decreasing flow of information with increasing distance. This implies that any cell, individual or agent of the system reacts upon, and adjusts its behaviour in response to, the occurrences of the near-by surroundings mainly. All elements of the system will, in time, be influenced by all changes of the system but information from far away will be delayed as well as damped. As regards the actual decision problem concerning the choice of tree species, site quality largely determines which tree species are attractive. However, the best way of using a specific site also depends on the use of the surroundings, partly because they influence the growth conditions of the site but also because the costs of activities related to a specific use may be scale dependent (e.g., costs of fencing or harvesting), or the recreational benefits may be related to the distance to cities, roads, parking lots or areas with a different land use (yielding a measure of the structural diversity of the forest). In conclusion, those characteristics that determine the value of a specific use at a given site are the properties of the site and the use of some specified neighbourhood. Accordingly, a cellular automaton implemented on a regular two-dimensional lattice constitutes a natural model of spatial interrelationships and, when supported by appropriate guiding functions, it may be able to converge towards globally near-optimal solutions. However, the number of attractors is likely to be high and therefore the probability of identifying the global optimum is indeed low. As mentioned above, a few recent studies have applied cellular automata for geographic or economic problems but, to the best of our knowledge, all these applications can be characterised as explorative simulation studies. Economic optimisation of the configuration as a whole has, so far, not been attempted.

2. A Spatial Economic Model

2.1 Vollerup afforestation area

According to the 1989 Zoning Act, the Danish county councils were instructed to produce land-zoning plans for afforestation based on existing structural plans for agriculture and nature conservation. The Vollerup afforestation area was established as a result of these plans. This area is situated close to the town of Kalundborg and was mainly designated for afforestation due to recreational interests, but also in order to protect groundwater resources. In addition, the area was chosen as an experiment by the Danish Forest and Nature Agency. The total forested area is 180 ha. Below, the cellular automaton is described as regards i) its lattice structure, ii) the possible states of the cells, iii) the neighbourhood, and iv) the transition rules (economic valuation).

2.2 Lattice structure

In this study the area including the forest is defined as a 336×504 ($m \times n$) lattice of square sites. In approximately 40 per cent of the area the land use is fixed once and for all, i.e. the use of areas occupied by roads, villages, farm land, existing forests, and the sea cannot be changed by the self-organisation process. Each of the remaining sites may be used in four different ways.

2.3 Cell states

The four different land use alternatives are the elements of the land use set u. The solution space of the automaton, $U = u^{mn}$, is a set of land use configurations which includes all possible states of the automaton. As each of the approximately 100,000 active cells can be used in four different ways, the total number of different land use setups is approximately $4^{100,000}$. Accordingly, we are facing a tremendous combinatorial problem. As regards the four elements of u, we have chosen to distinguish between pastures and three different tree species: Beech (*Fagus sylvatica* L.), Norway spruce (*Picea abies* (L.) Karst.), and Oak (*Quercus robur* L.). The pastures are assumed to be used for forage production and are intended as an open land alternative in the vicinity of inhabited areas mainly.

2.4 Neighbourhood

For description of the neighbourhood of a cell we have defined three neighbourhood functions Φ_1, Φ_2 and Φ_3. Φ_1 is the number of immediate neighbours (0-8) with the same use as the cell in question. Assuming scale-dependent fixed costs, the higher this number the lower the fixed costs associated with the current use of the cell. Φ_2 describes the diversity of land use in the immediate neighbourhood, structural variation being assumed to increase both the biodiversity and the aesthetic value of the landscape. The diversity is measured by means of a modified Simpson index where the population includes the current cell and its eight immediate neighbours. However, the size of the population may be less than nine cells if some of the adjacent cells are, e.g. roads or farms. Therefore, the diversity in each cell is measured relative to the maximum Simpson index, given the size of the population. Φ_3 is the geographical distance between the current cell i and the nearest inhabited area. Close to inhabited areas some land uses are assumed to be more appropriate than others due to their aesthetical characteristics.

2.5 Transition rules

Our objective is to maximise the economic value of the land use setup within the area evaluated by the economic model. As the objective function may include production of timber as well as non-market values, the calculated value of a specific land use candidate in a specific cell can be interpreted as a fitness value which depends on the site itself as well as the state and structure of the surroundings. The overall problem is formulated as a multiple objective problem where the objective function is the sum of soil expectation value (SEV), weighted scale-dependent costs, weighted deviation from a target diversity and weighted aesthetic value. The global objective function may be specified as:

$$V^* = \max_U \left\{ \sum_{i=1}^{M} \mathrm{v}(\boldsymbol{x}_i) \right\} \qquad (1)$$

where V^* is the maximum fitness value of the whole land use setup (U), $M = mn$ is the total number of active cells, $\boldsymbol{x}_i$ is an information vector describing the characteristics of cell i and its neighbourhood, and v is an economic model used for calculation of the economic fitness value, i.e. the transition valuation function.

However, in cellular automata decisions are not taken at the global level. Therefore, we define a local objective function which specifies how to choose between land use candidates in each individual cell at each iteration step.

$$V_{it}^{*} = \max_{u} \left\{ \mathrm{v}(\boldsymbol{x}_{it,1}),..,\mathrm{v}(\boldsymbol{x}_{it,o}) \right\} \tag{2}$$

where V_{it}^{*} is the maximum current fitness value of any land use in cell i at iteration t, o is the number of different land use candidates and $\boldsymbol{x}_{it,l}$ is the information vector of cell i at iteration t for land use l.

In general terms the economic model used for calculation of the fitness value, $\mathrm{v}_{it} = \mathrm{v}(\boldsymbol{x}_{it})$, is specified as:

$$\begin{aligned}
&\mathrm{v}_{it} = \mathrm{v}(\boldsymbol{x}_{it}) = \alpha_1 \mathrm{SEV}(u_{it}, WRC_i) - \alpha_2 \mathrm{F}(u_{it}) + \alpha_3 \mathrm{DIV}(u_{it}) + \alpha_4 \mathrm{DIST}(u_{it}) \quad , \\
&\mathrm{SEV}(u_{it}, WRC_i) = a(u_{it}) \cdot WRC_i + b(u_{it}) \quad , \\
&\mathrm{F}(u_{it}) = \omega_1(u_{it}) - \omega_2(u_{it}) \cdot \Phi_1(i) \quad , \\
&\mathrm{DIV}(u_{it}) = \Phi_2(i) - \omega_3(u_{it}) \quad , \\
&\mathrm{DIST}(u_{it}) = \omega_4(u_{it}) / \Phi_3(i) \quad ,
\end{aligned} \tag{3}$$

where α_1, .., α_4 are weights, $a(u_{it})$, $b(u_{it})$, and $\omega_1(u_{it})$, .., $\omega_4(u_{it})$ are land use dependent parameters, u_{it} is the land use in cell i at iteration t, SEV is the soil expectation value, F is a fixed cost function, DIV is a diversity function, and WRC_i is the water retention capacity in cell i. In the case study the annual discount rate is 1.5 per cent. The estimated functional relationship between WRC and SEV for each tree species is based on Holstener-Jørgensen (1958), Møller (1933), and Danish Forest Society et al. (1995). The annual rent for grazing (pastures) is assumed to be DKK 500 and 750 per year for the lowest and highest estimated WRC values, respectively. Geographical data on WRC are estimated on the basis of Granat (1995). WRC in each cell is adjusted according to expected rooting depth which is restricted in considerable parts of the case area by ironstone hardpans or the level of the groundwater table.

The fixed costs are scale dependent, i.e. it is assumed that two adjacent cells with similar land use have lower fixed costs than two non-adjacent cells. Moreover, the fixed costs are assumed to decrease linearly when the number of neighbours with the same land use increases .

In (3) the two components DIV and DIST are implicit non-market value functions, describing the variation and aesthetical value of a given land use configuration. Other types of implicit non-market value functions could be tested. Some examples may include: i) distance to farms, summer cottages, and roads, ii) size of groups of cells with a specific land use, and iii) distance between a given cell and cells with a specific land use located in a particular direction. The last example is intended to be used, e.g. when simulating the risk of windthrow for various tree species as a function of distance to open land.

3. Spatial Optimisation Procedure

We adopt features of evolutionary strategies for maximising the objective function (1). Evaluation of a given cell can be seen as a reproduction process where the outcome of the process depends on the current fitness of the land use alternatives. If all cells are evaluated in all iterations, new structural variation normally will not arise. Therefore, only part of the cells should be evaluated. Moreover, to add new variation mutations can be included, i.e. random changes which are not advantageous at the current stage, but may improve the configuration in the long run. The probability of evaluating a given cell, and hence accepting the potential improvement of the current land use setup, may be high in the beginning of the process and gradually decrease towards zero. Mutations may diversify the setup and lead to improvements in the long run. The probability of such mutations may be high in the initial iterations and approach zero in the final ones.

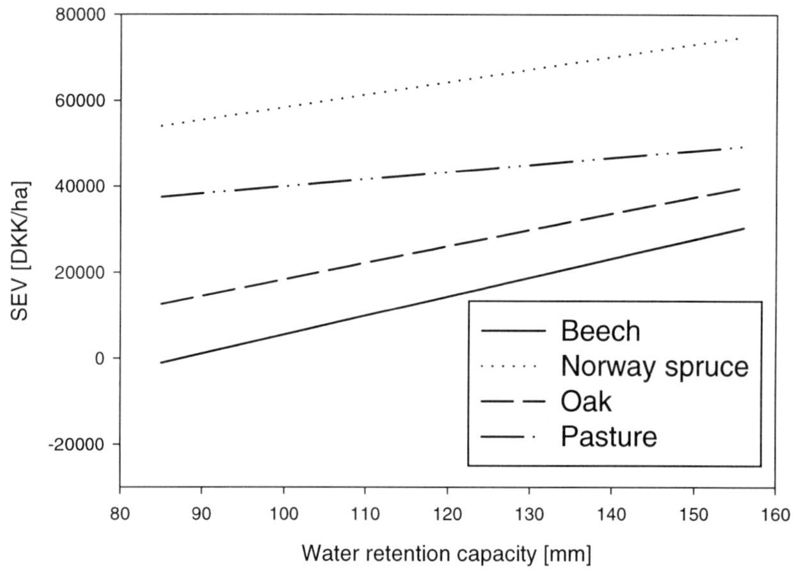

Figure 1. SEV as a function of water retention capacity.

4. Preliminary Results

The relationships between SEV and WRC are presented in Figure 1. If we maximimise SEV and disregard fixed costs, diversity and distance to inhabited areas ($\alpha_2 = \alpha_3 = \alpha_4 = 0$), the decision problem is simple. Norway spruce should be chosen in all cells. The problem is so to say completely separable, i.e. the cells are not interacting. However, realistic problem setups are generally nonseparable as the decision in one cell will affect the value of the land use in neighbouring cells. For example, the diversity index Φ_2 is affected by the state of each individual cell in the population, and the fixed costs decrease when the number of neighbouring cells with a similar land use (Φ_1) decreases. The cells are indeed interacting, just like the agents in any market.

Figure 2 presents an example showing the development of the objective function and the average diversity value as the system evolves. In this particular case the diversity weight (α_3) is ten times the standard weight. The initial solution is chosen randomly, i.e. the diversity is likely to be close to the average (0.5). The average diversity value is 0.64 at iteration 2 and converges towards 0.81. For the optimum solution the area can be expected to be divided almost equally between the four land use alternatives. The observed optimal shares are 24.7 per cent pasture, 19.4 per cent beech, 27.2 per cent Norway spruce, and 21.1 per cent oak (Figure 3). Even though the weight on diversity has been increased, Norway spruce dominates due to the high SEV value of this species. It is found that after only seven iterations the objective value, the diversity value and the shares of the area are almostsimilar to the optimal values.

In Figure 4 a second example is presented. In this case the weight on fixed costs (α_2) is increased by a factor ten. The optimisation procedure converges after 15 iterations. Initially, the average diversity index is 0.66 but gradually it decreases towards 0.24. Aggregation is clearly reflected in this result. Norway spruce increases its share of the area and reaches 36.2 per cent in the final solution (Figure 5). The shares of pasture and oak remain fairly constant as the system evolves but the share of beech decreases significantly. In Figure 6 a similar example is shown where the emphasis on fixed costs forces the system to move towards a structurally ordered state with a high degree of aggregation.

5. Discussion and Conclusion

Preliminary results indicate that two-dimensional cellular automata are promising tools for solving high dimensional nonlinear spatial optimisation problems in multiple-use forest management planning. It is also indicated that two-dimensional automata combined with probabilistic (evolutionary) strategies suit the optimisation of land use in afforestation projects.

There is a profound need for developing a tool for optimisation of spatial multiple-objective forest management problems. The cellular automaton is particularly interesting since it allows the decision-maker to watch the land use setup evolving from disordered initial stages towards a structure which reflects the growth conditions as well as the characteristics of the economic model applied. For example, it can be observed where the dynamic stability of the solution is high and where it is not. Moreover, by varying the parameters of the economic model the sensitivity of the solution in various parts of the area can be investigated. Insights of this kind may be important for the final decision.

The concepts of self-organisation are indeed linked to evolutionary economics which has emerged in recent years and contrasts with neoclassical economy. In evolutionary economics it is believed that the dynamics of complex economic systems can be modelled as an evolution relying on the concepts of evolutionary biology. Economics describes market systems which consist of individual agents. It is assumed that these individuals are rational agents, maximising their own utility, and that they are interacting. Evolutionary theory deals with the interaction of self-interested individuals or organisms who are 'trying to increase fitness' and produce as much offspring as possible. In this paper we have presented a forest management planning application of the idea characterising evolutionary economics.

Future research should emphasise the essential issues of multiple-objective optimisation, e.g. assessing appropriate weighting schemes and solving the problem of identifying Pareto efficient solutions. This paper presented four production functions of both timber and non-market outputs. Refinement and further development of these functions may increase the model's applicability. Moreover, the probabilistic elements used in the optimisation procedure may be changed into adaptive functions, and by requiring the value of potential changes of the land use configuration to exceed some adjustable threshold the efficiency of the process may be improved. Finally, large-scale mutations - simulating catastrophic changes - may help the process to move from an inferior local optimum towards a globally near-optimal solution.

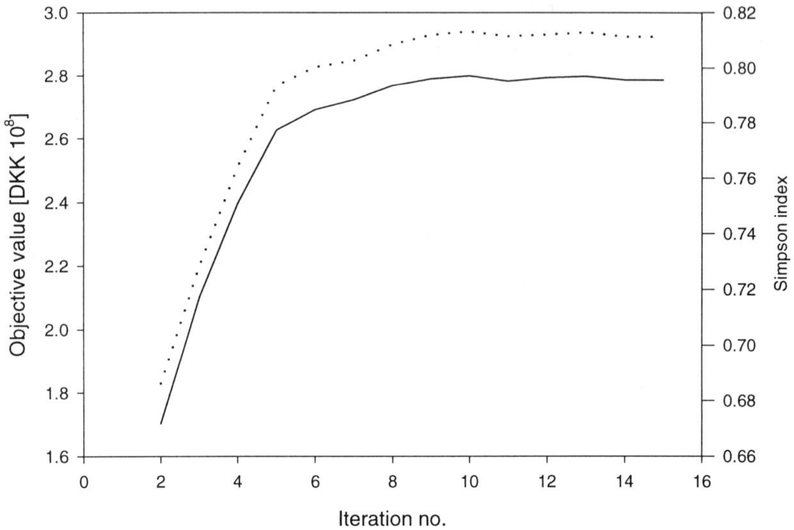

Figure 2. The objective value (solid) and Simpson index (dotted) as functions of iteration number. SEV, fixed costs, biodiversity and distance to settlements are weighted 1, 1, 1000000, and 500, respectively.

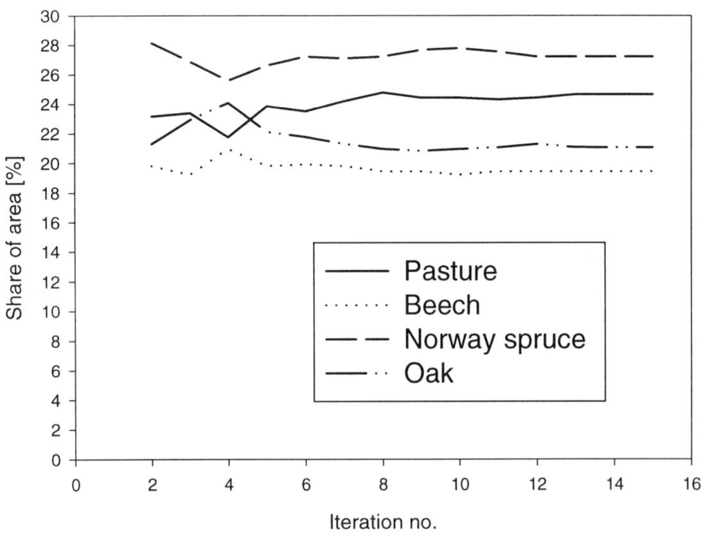

Figure 3. The share of area using the weights 1, 10,1000000, 500 for SEV, fixed costs, biodiversity, and distance to settlements.

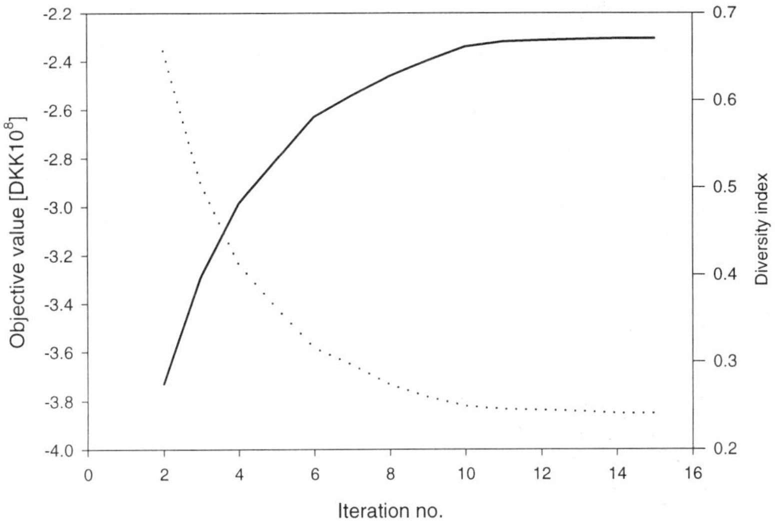

Figure 4. The objective function (solid) and the Simpson index (dotted) as functions of iteration. SEV, fixed costs, biodiversity, and distance to settlements are weighted 1, 10, 100000, and 500, respectively.

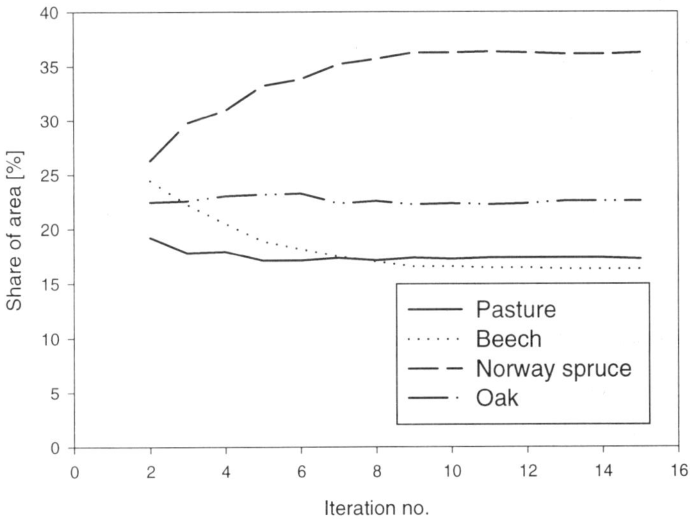

Figure 5. The share of area using the weights 1, 10, 100000, and 500 for SEV, fixed costs, biodiversity and distance to settlements, respectively.

6a.

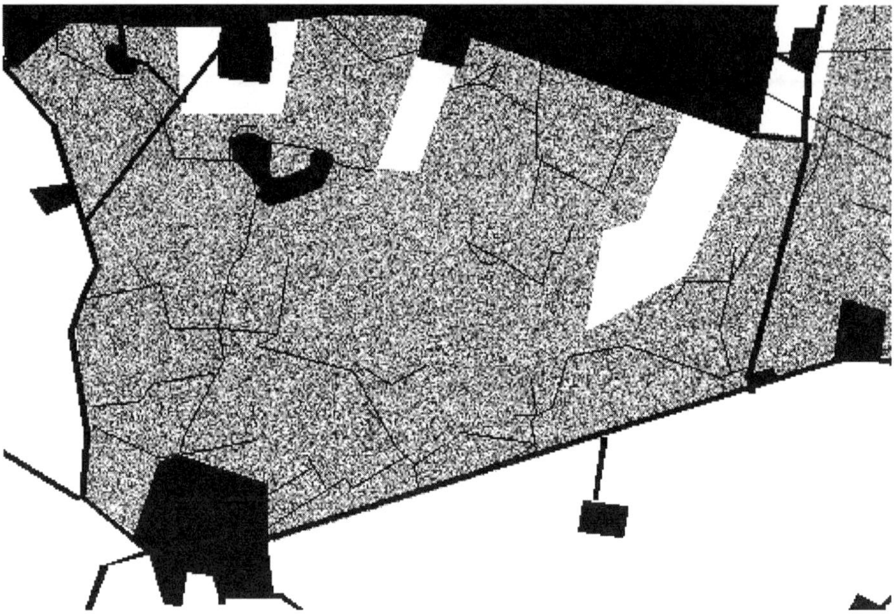

6b.

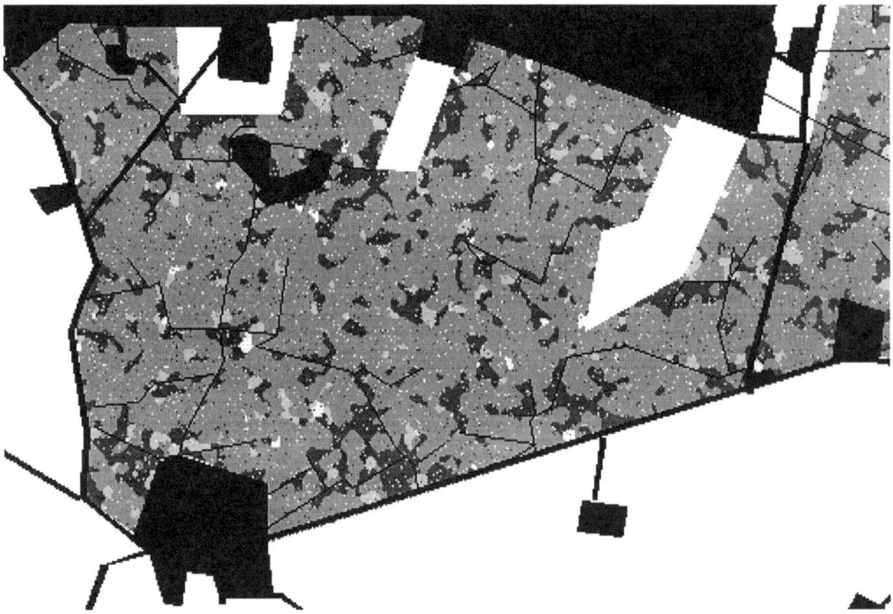

Figure 6 Self-organisation: Example showing the development from a random initial land use setup (a) towards a structurally ordered state (b)

References

BALMANN, A. 1997. Farm-based modelling of regional structural change: A cellular automata approach. European Review of Agricultural Economics 24:85-108.

BARE, B.B. AND MENDOZA, G. 1988. Multiple objective forest land management planning: An illustration. European Journal of Operational Research 14:44-55.

BETTINGER,P., SESSIONS, J. AND BOSTON, K. 1997. Using tabu search to schedule timber harvests subject to spatial wildlife goals for big game. Ecological Modelling 94:111-123.

CLEMENTS, S.E., DALLAIN, P. L. AND JAMNICK, M. S. 1990. An operational, spatially constrained harvest scheduling model. Canadian Journal of Forest Research 20:1438-1447.

DANISH FOREST SOCIETY, Danish Land Development Service, Forest and Nature Agency, 1995. [Forest Economy Tables]. (Unpublished).

DIRICKX, Y.M.I. AND JENNERGREN, L.P. 1979. System Analysis by Multilevel Methods - with Application to Economics and Management. International Series on Applied System Analysis. John Wiley & Sons. Toronto.

ENGELEN, G. 1988. The theory of self-organization and modelling complex urban systems. European Journal of Operations Research 37:42-57.

GRANAT,H. 1995. [Soil classification for the Vollerup afforestation area]. Report no. 21. The Forest and Nature Agency, Ministry of the Environment and Energy. (In Danish).

HOF, J.G. AND JOYCE, L.A. 1992. Spatial optimization for wildlife and timber in managed forest ecosystems. Forest Science 38:489-508.

HOLSTENER-JØRGENSEN, H. 1958. [Physical soil-investigations in Danish beech-stands]. The Danish Forest Experiment Station. Report no. 25:93-224 (In Danish, Summary in English).

KEENAN,D.C. AND O'BRIEN, M.J. 1993. Competition, collusion, and chaos. Journal of Economic Dynamics and Control 17:327-353.

LOCKWOOD, C. AND MOORE, T. 1993. Harvest scheduling with spatial constraints: A simulated annealing approach. Canadian Journal of Forest Research 23:468-478.

MØLLER, C.M. 1933. [Site class tables and growth tables for beech, oak and spruce in Denmark]. Dansk Skovforenings Tidsskrift 18:457-513. (In Danish).

NELSON,J.D. AND FINN, S. T. 1991. The influence of cut block size and adjacency rules on harvest levels and road networks. Canadian Journal of Forest Research 21:595-600.

NELSON,J., BRODIE, J.D. AND SESSIONS, J. 1991. Integrating short-term, area based logging plans with long-term harvest schedules. Forest Science 37:101-122.

O'HARA, A.J., FAALAND, B.H. AND BARE, B.B. 1989. Spatially constrained timber harvest scheduling. Canadian Journal of Forest Research 19:715-724.

PRASAD, K. 1988. Parallel distributed processing models for economic systems and games. Computer Science in Economics and Management 1:163-174.

ROISE, J.P. 1990. Multicriteria nonlinear programming for optimal spatial allocation of stands. Forest Science 36:487-501.

TARP, P. AND HELLES, F. 1997. Spatial optimization by simulated annealing and linear programming. Scandinavian Journal of Forest Research 12:390-402.

THOMPSON, W., HALME, M., BROWN, S., VERTINSKY, I. AND SCHREIER, H. 1994. Timber harvest scheduling subject to wildlife and adjacency constraints. In Sessions, J. and Brodie, J.D. (eds.): Management Systems for a Global Economy with Global Resource Concerns. Proceedings of the 1994 Symposium on System Analysis in Forest Resources, California.

TORRES-ROJO, J.M. 1989. The use of relaxation to solve harvest scheduling problems with flow, wildlife habitat, and adjacency constraints. Ph.D. dissertation, Department of Forest Resources, Oregon State University, Corvallis OR.

VON NEUMANN, J. 1966. Theory of Self Reproducing Automata. University of Illinois Press. Urbana.

WEINTRAUB, A., BARAHONA, F. AND EPSTEIN, R. 1994. A column generation algorithm for solving general forest planning problems with adjacency constraints. Forest Science 40:142-161.

YOSHIMOTO, A., BRODIE, J.D. AND SESSIONS, J. 1994. A new heuristic to solve spatially constrained long-term harvest scheduling problems. Forest Science 40:365-396.

Chapter VI

Cost Efficiency

A Method for Comparison of Efficiency Scores: A Case Study of Danish Dairy Farms

Mette Asmild,
Jens Leth Hougaard and
Dorte Kronborg

Abstract

This paper suggests the use of a particular parametric test when comparing the efficiency scores resulting from efficiency analysis. As an illustrative case Data Envelopment Analysis (DEA) is run on a set of Danish dairy farm data. Assuming that the efficiency scores resulting from DEA are right censored log-normally distributed, the influence of potential external factors affecting the efficiency can be investigated. It is found that efficiency increases with size and more surprisingly, that farms with jersey cows tend to be more efficient than non-jersey farms.

Keywords: Cost efficiency, DEA, censored log-normal distribution, dairy farms

1. Introduction

The analysis of efficiency is an important tool for managerial decision-making in general. Therefore efficiency analyses have been applied in numerous fields including dairy farms as in the present case study (e.g. Hallam and Machado 1996, Lund at al. 1993, Bravo-Ureta and Rieger 1990, 1991). One of the most popular tools for efficiency measurement is the by now well-known Data Envelopment Analysis (DEA) (e.g. Charnes et al. 1978, Charnes et al. 1995). This method is highly operational since it relies on linear programming and is able to handle multiple inputs and outputs. Furthermore, DEA has the advantage of letting data determine the production possibility set with few a priori assumptions on the functional form (it is non-parametric).

The aim of this paper is to introduce a statistical model for the efficiency scores, which can be used to evaluate the influence of potential external factors on the obtained efficiency results. The statistical model used in the present study accounts for the inherent problem of dealing with multiple (fully) efficient units. In the case study the right censored log-normal distribution appears to be an appropriate model for the efficiency estimates. The proposed model differs from what is usually assumed when discussing the statistical foundation of DEA.

In the case study of Danish dairy farms the results of Lund et al. (1993) are re-examined using a slightly different model and a considerably larger data set. The above mentioned statistical model is used to evaluate the inference of size and whether the livestock consists of jersey or non-jersey cows. It should be noted that even though this case study is not directly related to the field of natural resources the methodology presented and developed is easily applied to a wide range of scenarios including natural resources.

The paper is organised as follows. First a brief review of the DEA methodology is given. In the next section the statistical model is presented and discussed. Hereafter, the data set used in the case study is presented and the following section presents the efficiency scores resulting from the empirical analysis as well as the estimated statistical distributions. The paper closes with some concluding remarks on the applicability of the proposed statistical model.

2. Methods

2.1 The DEA methodology

The cost efficiency of the Danish dairy farms is measured using Data Envelopment Analysis (DEA) (e.g. Charnes et al. (1978), Charnes et al. (1995)). DEA is a non-parametric method, which combines the Farrell efficiency index with a mathematical description of the production possibility set. Since we are analysing cost efficiency, we apply the input oriented version of DEA, using Farrell's input efficiency index.

The Farrell input efficiency index E^F is formally defined as

$$E^F((x,y),T)=\min\{E \mid (Ex,y)\in T\},$$

where (x,y) is an input-output combination and T is the production possibility set.

The production possibility set T is constructed from n observed input-output combinations $(x^j,y^j)\in \Re_0^{m+p}$, j =1, ... ,n , combined with assumptions about free disposability, convexity and returns to scale, i.e.

$$T(k)=\{(x,y)\in \Re_0^{m+p} \mid \exists \lambda \in \Re_0^n : x \geq \sum_{j=1}^{n} \lambda^j x^j, y \leq \sum_{j=1}^{n} \lambda^j y^j, \lambda \in \Lambda(k)\},$$

where k=crs and k=vrs indicate constant returns to scale and variable returns to scale respectively, with

$$\Lambda(crs)=\Re_0^n, \Lambda(vrs)=\left\{\lambda \in \Re_0^n \mid \sum_{j=1}^{n} \lambda^j = 1\right\}.$$

Combining the above definition of the production possibility set with Farrell's input efficiency index results in the following linear programming problem for calculating the input efficiency score (DEA score) of each observation (x', y')

$$E^F((x',y'),T(k))=\min\left\{E \mid Ex' \geq \sum_{j=1}^{n} \lambda^j x^j, y' \leq \sum_{j=1}^{n} \lambda^j y^j, \lambda \in \Lambda(k), E \geq 0\right\}.$$

These input efficiency scores measure how much all inputs can be reduced proportionately within the production possibility set and take, by definition, values between 0 and 1 (0 and 100 %), where a score of 1 indicates a (fully) efficient unit.

2.2 The statistical model

Several authors (e.g. Banker 1993 and Simar 1996) have approached the comparison of efficiency scores by means of parametric statistical methods. In one of the earliest attempts to establish a statistical foundation Banker suggests two feasible methods, based on the assumption that the probability density function for the efficiency scores is a monotone decreasing function, either a half-normal or an exponential density. Also other authors tend to use a truncated normal distribution as probability model for efficiency scores. The family of truncated distributions includes the half-normal distribution suggested by Banker (1993) and arises when one samples from an incomplete population or equivalently, from a conditional distribution. However, the assumption of a truncated continuous distribution is contradictory to empirical results in general, including those obtained in the present study. Also from purely methodological reasons the DEA method implies that we do have a strictly positive probability for observing efficiency scores of precisely 100 %. In the present context a truncated continuous distribution can only be used to fit an empirical distribution for efficiency scores given that the units considered not are fully efficient. However, comparison of efficiency scores without accounting for the fully efficient units, and thereby not utilise all relevant information, can result in meaningless conclusions. By contrast, a right censored distribution accounts for this type of distribution.

In the present case study the statistical analysis is based on the assumption that the efficiency scores can be described by a censored log-normal distribution with log (100) as right censoring point. This assumption is based on empirical evidence as will be seen in the sequel.

The two-parameter log-normal distribution is the distribution of a stochastic variable X if log X is normally distributed. The density function is

$$f(x)=\frac{1}{\sqrt{2\pi\sigma^2}}\frac{1}{x}e^{-(\log x-\mu)^2/2\sigma^2}, 0<x<\infty,$$

where e^{μ} is a scale parameter (the median) and σ^2 is a parameter describing the shape of the distribution. Estimation of parameters in the censored log-normal distribution is performed as maximum likelihood estimation applying the EM-algorithm (Dempster et al. 1977) in the standard way for incomplete data. Moreover, 95% confidence limits are calculated by use of the profile likelihood technique (e.g. Azzalini 1996).

Comparison of efficiency scores between groups of farms is performed by likelihood ratio tests for the hypotheses of equal μ's, equivalently e^{μ}'s, over groups. It may be argued, as in stochastic frontier models (Meeusen and van den Broeck 1977, Aigner et al. 1977), that due to measurement errors on the output variable, the calculated DEA inefficiencies may be 'pseudo'-inefficiencies and composed of both real inefficiency and a noise term. However, as long as the noise term is assumed to have mean zero and the components act additively, the mean 'pseudo'-inefficiency and the mean 'real'-inefficiency are the same. So, when the interest is comparison of efficiency among subgroups, the obtained DEA scores can be used for inference purposes.

Alternatively, comparisons of efficiency scores between different groups may be performed using non-parametric tests as, e.g., the Mann-Whitney test (e.g. Siegel and Castellan 1988). The counterpart within the framework of non-parametric statistics to the problem of multiple efficient units can be seen as the problem of handling ties. An advantage of the parametric method, however, is that the actual values of the efficiency scores are used instead of the ranking. This implies that all the information available is utilised, and thereby that one is getting more powerful and precise results.

Finally, note that standard maximum likelihood theory is based on the assumption that the stochastic variables (here efficiency scores) are independent and identically distributed. The definition of the efficiency scores implies that the independence assumption almost automatically is violated. In turn this implies, also mentioned by Simar (1996), that one cannot rely on standard methods for inference. However, dependencies between inefficiencies vanish asymptotically for large number of decision-making units (Banker 1993). In the case described below the number of units is relatively large and, it turns out (Figure 1) that the estimated probability density function describes the empirical density well, indicating a high reliability of the estimated parameters. Still the calculated likelihood ratio test statistics and corresponding p-values must be interpreted cautiously.

3. A Case Study of Danish Dairy Farms

The data set originates from the Danish Agricultural Advisory Office. Here accounts for about 13,000 Danish farms are registered on a yearly basis. In the present study, observations from 1994 are used. To ensure a homogeneous sample we concentrate on full-time farmers that are mainly milk producers. After removing a few defective observations a sample of 1,810 farms remains. Based on the analysis in Lund and Hansen (1993) we have calculated from a large number of registered economic variables the principal non-allocative cost components (measured in Danish kroner).

The five input variables used in the empirical analysis are:

1. Building costs, defined as the book depreciation of the building investments plus the costs of maintenance.
2. Equipment costs, defined as the sum of the book deprecation of the inventory, the maintenance costs and the contract operations costs. The latter is included because it substitutes investment in inventory.
3. Capital costs, calculated as 4 per cent of all capital locked up in land, production buildings, livestock, inventory and stock, which is a proxy for the opportunity cost of the capital locked up in the production assets.
4. Labour costs, the wages actually paid to workers plus a calculated payment of family labour. The payment of family labour is an estimate based on the difference between wages obtained from work done outside the farm and an average wage for farmers. See Luijt and Hillebrand (1991) for a related approach to the pricing of family labour.
5. Miscellaneous costs, calculated as the sum of costs regarding insurance, energy, water etc. plus the allocative costs, i.e. fodder, seed grain and fertilisers.

3.1 The single output variable used is gross returns.

A multiple regression analysis with gross returns as dependent variable and the five input variables as independent variables shows that all the inputs have significant effect and have the expected (positive) signs.

Note that the above model differs from the variables used in the previous studies by Lund et al. (1993) and Lund and Hansen (1993) on three main points:

- Here equipment and contract operations are aggregated as these are considered to be perfect substitutes. Treating the variables separately may result in cases where units become efficient by disregarding one of these two dimensions.

- It is necessary to include the variable or allocative costs in an analysis of the non-allocative costs, because they affect the output. Otherwise, inefficiency in the allocative costs would look like inefficiency in the non-allocative costs.

- When including the allocative costs in the analysis gross returns must be preferred as output variable instead of gross margins, to avoid that allocative costs affect both cost and revenue.

A short description of the variables can be seen in Table 1.

Variables	Mean	Min.	Max.	Std.dev.
Building	50478.11	1127.00	256426.00	29303.2
Equipment	182319.91	23339.00	699624.00	88156.3
Labour	292968.69	0	737847.00	88512.7
Capital	108015.35	18134.80	494672.96	53075.8
Miscellaneous	464164.31	93353.00	1632762.00	2065219.9
Gross Returns	1097293.93	157640.00	4022522.00	502106.1

Table 1 Descriptive statistics for input and output variables

Note that due to the definition of labour cost the minimum value can be 0. This value is obtained for a single farm and can be caused by a farm with no employees and a family having external salaries exceeding the average wage for farmers.

For further analysis of the DEA results the efficiency scores of the farms are divided into six groups according to the size of the farm (less than 30 cows, 30-60 cows and more that 60 cows) and whether its livestock consists of jersey or non-jersey cows. It should be noted that data in each subgroup can be described by the same technology as the whole sample.

4. Results

To assign an efficiency score to each farm, a DEA is performed with the above-described variables as input and output variables. Efficiencies are calculated assuming a constant return to scale technology and a variable return to scale technology, respectively.

The obtained efficiency results are summarised in Table 2.

CRS							
Race	Number of cows	N	No. efficient	Min.	Lower quartile	Median	Upper quartile
Non-jersey	≤ 30	257	9 (3.5%)	48.89	71.43	77.03	85.96
	30 < - ≤ 60	941	16 (1.7%)	47.15	74.03	79.46	85.59
	> 60	363	8 (2.2%)	46.52	78.25	82.86	87.57
Jersey	≤ 30	42	2 (4.8%)	47.88	63.66	78.71	87.08
	30 < - ≤ 60	148	5 (3.4%)	56.13	77.33	82.43	89.00
	> 60	59	4 (6.8%)	66.73	77.78	83.39	89.08
Total		1810	44 (2.4%)	46.52	74.81	80.46	86.49
VRS							
Race	Number of cows	N	No. efficient	Min.	Lower quartile	Median	Upper quartile
Non-jersey	≤ 30	257	20 (7.8%)	57.08	76.42	81.61	90.44
	30 < - ≤ 60	941	22 (2.3%)	47.22	74.94	80.31	86.17
	> 60	363	20 (5.5%)	47.75	79.56	84.61	90.50
Jersey	≤ 30	42	9 (21.4%)	63.62	78.74	85.12	98.83
	30 < - ≤ 60	148	6 (4.1 %)	61.54	78.08	83.46	89.76
	> 60	59	8 (13.6%)	67.03	78.07	84.56	91.15
Total		1810	85 (4.7%)	47.22	76.44	82.12	88.22

Table 2 Descriptive statistics for efficiency scores with different technological assumptions

First we notice that there is a large number of inefficient farms. For the crs technology only 44 out of a total of 1810 farms are best performers. Assuming the vrs technology, twice as many farms turn out to be fully efficient, but this is still a rather small ratio of the total sample. Per definition the vrs assumption gives an overall higher efficiency score than the crs assumption. The median efficiency is 80.46 for crs versus 82.12 for vrs.

Moreover, it can be noticed that the efficiency scores obtained assuming crs generally are quite close to the vrs scores. This indicates that an assumption of constant returns to scale seems reasonable.

As seen from the percentiles in Table 2 the distribution of efficiency scores is characterised by being asymmetric with a heavy upper tail. This type of skewness of the empirical distribution is present in all groups and by both types of technology. In Figure 1 distributions of efficiency scores are shown graphically for farms with between 30 and 60 cows.

The histograms in Figure 1 support the tendency of a skew distribution mentioned above. The log-normal distribution is well suited to describe this type of distributions. Because of the strictly positive probability mass at the upper limit of the efficiency scores, a right censoring point, log(100), is built into the log-normal distribution. Within each group the parameters in the censored log-normal distribution are estimated for both assumptions about returns to scale. The solid lines in Figure 1 are the estimated probability density functions. The estimated means $\hat{\mu}$, medians $e^{\hat{\mu}}$ and corresponding 95% confidence limits are shown in Table 3.

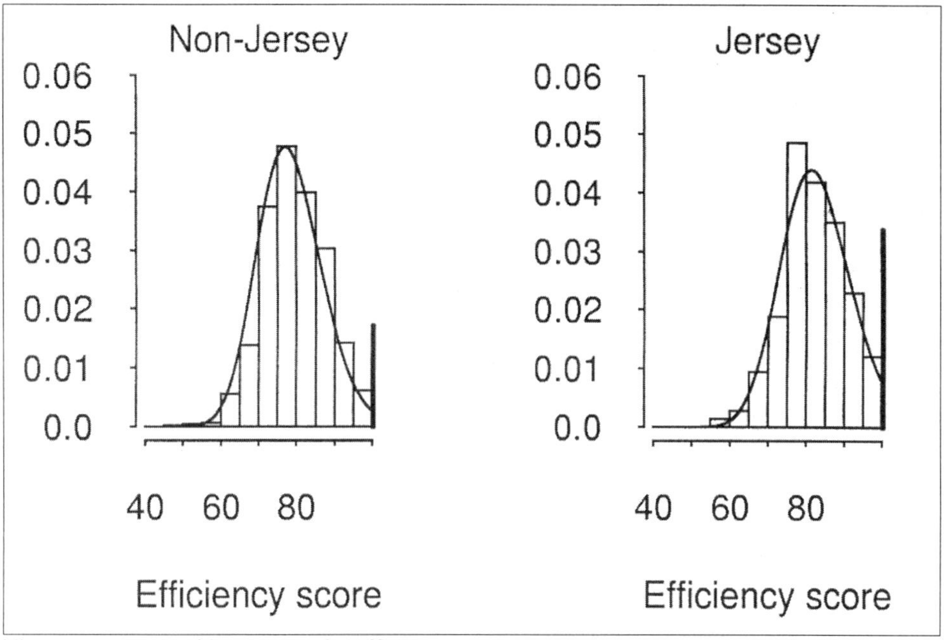

Figur 1 Normalised histograms for efficiency scores assuming crs technology, together with density function for the corresponding estimated censored log-normal distributions. Farms with 30-60 cows

CRS						
Race	**Number of cows**	**N**	**Log scale** $\hat{\mu}$	**95% conf. Limits**	**Perc. Scale** $e^{\hat{\mu}}$	**95% conf. limits**
Non-jersey	≤ 30	257	4.353	(4.336,4.370)	77.69	(76.39,79.00)
	$30 < - \leq 60$	941	4.376	(4.369,4.383)	79.51	(78.96,80.05)
	> 60	363	4.418	(4.408,4.428)	82.94	(82.13,88.75)
Jersey	≤ 30	42	4.335	(4.277,4.393)	76.29	(71.99,80.90)
	$30 < - \leq 60$	148	4.411	(4.393,4.429)	82.35	(80.88,83.82)
	> 60	59	4.426	(4.400,4.452)	83.58	(81.42,85.78)
VRS						
Race	**Number of cows**	**N**	**Log scale** $\hat{\mu}$	**95% conf. limits**	**Perc. Scale** $e^{\hat{\mu}}$	**95% conf. limits**
Non-jersey	≤ 30	257	4.418	(4.402,4.430)	82.91	(81.64,83.94)
	$30 < - \leq 60$	941	4.387	(4.380,4.394)	80.39	(79.84,80.93)
	> 60	363	4.441	(4.430,4.452)	84.86	(83.95,85.76)
Jersey	≤ 30	42	4.468	(4.420,4.521)	87.22	(83.07,91.93)
	$30 < - \leq 60$	148	4.425	(4.408,4.442)	83.48	(82.08,84.90)
	> 60	59	4.450	(4.419,4.481)	85.60	(83.04,88.29)

Table 3 Estimated means $\hat{\mu}$, medians $e^{\hat{\mu}}$ and 95 % confidence limits

Since an assumption of increasing or at least constant returns to scale, seems most appropriate from an economic point of view, we consider the crs case first.

For non-jersey farms as well as jersey farms we observe significant differences in the mean efficiency among the size groups. The likelihood ratio test statistic is $-2 \log Q = 63.6$ for comparison of the three means for the non-jersey farms. Evaluated in the χ^2 distribution with two degrees of freedom, this corresponds to p≈0.0000. Similarly, for jersey farms the test statistic is –2 log Q =7.32, resulting in a low p-value, p=2.5 %. Both for jersey and non-jersey farms the efficiency tends to increase with increasing size.

There seems to be a difference between the mean efficiency of jersey versus non-jersey farms independent of size. However, a closer inspection reveals that this difference is not significant for the larger farms with more than 60 cows (-2logQ=0.347, df=1, p=55.6 %), nor for the smallest farms with less than 30 cows (-2logQ=0.356, df=2, p=55.1 %). For the majority of farms, the group with 30-60 cows, jersey farms are, however, significantly more efficient than non-jersey farms. (-2logQ=12.92, df=1, p=0.03 %).

The fact that farms with jersey cows tend to be more efficient than non-jersey farms is supported by the results for the variable returns to scale technology in Table 3. Concerning the variation between size groups the mean efficiency does not differ significantly within the group of jersey farms (-2logQ=4.122, df=2, p=12.7%), whereas there is a significant difference within the group of non-jersey farms (-2logQ=149.79, df=2, p≈0.0000). This difference does not exhibit the expected pattern, since the smallest farms are more efficient than the middle group. Rather than being a consequence of a different utilisation of the capacity costs, we believe this finding to be caused by the DEA method itself. In a DEA model with varying returns to scale, the farms with the smallest consumption of the various inputs are efficient by definition. Since this group consists of a relatively small number of farms, these efficient farms will have a fairly large impact on the average efficiency.

To summarise, the results of the DEA showed an average efficiency around 80 % which is in line with earlier results (Lund et al. 1993, Hallam and Machado 1996), despite the fact that the present study is based on a different and much larger data set. The results indicate a considerable input reduction potential in dairy production. However, the specific value of the average efficiency should be interpreted somewhat cautiously since it depends critically on the choice of model as well as the size of the sample and the number of input and outputs. Furthermore, our result confirms earlier findings as to economy of scale (Hallam and Machado 1996, Lund et al. 1993, Kalaitzandonakes et al. 1992, Weersink and Tauer 1991, Weersink et al. 1990). A new result, on the other hand, is that jersey farms tend to be more efficient than non-jersey farms. This may be caused by a real difference in efficiency, but could also be a result of some underlying unmodelled production aspect. A further examination of possible explanatory variables has shown that neither the age of the farmer, nor whether the farm is an organic farm or not could explain the differences in efficiency.

5. Concluding remarks

The proposed parametric model for efficiency scores - the censored log-normal distribution - was found to give a satisfactory description of the obtained empirical distribution. Application of this model when comparing efficiency scores for groups defined by external factors gives a natural method for handling the fact that these distributions often contains several fully efficient units. It may be argued that the problem diminishes for large data sets. However, the data set analysed in the present paper is large (1810 units) and still a relative large number of fully efficient units is observed even assuming constant return to scale. Moreover, even for large data sets the fully efficient units may be unequal distributed over the groups compared and consequently must be accounted for when comparing efficiency scores.

References

AIGNER, D.J., LOVELL, C.A. AND SCHMIDT, P. 1977. Formulation and Estimation of Stochastic Frontier Production Function Models. Journal of Econometrics 6:21-37.

AZZALINI, A. 1996. Statistical Inference: Based on the Likelihood. Chapman-Hall.

BANKER, R.D. 1993. Maximum Likelihood, Consistency and Data Envelopment Analysis: A Statistical Foundation. Management Science 39:1265-1273.

BRAVO-URETA, B.E. AND RIEGER, L. 1990. Alternative Production Frontier Methodologies and Dairy Farm Efficiency. Journal of Agricultural Economics. 41:215-226.

BRAVO-URETA, B.E. AND RIEGER, L. 1991. Dairy Farm Efficiency Measurement using Stochastic Frontiers and Neo-classical Duality. American Journal of Agricultural Economics 72:421-428.

CHARNES, A., COOPER, W.W. AND RHODES, E. 1978. Measuring the efficiency of decision-making units. European Journal of Operational Research 2:429-444.

CHARNES, A., COOPER, W.W., LEWIN, A.Y. AND SEIFORD, L.M.(ED.) 1995. Data Envelopment Analysis: Theory, Methodology and Applications. Kluwer Academic Publishers, Dordrecht.

DEMPSTER, A.P., LAIRD, N.L. AND RUBIN, D.B. 1977. Maximum Likelihood from Incomplete Data via the EM Algorithm (with discussion). Journal of the Royal Statistical Society B. 39:1-38.

HALLAM, D. AND MACHADO, F. 1996. Efficiency Analysis with Panel Data: A Study of Portuguese Dairy Farms. European Review of Agricultural Economics 23:79-93.

KALAITZANDONAKES, N.G., WU, S. AND MA, J. 1992. The Relationship between Technical Efficiency and Firm Size Revisited. Canadian Journal of Agricultural Economics 40:427-442.

LUIJT, J. AND HILLEBRAND, J.H.A. 1991. Fixed Factors, Family Farm Income and the Continuity of Dutch Dairy Farms. European Review of Agricultural Economics 19:265-282.

LUND, M. AND HANSEN, L.C.E. 1993. Analysing Agricultural Capacity Costs (in Danish). Statens Jordbrugsøkonomiske Institut, Denmark. Report no. 71.

LUND, M., JACOBSEN, B.H. AND HANSEN, L.C.E. 1993. Reducing Non-allocative Costs on Danish Dairy Farms: Application of Nonparametric Methods. European Review of Agricultural Economics 20:327-341.

MEEUSEN, W. AND VAN DEN BROECK, J. 1977. Efficiency Estimation from Cobb-Douglas Production Function with Composed Error. International Economic Review 8:435-444.

SIEGEL, S. AND CASTELLAN, J.M. 1988. Non-Parametric Statistics. McGraw-Hill, New York.

SIMAR, L. 1996. Aspects of Statistical Analysis in DEA-Type Frontier Models. Journal of Productivity Analysis 7:177-185.

WEERSINK, A. AND TAUER, L.W. 1991. Causality between Dairy Farm Size and Productivity. American Journal of Agricultural Economics 73:1138-1145.

WEERSINK, A., TURVEY, C.G. AND GODAH, A. 1990. Decomposition Measures of Technical Efficiency for Ontario Dairy Farms. Canadian Journal of Agricultural Economics 38:439-456.

Biological Insecticides contra Insect Growth Regulators in Polish Forestry - a Financial Cost-Benefit Analysis

Peter Tarp
Finn Helles
Stanislaw Zajac

Abstract

Strategic decision-making within integrated forest pest management is often mistakenly based on a short-term treatment cost evaluation. A modelling framework is presented having a long-term perspective focused on the environmental impact associated with two groups of control agents - insect growth regulators and biological insecticides. The analysis is linked to the dynamical eco-physiological processes in a pine forest soil, based on the decomposing role of the soil fauna biomass in the ecological system. A considerable part of the soil fauna biomass depends on chitin formation and deposition which is halted by the presence of an effective insect growth regulator on the litter. The persistency of Dimilin (an insect growth regulator) on leaf surfaces has been reported to be up to 12 months, as opposed to biological insecticides having a short persistency of up to two weeks. Therefore, a subsequent mortality of elements of the soil fauna may block the release of nutrients associated with decomposition of litter and lead to a reduction of the growth and yield of the stand. A sensitivity analysis of the incremental cost is performed by use of a long-term dynamic linear programming model for the state forest area of Poland, leading to a financial cost-benefit analysis. Based on historical Polish data on treatment of defoliating forest pests it is found that a possible yield reduction of approximately six per cent of the treated forest area seems to make a biological insecticide more profitable than an insect growth regulator, in despite of the annual treatment cost per ha of the former being considerably lower.

Keywords: Integrated forest pest management, economic result evaluation, biological pesticides, insect growth regulators, linear programming

1. Introduction

Strategic decision-making regarding the choice of control agents within integrated forest pest management should be performed on a marginal basis, but there is a tendency of mistakenly using a short-term perspective with focus on the annual procurement cost pertaining to a specific strategy. Focus should be on the objective of the efforts, viz. to protect and improve the productive capacity of the forest resources, and therefore a long-term perspective is necessary with due observation of the pest management's impact on the forest resources. In this paper we provide an appropriate modelling framework for strategic decision-making within integrated forest pest management, exemplified by an evaluation of the use of insect growth regulators compared to the use of biological insecticides.

Forest pest management and control is a very important issue in Polish forestry. The forest area is dominated by coniferous species occupying 79.1 per cent of the state forest area, and Scots pine (*Pinus silvestris* L.) accounts for 70.9 per cent. About 82 per cent of the entire forest area of 8.7 million ha is state owned (Strykowski et al. 1993, Central Statistical Office 1994).

The major forest pests during the last decades have been lepidopteran and some hymenopteran species defoliating coniferous forests, Nun Moth (*Lymantria monacha* L.) being the most important pest species (Kolk 1992). The average length of epidemic infestations by Nun Moth since 1950 has been about seven years, and the average size of the annual treatment area during the infestations amounted to 600,000 ha, corresponding to 12.7 per cent of the present state forest area of Scots pine.

Defoliating forest pests were treated with pyrethroids from the 1980s until 1994, where collaboration with foreign institutions and financing by international donors with emphasis on environmental concern led to increased use of insect growth regulators and biological insecticides. In 1994, approximately 70 per cent of the infested forest area was treated with the insect growth regulator Dimilin 480 SC, about 20 per cent with the biological *Bacillus thuringiensis* insecticide Foray 48 B, and the remaining 10 per cent with pyrethroids (Glowacka 1994).

The 1994 forest pest treatment campaign was followed by a combination of ecological and economic monitoring, studies, evaluation, and research with the aim to compare the former wide use of pyrethroids with more environmentally friendly alternatives. A preliminary long-term evaluation of pyrethroids compared to insect growth regulators and biological insecticides indicate that the latter groups of control agents may be more profitable than the use of pyrethroids (Tarp and Helles 1994). The presentation here is concerned with a financial cost-benefit analysis comparing the use of an insect growth regulator and *Bacillus thuringiensis*.

2. Material and Methods

2.1 Characteristics of an insect growth regulator compared to *Bacillus thuringiensis*

Dimilin is the trade name of the selective insecticide diflubenzuron. It is a synthetic growth regulator which, when ingested, interferes with chitin synthesis, causing the body wall of the larva to rupture during the moulting phase. Protein synthesis is unaffected, but its deposition in the cuticle can be disturbed (Duphar 1995).

Diflubenzuron has potential for persistence or accumulation in highly acidic conditions (pH < 4.0). No degradation was observed after 56 days under such conditions (Ivie et al. 1980). The persistency of diflubenzuron outdoors has been reported to be one year or more. This long-term effect gives cause for serious concern because food chains can be interrupted, which means that hidden, unwanted side-effects can not be excluded (Hunter and Skatulla 1994). The stability of diflubenzuron on leaf surfaces is high, with as much as 90 per cent remaining 30-60 days after treatment. About 20-25 per cent has been reported to remain on pine needles after 62-63 days (Willcox and Coffey 1978, Mutanen et al. 1988, Van Den Berg 1986).

Springtails and mites in soil were incidentally affected in different types of forests after treatment with diflubenzuron (USDA 1989, Duphar 1995). Chitin-producing organisms present in the aquatic ecosystem and in soil during treatment may be affected (Willcox and Coffey 1978). Some non-target moulting insects, especially canopy feeding macrolepidoptera and non-lepidopteran mandibulate herbivorous insects may be impacted by diflubenzuron. Sucking herbivorous insects, microlepidoptera, and predacious insects would be less affected (Martinat et al. 1988).

Bacillus thuringiensis, commonly known as Bt, is a naturally-occurring, spore-forming, crystal-producing, rod-shaped bacterium. The variety registered for the control of various agricultural and forest Lepidoptera is Bt Berliner, variety *Kurstaki*, of various strains. Commercial formulations of Bt contain the spores and crystals as their primary ingredients. When ingested by larvae, the crystals dissolve in the alkaline gut, releasing a toxic protein. The protein breaks down the lining of the gut, causing the insect to cease feeding. The bacterium then invades the internal tissues of the larvae, multiplies rapidly and causes organ disintegration. Death of larvae is caused by a combination of infection and starvation, usually within 7 to 10 days (USDA 1985, Dubois and Lewis 1981).

The intake of Bt by larvae leads to a rapid paralysis of intestines followed by death. In contrast to diflubenzuron, it is effective already after one to three days, where the caterpillars stop eating and die. The persistency of Bt is about 7-14 days (USDA 1989, Hunter and Skatulla 1994, Novo Nordisk 1994).

2.2 Economic model description

The analysis presented here is based on the national forest inventory of Poland, prepared by the General Directorate of State Forests. A dynamic linear programming optimisation model has been developed (see a general outline in Figure 1) with the objective of maximising the net present value of the state forest area, applying a real discount rate of 2 per cent per annum.

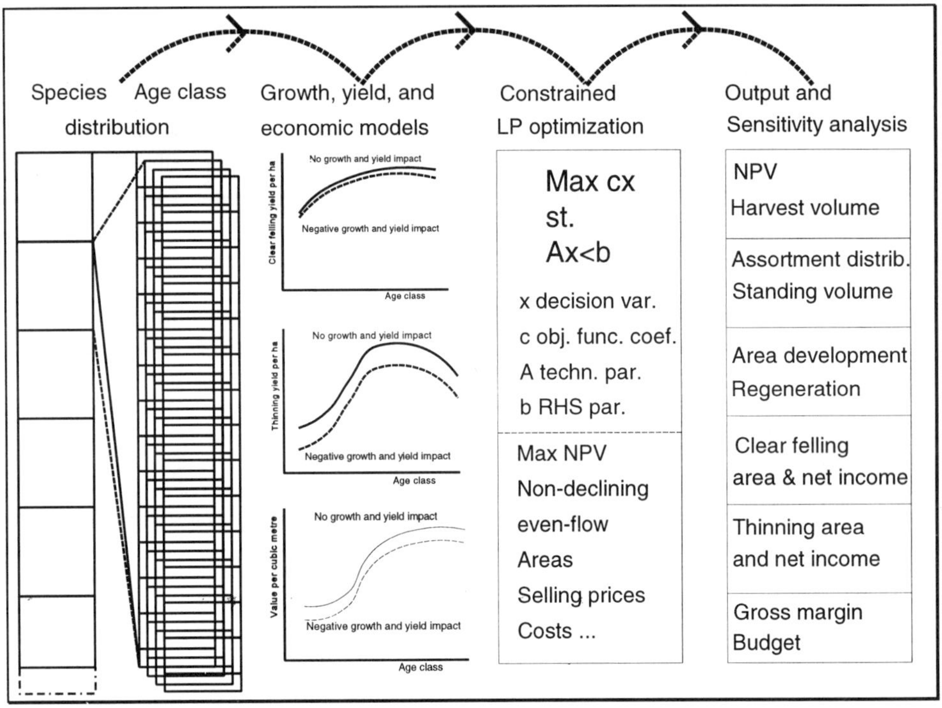

Figure 1 General outline of LP model

The mathematical formulation of the LP programme is given below:

$$\text{MAX} \sum_{t=1}^{N} a^t \mathbf{V}'\mathbf{h}_t + a^{N+1}\mathbf{r}'\mathbf{x}_{N+1}$$

s.t.

$$\mathbf{x}_{t+1} = \mathbf{R}\mathbf{x}_t - \mathbf{S}\mathbf{h}_t, \; t=1,2,...N$$

$$\mathbf{0}<\mathbf{h}_t<\mathbf{x}_t, \; t=1,2,...,N$$

$$(1-b)\mathbf{v}'\mathbf{h}_{t-1}<\mathbf{v}'\mathbf{h}_t<(1+b)\mathbf{v}'\mathbf{h}_{t-1}, \; t=2,3,...,N,$$

where a^t is the discount factor, t denotes time periods, $\mathbf{V}'\mathbf{h}_t$ is a revenue function, $\mathbf{r}'$ is the vector of per ha NPVs for areas in age classes 1,2,...,k at time N+1, $\mathbf{x}_{t+1}$ denotes the area of the age classes at time t+1, $\mathbf{Rx}_t$ denotes regeneration areas, and $\mathbf{Sh}_t$ denotes clear felling areas, $\mathbf{0}$ is a k-dimensional column of zeros, b is the maximum proportional decrease of sequential volume flow, and $\mathbf{v}'$ is a vector of per ha clear felling volumes. Thus, **R** and **S** are matrices. A detailed description of the model structure is given by Reed and Errico (1989), Tarp (1994) and Tarp and Helles (1995).

The model is based on a stratification into seven species groups, an age-class width of 10 years, a planning time frame of 100 years, and consists of approximately 7,000 rows, 7,000 columns, and 23,000 elements.

The main parameters in the LP model from an economic point of view are: i) the yield at clear felling, ii) the yield of thinnings, and iii) the stumpage price. These three parameters are the most sensitive and direct the search for the optimal solution (maximum NPV), which is constrained by non-declining even-flow constraints (cf. Figure 1).

The effect of a growth reduction on the diameter increment is simulated external to the LP model and is based on the assumption of a fixed height-standing volume relationship. The simulation leads to determination of the age-DBH (diameter at breast height) and age-stumpage price functions for an average stand, i.e. expressing the aggregate outcome of harvesting in infested (treated) and non-infested (non-treated) stands.

Two models (1 and 2) have been developed and applied. Model 1 is based on the entire Polish state forest area and Model 2 covers specifically the pine state forest area. The difference between Model 1 and 2 is described in more detail in Section 3.

The result of a DBH and stumpage price development simulation assuming a yield reduction of 10 per cent is shown in Figure 2A for Model 2. The dotted lines show the deviations from the normal DBH-age class and stumpage price-age class development. The reason for this result is ascribed to the aggregation level of the model, where a given growth reduction in some cases will cause a reduction of the DBH (and hence of the stumpage price) assumed at a given stand age and in other cases not (this procedure is applied in Model 2).

In Model 1 the effect of a growth reduction is taken into account as a reduction factor (RF) for each age-class, i.e. the effect on the stumpage price is simulated so that it is explicitly expressed over the full range of age classes. The reduction factors assuming a 10 per cent yield reduction are: for age classes 3-5: 0.98, for 6-8: 0.985, and for age classes 9 and above 0.99, as shown in Figure 2B. The reduction factor decreases with increasing age due to the high slope of the stumpage price function in the young-medium age classes (see, e.g. Figure 2B). The difference between the two stumpage price functions is hardly visible because the stumpage price functions are expressed for an average stand.

The possible negative impact on the thinning and clear felling yield from an average stand is simulated by use of formula 1:

$$Y_r(p) = Y_n(p) - Y_n(p) \times YRF(p) \times PIA \times LOE \quad (1)$$

where
$Y_r(p)$ = reduced yield in period p,
$Y_n(p)$ = normal yield in period p,
YRF(p) = yield reduction factor in infested stand in period p,
PIA = pest infestation area as a proportion of the total area,
LOE = length of epidemic in years.

The yield reduction factor is assumed to be constant over the entire planning horison, i.e. the effect is assumed to be immediate and persistent when simulating the impact of the control agent on the thinning yield. But the simulation of the clear felling yield comprises, however, a gradual increase of the yield reduction factor over a period of about one rotation (100 years), where the maximum yield reduction is encountered. YRF(p) is increasing ten-fold from period 1 to period 10 when the clear felling yield is simulated, according to an assumed linear relationship. The maximum yield reduction factors applied for simulating the thinning and clear felling yields are equal, i.e. if the thinning yield in an infested stand is assumed to be reduced by, e.g. 5 per cent, then the clear felling yield in an infested stand is assumed to be reduced by 1 per cent in period 2 (2/10 x 5 per cent) and 5 per cent in period 10 (10/10 x 5 per cent).

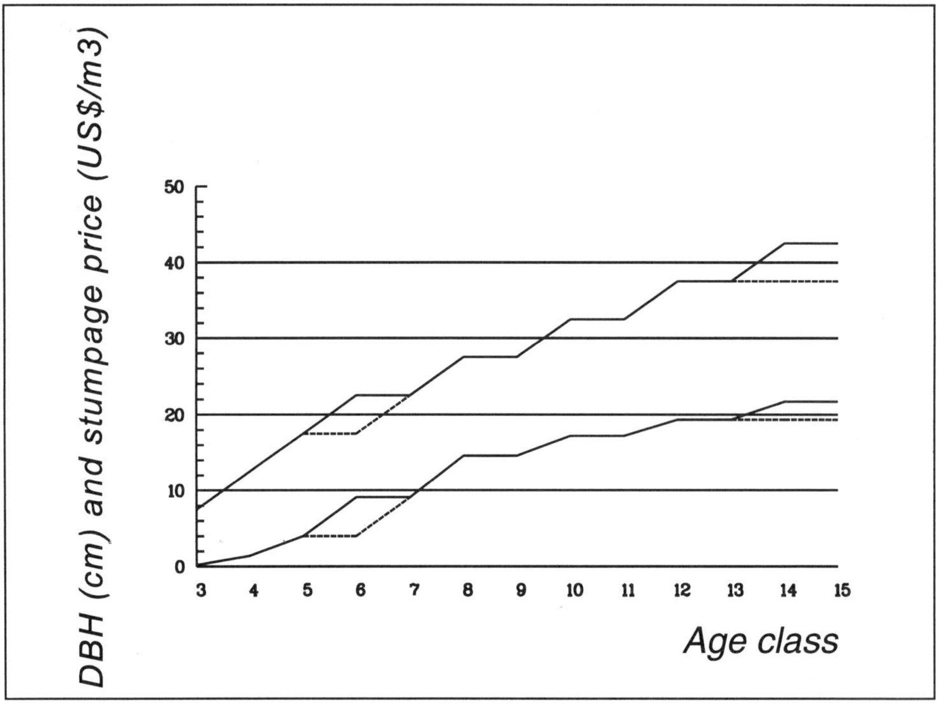

Figure 2A Result of DBH-age class and stumpage price-age class simulation, Model 2. (Dotted lines denote the development after a yield reduction of 10 per cent)

For illustration the numerical inputs to the LP model from the simulation of yield and stumpage price impacts are shown in Table 1. The relatively small impact on the yield is attributed to the size of the infestation area which amounts to 12.7 per cent of the entire pine forest area. The consequences of a yield reduction is transformed into the impact on an average stand.

The cost of treatment as observed in the 1994 forest pest campaign with diflubenzuron and Bt is presented by Hunter and Skatulla (1994) as shown in Table 2. The economic effect of a yield reduction is analysed below with due consideration of the effect on the diameter growth of the stand and thus of the effect on the stumpage price per cubic metre.

Table 1 Numerical inputs for an average pine stand to the LP model for economic analysis

Age Class	Clear felling volume per ha No impact, basic model Fourth period	10 per cent yield redution	Thinning volume per ha No impact, basic model First period	10 percent yield reduction
3			9.9	9.8
4			26.3	26.1
5			31.0	30.9
6			32.9	32.7
7	264	263	32.9	32.7
8	275	274	33.6	33.4
9	282	281	32.9	32.7
10	290	289	32.9	32.7
11	293	292	32.5	32.3
12	296	295	29.2	29.1
13	298	297	28.5	28.3
14	298	297	23.7	23.6
15	298	297		

Table 2 Treatment cost per ha with use of diflubenzuron and *Bacillus thuringiensis*

	Diflubenzuron	Bacillus thuringiensis US$ per ha
Threat monitoring	0.39	0.39
Active ingredient	9.58	13.60
Solvent	0.98	-
Aircraft hire	2.79	2.79
Application equipment	0.24	0.24
Airfield services	0.20	0.20
Spraying costs	3.31	3.31
Follow-up	1.10	1.10
Total	18.59	21.63

Figure 2B Result of DBH and stumpage price simulation for an average pine stand, Model 1

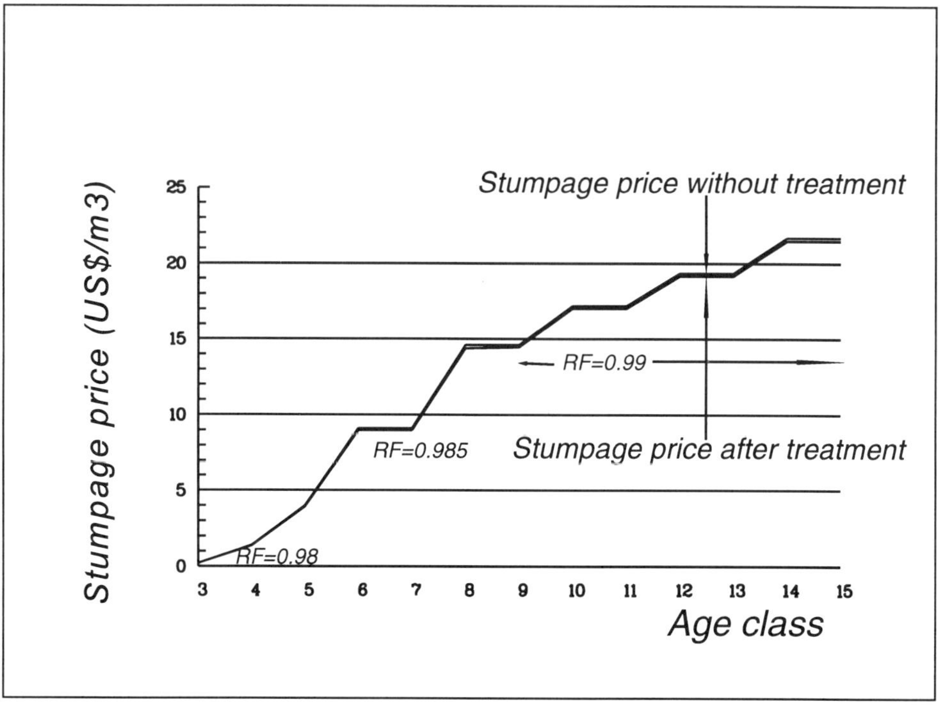

2.3 Eco-physiological and biological model description

An economic evaluation of the effect of defoliating forest pests is usually linked to the relationship between defoliation and subsequent growth and yield of stands, e.g. based on eco-physiological characteristics as illustrated by Flower-Ellis and Persson (1980), Linder and Troeng (1980), Linder and Lohammar (1981), and Troeng and Linder (1982a, b). However, this presentation is focused on the effect of changes of the eco-physiological processes dependent on the activity of the decomposing soil fauna biomass.

The body of arthropods is covered by an often thick cuticula. Therefore, these animals need to change the body cover during growth. Mites and beetles most often go through larva stages and a moulting stage where chitin formation is required (Bejer-Petersen 1966, 1968). The prospects for manipulation of the chitin formation and deposition processes are described by Fuhrman et al. (1992) and Spindler-Barth (1992). The egg shell of nematodes contains up to 30 per cent chitin (Bird 1976, Spiegel and Cohn 1985, Gooday et al. 1988). Blocking of egg production in nematodes by inhibition of a certain enzyme is reported by Ginger (1991). So, the presence of effective diflubenzuron on the food substance (pine needles) may also halt the regeneration of nematodes.

The fauna biomass in a pine forest in Central Sweden has been estimated by Persson et al. (1980). The contribution to the annual mean biomass and annual respiration of the main taxonomic groups within the soil fauna in a pine forest is shown in Figure 3.

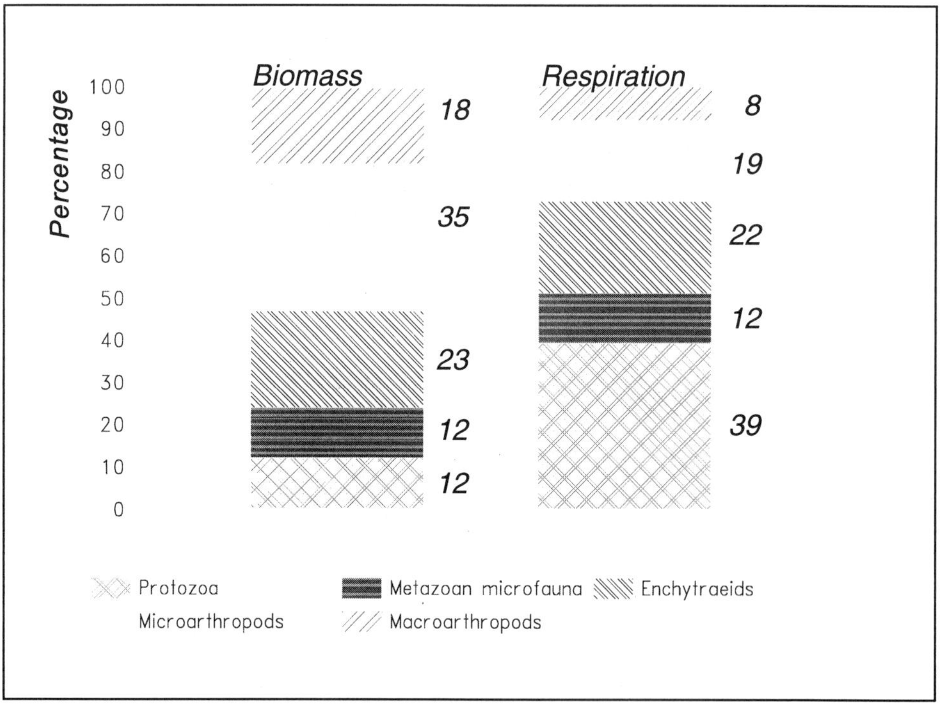

Figure 3 Contribution to the annual mean biomass and annual respiration of the main taxonomic groups within the soil fauna. Source: Persson et al. (1980)

It is seen that the non-bacteria biomass is dominated by microarthropods and the non-bacteria respiration is dominated by enchytraeids. Both the non-bacteria biomass and the non-bacteria respiration are dominated by arthropods.

A classification of the soil fauna according to feeding pattern is presented by Persson et al. (1980). The contribution to the annual mean biomass and annual respiration of these main trophic groups within the soil fauna is shown in Figure 4. It is seen that 14-19 per cent of the biomass and respiration is accounted for by saprovores.

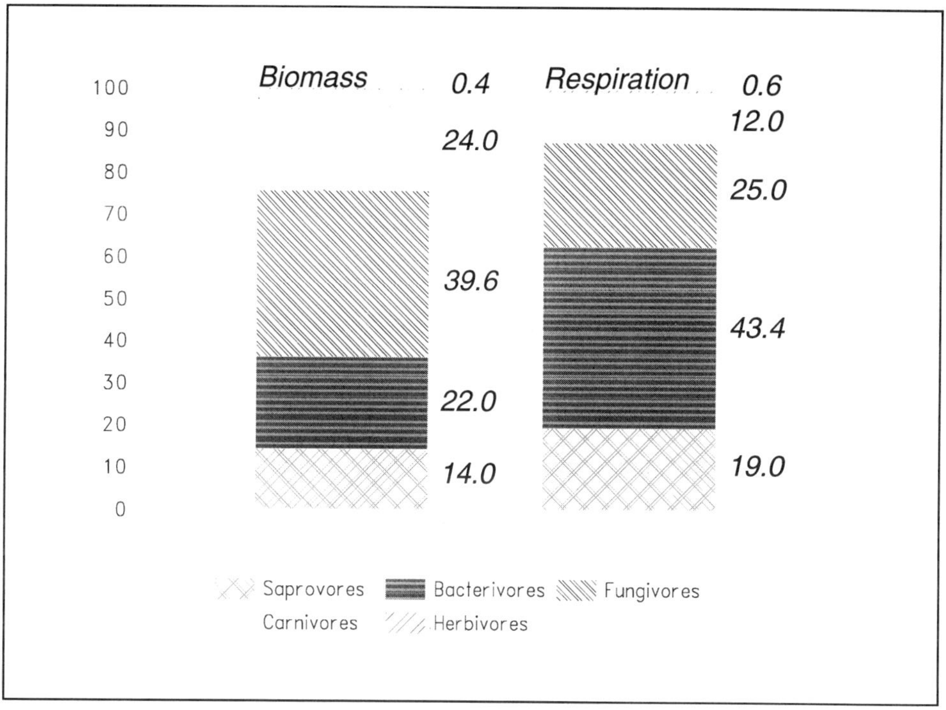

Figure 4 Contribution to the annual mean biomass and annual respiration of the main trophic groups within the soil fauna. Source: Persson et al. (1980)

The needle biomass in pine stands has been estimated at 2.2-2.8 tonnes dry weight per ha. Based on the presence of three age classes of needles, the normal annual needle litterfall is estimated at 733-933 kg dry weight per ha, corresponding to up to about 100 g per square metre (Larsson and Tenow 1980). Flower-Ellis and Olsson (1978) have estimated the annual mean value of total above-ground litterfall from pine trees at 140 g dry weight per square metre, corresponding to 1,400 kg per ha. Bosatta et al. (1980) have estimated the annual litterfall in an old pine stand in Sweden at 1,900 kg per ha.

Generally, the presence of forest cover compared to bare land decreases the pH value (Møller 1977). The soil fauna plays an important role for transformation of leaves and needles. The effect of the fauna is more profound on rich and productive sites compared to poor sites (Møller 1977). The importance of soil animals on decomposition of pine needle litter is described by Berg et al. (1980).

The pH in the litter layer of a 120 year old pine stand has been estimated at 3.5 by Bringmark (1980). The pH of the soil of a similar pine stand in Central Sweden was estimated by Persson et al.

(1980). The pH of the litter layer was 4.4, 4.0 of the humus layer, and 4.9 of the so-called enrichment layer - 10-40 cm below the surface.

The following economic analysis is focused on the effect of changes in the eco-physiological chain as illustrated in Figure 5 where the quantitative data refer to Figures 3 and 4. The relationship between the decomposer biomass and a forest stand is illustrated in Figure 6.

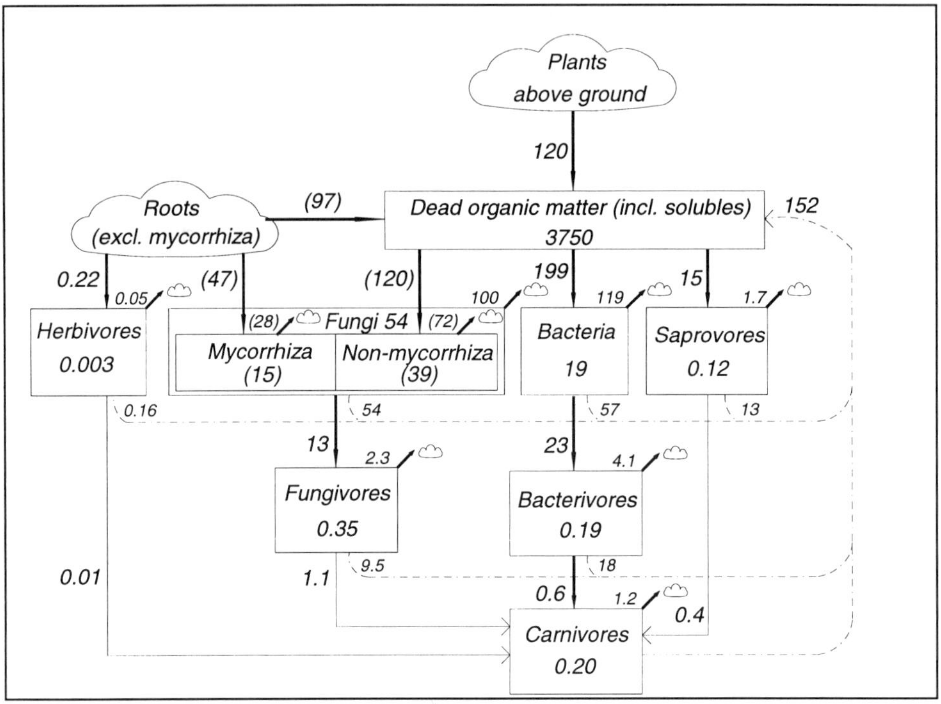

Figure 5 Carbon flow diagram of the soil (to a depth of 30 cm in the mineral soil) in a 120-year-old Scots pine stand in Central Sweden. Boxes denote annual mean biomass (g C per square metre) and arrows denote flows (g C per square metre per year), of which arrows with small clouds indicate respiration. Broken lines ending in an arrow denote transfer of dead (non-predated) organisms and faeces to the pool of dead organic matter. All values are uncertain and in particular those influenced by the assumptions of mycorrhiza (within parentheses). Source: Persson et al. (1980, p. 453)

The length of former infestation periods in Poland during the last 45 years has been approximately seven years. The approximate length of non-infested periods has been 2.5 years (Hunter and Skatulla 1994). Thus, on average it may be expected that an important infestation will occur once every decade.

If the persistency of diflubenzuron is assumed to be 12 months, the effect on the soil fauna may be observed during the whole treatment period plus one year after the last treatment.

If it is assumed that the effect in terms of avoiding defoliation is equal to use of diflubenzuron and Bt, these two options may be compared in economic terms, envisioning the persistency of Bt being only up to about two weeks. So, the effect of Bt on non-target soil fauna elements may be nil, because the litterfall is assumed to take place during a longer period, viz. from May until November (Flower-Ellis and Persson 1980).

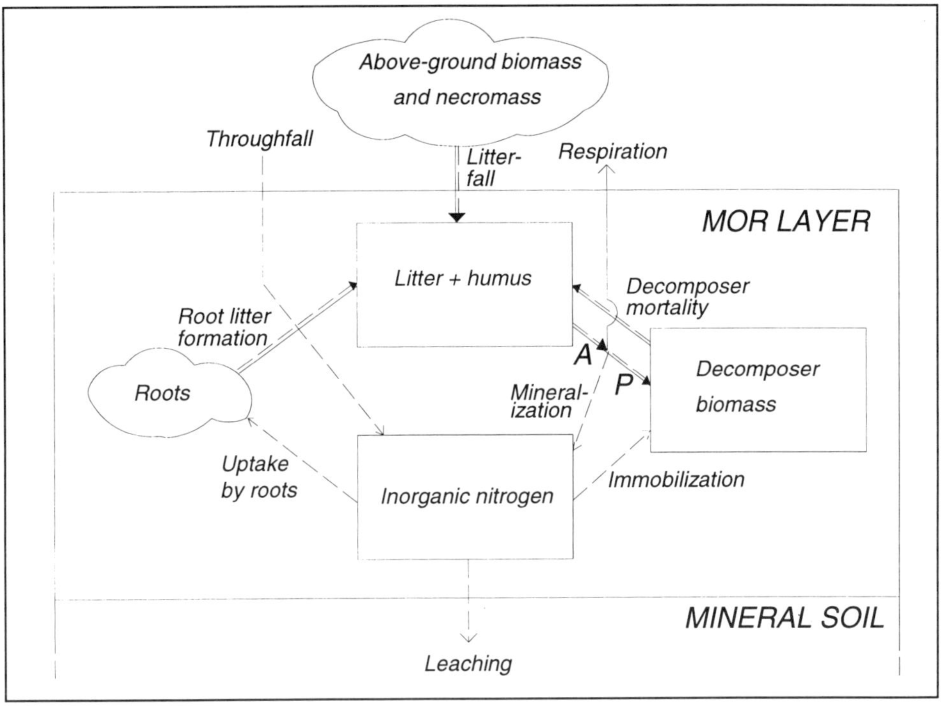

Figure 6 A nitrogen transfer model, represented by some components and processes of carbon (solid line) and nitrogen (broken line). A = assimilation (decomposition), P = production (incorporation into decomposer biomass). Source: Bosatta et al. (1980)

3. Results

Model 1 - for the entire Polish state forest area: The NPV is estimated at US$ 32.6 billion, corresponding to US$ 5,000 per ha, assuming a discount rate of 2 per cent and no pest infestation. The long-term non-declining even-flow harvest is estimated at an annual harvest of 37.2 million cubic metres, of which pine accounts for 27.6 million. A yield reduction for pine of 5 per cent reduces the NPV by US$ 53 million and the long-term sustainable pine harvest by 102,000 cubic metres annually. A yield reduction of 10 per cent leads to a decrease of the NPV of US$ 252 million and a reduction of the annual pine harvest of 128,000 cubic metres.

Model 2 - specifically for the pine state forest area: The NPV is estimated at US$ 15.9 billion, corresponding to US$ 3,300 per ha, assuming a discount rate of 2 per cent and no pest infestation. The long-term non-declining even-flow harvest of pine has been estimated at 26.6 million cubic metres annually. A yield reduction for pine of 5 per cent reduces the NPV by US$ 21 million and the long-term sustainable pine harvest by 61,000 cubic metres annually. A yield reduction of 10 per cent leads to a decrease of the NPV of US$ 444 million and a reduction of the annual pine harvest of 108,000 cubic metres.

Both models are most sensitive to the yield functions (clear felling and thinning) and the stumpage price function. The thinning yield function is affected immediately in the first planning period of 10 years with respect to the full effect of a growth reduction, whereas the clear felling yield function is gradually affected so that the full effect of a growth reduction is found after approximately one rotation of pine (cf. the simulated yield data in Table 1). The stumpage price effect of a growth reduction is assumed to last over the entire planning horison, which is the result of a compromise linked to the model structure; but the combined effect of the yield and stumpage price functions is assumed to reflect the real situation reasonably well, viz. a situation with decreasing yield and lower stumpage prices. So, the thinning yield function is critical in the short term and the clear felling yield function is critical in the longterm. The stumpage price function is not critical with respect to specific time periods.

The impact of the control agents on the soil fauna depends on the ability of the ecosystem to restore equilibrium after a proportion of the fauna has been killed. The dynamics of the soil fauna population is assumed to act as a buffer mechanism with certain thresholds of the critical biomass, enabling effective resistance to a reduction of a given population. A high treatment level (high application rate) is assumed to comprise a higher risk of reaching a critical biomass of important soil fauna groups. It is not possible based on the references cited here to depict the relationship between treatment level and decomposition of needle litter, but it is possible to analyse the economic effect associated with a growth reduction caused by a biomass reduction exceeding the above thresholds where the ecosystem cannot restore the activity (measured by respiration, biomass, or size of population) of the soil fauna biomass to the level before treatment. So, it is feasible to describe some possible consequences of impacts on the soil fauna biomass and then to analyse the economic consequences of a hypothetical reduction of the soil fauna activity, which is assumed to be functionally related to the growth and yield functions of a pine forest stand.

The differences between the results applying Model 1 and Model 2 are mainly caused by the effect of the growth reduction on the stumpage price function as described above. A change of the stumpage price function affects the economic optimal rotation age, the NPV, and the long-term sustainable harvest level. It is noted that the differences between the results of Modelles 1 and 2 are not biased (one-sided), so the application of the two models serves the purpose of sensitivity analysis.

The incremental cost of treatment with Bt compared to treatment with diflubenzuron has been estimated at about US$ 3 per ha per year (see Table 2). Based on the assumption that an area of about 600,000 ha per year is treated every year during an infestation, and the length of an infestation being 7 years, the corresponding NPV with use of a real discount rate of 2 per cent per annum amounts to US$ 63 million ((US$ 3/ha x 600,000 ha/year x 7 years/10 years)/(10 years x 2 per cent per year)) = US$ 1.26 million/year capitalised at 2 per cent = US$ 1.26 million/(year x 2 per cent per year) = US$ 63 million).

Due to the longer persistency of diflubenzuron compared to Bt it is pertinent to analyse the economic impact of a growth and yield reduction caused by the presence of effective diflubenzuron on the litter - through mortality of part of the decomposer soil fauna biomass, blocking the ecosystem cycle as illustrated in Figures 3 and 4. Based on an assumption of equal efficacy of the two agents with respect to control of defoliating pests, a sensitivity analysis is performed and shown in Table 3.

Table 3 Sensitivity analysis of variation of the impact of diflubenzuron on the decomposition of pine litter and subsequent effect on the yield

Decrease of yield (per cent)	Incremental NPV (US$ million)		Incremental annual pine harvest (1,000 cubic metres)	
	Model 1	Model 2	Model 1	Model 2
30	682	534	572	326
20	516	487	313	276
10	252	444	128	108
5	53	21	102	61

The economic consequences associated with a reduction of the yield are illustrated in Figure 7 together with the estimated NPV of the incremental NPV of treatment with Bt compared with diflubenzuron. The difference in form and development of the two curves is attributed to the different procedures in the two models applied for modelling the effect of a growth reduction on the stumpage prices. It is seen that a more smooth development is expressed by use of Model 1 compared to Model 2, probably because the stumpage price impact takes effect explicitly over the full range of age classes in Model 1. So, it seems that use of a function which distributes the effect on the stumpage price over the age classes is preferable to the use of piece-wise linear or discrete stumpage price functions, which may cause considerable deviations from the real stumpage price function due to aggregation characteristics.

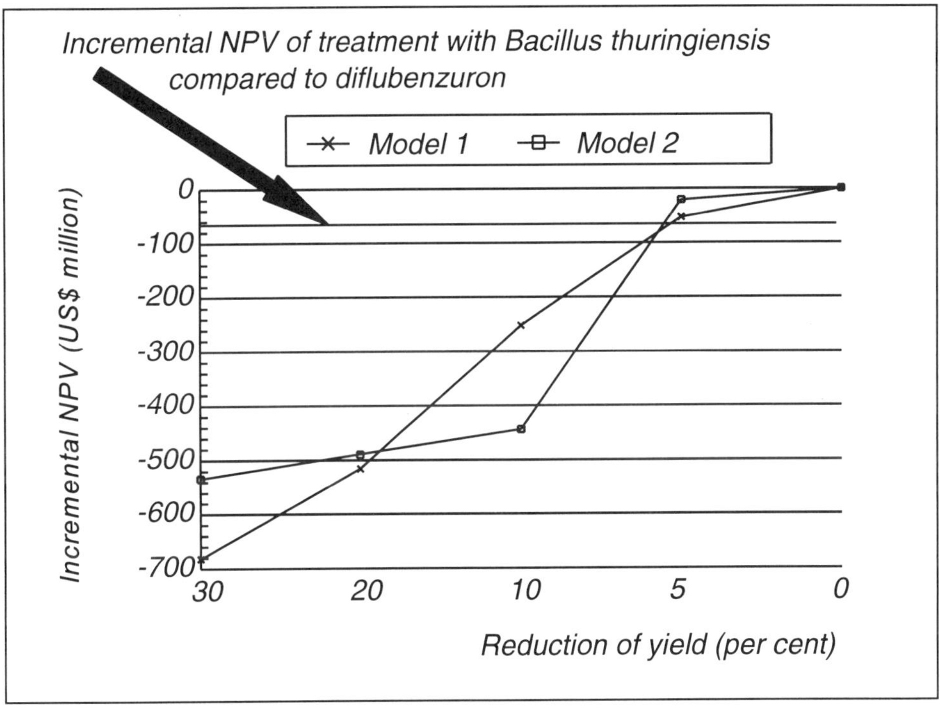

Figure 7 Sensitivity analysis of the impact of diflubenzuron on decomposition of needles and subsequent yield of pine for Model 1 and Model 2

It is seen that the estimated break-even point is found with a yield reduction of approximately 6 per cent by use of both models. That is if the yield reduction caused by the use of diflubenzuron exceeds approximately 6 per cent during the treatment period, then Bt is a more profitable treatment option. These results have been confirmed by stand level analyses and dual analyses of the national models.

4. Discussion and Conclusion

The effect of killing a proportion of the decomposing fauna species may have an effect on the ecological cycle which is much higher than shown in Figures 3 and 4, because the inhibition of the activity of a certain link in this cycle may impede the functioning of the cycle as a whole. Therefore, even if the respiratory activity of decomposing organisms is in the magnitude of, e.g. less than 5 per cent, mortality of this group may hinder the functioning of the processes further down, e.g. the decomposition accomplished by bacteria.

The major point to the defibrination performed by part of the fauna, making the debris accessible for other smaller animals or fungi. The species predominantly involved in these processes are probably Enchytraeidae (brushworms), Nematoda (roundworms), Diplopoda (millipedes), Elaterid larvae (lengthy, hard, and flat beetles), Diptera larvae (two-winged), and phthiracarid mites, Collembola (springtails), Protura (primitive insects), Blattodea, Psocoptera, Coleoptera (beetles), Diptera Nematocera (mosquito larvae), Acari (mites), and Symphyla (arthropods resembling true insects and millipedes). All these animals depend on chitin formation and deposition, however, the importance of chitin fotmation for roundworms and brushworms is still not fully known and documented.

Brushworms, roundworms, spingtails, elaterid larvae, larvae of two-winged insects, mites, and mosquito larvae are known to have a positive impact on the decomposition in forest soil (Bejer-Petersen 1966).

Mites (Acarina) comprise the most numerous part of the microfauna in forest soils (1,000-10,000 animals per square metre). They probably play an insignificant metabolistic role in raw humus where they are most abundant, *but* their cutting-up activity is important (Bejer 1989). This activity strongly increases the surface area of the remains, hereby supporting the decomposing effect of other microorganisms. Persson et al. (1980) conclude that Acari (mites) is the most species-rich animal taxon in pine forest. Mites are classified as microarthropods. Acari and collembolans (springtails) dominated the microarthropod abundance almost totally. The mean abundance in September was 58,000 per square metre for Collembola, 930 for Protura (primitive insects), and 750,000 for Acari.

As the organic substance of litter is a carrier of nutrients, litter decomposition also means a release of nutrients (see Figure 6). The retention of litter in the upper soil layer is greater in soils without any animals present. Such a difference is assumed to affect the total nutrient turnover rate in a forest (Berg et al. 1980). It is indicated that mites and worms could have an influence on the decomposition of the slowly decomposed components like lignin and lignified tissues, and as such that they might have a larger effect on long-term litter turnover than is indicated by simple weight loss (Berg et al. 1980).

In the soil profile all organism groups seem to have biomass maxima in the upper horisons. The animal biomass in pine forest was found to decrease more abruptly with increasing soil depth than the microorganism biomass (Persson et al. 1980). Based on respiration studies, it seems that microorganisms account for 80-99 per cent and soil animals for 1-20 per cent of the decomposition in a pine forest. Thus, the quantitative dominance of the microorganisms is outstanding in most soils. *However, the soil organism community is an interacting system with many feed-backs, and organisms at higher trophic positions may exert important regulating functions without being characterised by high metabolic rates.* It is, therefore, important to distinguish between different trophic groups within the soil organism community in order to obtain a better understanding of what the metabolic activity really implies (Persson et al. 1980).

Based on Figures 3 and 4 it may be concluded that the effect of diflubenzuron being persistent on the needles probably is that the metazoan microfauna, the enchytraeidae, the microarthropods, and the macroarthropods are affected. Based on the estimates of biomass and respiration in a pine forest soil, the impact of these groups on the decomposition of litter may be assumed to be at least about one-third, i.e. these animal groups are assumed to account for at least this proportion of the decomposition of the litter.

The decomposition process is a complicated dynamical ecological mechanism. It may be assumed that some of the larger saprovores initiate this process. Later, smaller saprovores and possibly other trophic groups are involved. The smaller saprovores together with fungi probably account for a large proportion of the total decomposition, but the defibrination caused by the larger animals is assumed to be a requirement for the further decomposition of plant materials. Therefore, the killing of only part of the fauna in the forest soil may impede the decomposition process and eventually lead to obstruction of the nutrient cycle as illustrated in Figures 5 and 6. Thus, ecological disturbance caused by a negative impact on the biodiversity of the soil fauna may create growth and yield losses, which in the end is analogous to the effect of defoliation. The evidence presented above seems to give reason for considering the risk connected with persistency of an insect growth regulator causing a decrease of the growth and yield of the treated forest stand. The argumentation is strongly connected with the pH of pine forest soils being about 4 or lower. The pH value may even be lower on poor soils, i.e. in pine forests with a low site index. Therefore, if a possible negative impact on growth and yield due to treatment with an insect growth regulator is expected, then the risk of such a negative impact may be lower when a biological control agent is applied in pine forest with a medium to low site index.

Figure 7 shows that, with the present assumptions, the use of Bt is more profitable than the use of diflubenzuron if the effect through a decreased decomposition on the yield of a treated pine stand is higher than approximately a 6 per cent reduction of the normal yield. A yield reduction of approximately 6 per cent leads to an incremental NPV of the forest area of -US$ 63 million, which is equal to the incremental cost of using Bt compared with use of diflubenzuron. It is possible that an insect growth regulator exerts a negative impact on decomposition and subsequent growth and yield of a forest stand, but it is not possible to predict the magnitude of this impact with certainty based on the research referenced here. However, the sensitivity analysis indicates that a negative impact of an insect growth regulator causing a decrease of the yield of more than 6 per cent makes the use of Bt more profitable than the use of the insect growth regulator. The likelihood of a negative impact caused by an insect growth regulator is higher on poor forest sites where a relatively low pH is prevalent.

It is possible that the former use of broad-spectrum pyrethroids has had a similar effect on the soil fauna so that the existing growth and yield may be improved through a shift to the use of Bt. In this situation the argumentation is analogous to that presented above. If the decomposition of litter is improved due to the absence of insecticide, then it is likely that the growth and yield of the treated stands are increased.

The analyses indicate that a supposition of a forest pest control agent having a negative impact on the growth of the forest stand (through a negative impact on the soil fauna) should be taken seriously and may, also on economic grounds, lead to preference for the most environment-friendly control agent. This observation is particularly relevant where treatment is required on poor forest soils (pH of 4 or lower) with either Bt or an insect growth regulator. Development of models with a lower aggregation level, i.e. with more detailed economic data, will allow a more precise estimation of the trade-off associated with a growth reduction caused by the control agent. Also, a more detailed examination must be made of the soil fauna's role for the ecological processes related to decomposition of pine needle litter. The sensitivity of the soil fauna to remains of control agents on needle litter could be analysed through experiments, which are, however, beyond the limits of the research underlying this presentation.

Acknowledgements

The Danish Ministry of the Environment and Energy, the Environmental Protection Agency is thanked for funding the research underlying the findings presented here.

References

BEJER, B. 1989. Forstzoologi I. Forstentomologi. [Forest Zoology I. Forest Entomology.] Jordbrugsforlaget, Copenhagen. (In Danish).

BEJER-PETERSEN, B. 1966. Forstzoologi I. [Forest Zoology I.] DSR Forlag, Copenhagen. (In Danish).

BEJER-PETERSEN, B. 1968. Forstzoologi II. [Forest Zoology II.] DSR Forlag Copenhagen. (In Danish).

BERG, B., LOHM, U., LUNDKVIST, H. AND WIREN, A. 1980. Influence of Soil Animals on Decomposition of Scots Pine Needle Litter. In: Persson, T. (ed.). Structure and Function of Northern Coniferous Forests - an Ecosystem Study. Ecological Bulletin (Stockholm) 32:401-409.

BIRD, A.F. 1976. The development and organization of skeletal structures in nematodes. In: Croll, N.A. (ed.). The Organization of Nematodes. Academic Press, London:107-137.

BOSATTA, E., BRINGMARK, L. AND STAFF, H. 1980. Nitrogen Transformations in a Scots Pine Forest Mor - Model Analysis of Mineralization, Uptake by Roots and Leaching. In: Persson, T. (ed.). Structure and Function of Northern Coniferous Forests - an Ecosystem Study. Ecological Bulletin (Stockholm) 32:565-589.

BRINGMARK, L. 1980. Ion Leaching through a Podsol in a Scots Pine Stand. In: Persson, T. (ed.). Structure and Function of Northern Coniferous Forests - an Ecosystem Study. Ecological Bulletin (Stockholm) 32:341-361.

CENTRAL STATISTICAL OFFICE 1994. Forestry 1994. Warsaw. (In Polish).

DUBOIS, N.R. AND LEWIS, F.B. 1981. What is *Bacillus thuringiensis*. Journal of Arboriculture 7:233-240.

DUPHAR 1995. An Insecticide Interfering with Chitin Deposition. Dimilin. Trade mark of Duphar B.V. Technical information, 9th edition. Weesp, Holland.

FLOWER-ELLIS, J.G.K. AND OLSSON, L. 1978. Litterfall in an age series of Scots pine stands and its variation by components during the years 1973-1976. Swedish Coniferuous Forest Project, Technical Report 15.

FLOWER-ELLIS, J.G.K. AND PERSSON, H. 1980. Investigation of Structural Properties and Dynamics of Scots Pine. In: Persson, T. (ed.). Structure and Function of Northern Coniferous Forests - An Ecosystem Study. Ecol. Bull. (Stockholm) 32:125-138.

FUHRMAN, J.A., LANE, W.S., SMITH, R.F., PIESSENS, W.F. AND PERLER, F.B. 1992. Transmission-blocking antibodies recognize microfilarial chitinase in brugian lymphatic filariasis. Proceedings of the National Science USA 89:1548-1552.

GINGER, C.D. 1991. Filarial Worms: Targets for Drugs. Parasitology Today 7:262-264.

GLOWACKA, B. 1994. Assessment of the effects of control treatments on the nun moth (*Lymantria monacha* L.) populations. Ecological and economic evaluation of the impact of the nun moth control treatments on the forest environment (part 1.1). Phase 1. The Forest Research Institute, Department of Forest Protection, Warsaw.

GOODAY, G.W., BRYDON, L.J. AND CHAPPELL, L.H. 1988. Chitinase in female Onchocerca gibsoni and its inhibition by allosamidin. Molecular and Biochemical Parasitology 29:223-225.

HUNTER, I.R. AND SKATULLA, U. 1994. Monitoring and Assessment of Programme and Strategy for Remediation of Forest Infestation, Poland. Deutsche Gesellschaft für Technische Zusammenarbeit (GTZ) GmbH. PHARE Programme Services Contract No. 94-0583. Final Report. Eschborn.

IVIE, G.W., BULL, D.L. AND VERCH, J.A. 1980. Fate of diflubenzuron in water. Journal of Agriculture and Food Chemistry 28:330-337.

KOLK, A. 1992. Impact of bark beetles on forest management in Poland in 1981-1990. Journal of Applied Entomology 114:425-430.

LARSSON, S. AND TENOW, O. 1980. Needle-Eating Insects and Grazing Dynamics in a Mature Scots Pine Forest in Central Sweden. In: Persson, T. (ed.). Structure and Function of Northern Coniferous Forests - an Ecosystem Study. Ecological Bulletin (Stockholm) 32:269-306.

LINDER, S. AND LOHAMMAR, T. 1981. Amount and Quality of Information on CO_2-Exchange Required for Estimating Annual Carbon Balance of Coniferous Trees. Studia Forestalia Suecica 160:73-87.

LINDER, S. AND TROENG, E. 1980. Photosynthesis and Transpiration of 20-year-old Scots Pine. In: Persson, T. (ed.). Structure and Function of Northern Coniferous Forests - An Ecosystem Study. Ecological Bulletin 32:165-181.

MARTINAT, P.J, COFFEMAN, C.C., DODGE, K., COOPER, R.J. AND WHITMORE, R.C. 1988. Effect of diflubenzuron on the canopy arthropod community in a central Appalachian forest. Journal of Economic Entomology 81:261-267.

MUTANEN, R.M., SILTANEN, H.T. AND KUUKKA, V.P. 1988. Residues of diflubenzuron and two of its metabolites in a forest ecosystem after control of the pine looper moth, *Bupalus piniarius* L. Pesticide Science 23:131-140.

MØLLER, C.M. 1977. Vore skovtræarter og deres dyrkning. [Our Forest Tree Species and their Silviculture.] The Danish Forestry Society, Copenhagen. (In Danish).

NOVO NORDISK 1994. Foray 48B. Ultra Low Volume Suspension Concentrate for Forestry Application. Plant Protection Division, Novo Nordisk A/S, Bagsvaerd.

PERSSON, T., BÅÅTH, E., CLARHOLM, M., LUNDKVIST, H., SÖDERSTRÖM, B.E. AND SOHLENIUS, B. 1980. Trophic Structure Biomass Dynamics and Carbon Metabolism of Soil Organisms in a Scots Pine Forest. In: Persson, T. (ed.) Structure and Function of Northern Coniferous Forests - an Ecosystem Study. Ecological Bulletin 32:419-459.

REED, W.J. AND ERRICO, D. 1989. A New Look at Whole-Forest Modeling. Natural Resource Modelling 3:399-427.

SPIEGEL, Y. AND COHN, E. 1985. Chitin is present in gelatinous matrix of meloidogyne. Revue of Nematology 8:179-190.

SPINDLER-BARTH, M. 1992. Endocrine strategies for the control of ectoparasites and insect pests. Parasitology Research 78:89-95.

STRYKOWSKI, W., RATAJCZAK, E. AND PRZYBORSKI, W. 1993. Forest and Forest Products Country Profile. Poland. ECE/TIM/67.

TARP, P. 1994: Application of Operations Research Methods in Danish Forest Management Planning. Ph.D. thesis. The Royal Veterinary and Agricultural University, Copenhagen.

TARP, P. AND HELLES, F. 1994: Opportunity costs of biological forest pest management and control in Polish forestry. In: Sessions, J. and Brodie, J.D. (eds.). Management Systems for a Global Economy with Global Resource Concerns. Proceedings of the 1994 Symposium on Systems Analysis in Forest Resources, Asilomar Conference Center, Pacific Grove, California, September 6-9:122-133.

TARP, P. AND HELLES, F. 1995. Strategic Tree Species Analysis Based on Marginal Post-Tax Economic Optimization. Forest & Landscape Research 1:180-199.

TROENG, E. AND LINDER, S. 1982a. Gas exchange in a 20-year-old stand of Scots pine. I. Net photosynthesis of current and one-year-old shoots within and between seasons. Physiol. Plant. 54:7-14.

TROENG, E. AND LINDER, S. 1982b. Gas exchange in a 20-year-old stand of Scots pine. II. Variation in net photosynthesis and transpiration within and between trees. Physiol. Plant. 54: 15-23.

USDA 1985. Final environmental impact statement, land and resource management plan, Jefferson National Forest. Roanoke, VA. U.S. Department of Agriculture, Forest Service, Southern Region, Jefferson National Forest.

USDA 1989. Final Environmental Impact Statement. Appalachian Integrated Pest Management (AIPM). Gypsy Moth Demonstration Project. United States Department of Agriculture, Forest Service, Southern Region. Management Bulletin R8-MB 33. Pennsylvania.

VAN DEN BERG, G. 1986. Dissipation of diflubenzuron residues after application of Dimilin WP-25 in a forestry area in N. Carolina and some ecological effects, Duphar, B.V. Crop Protection Division Report #56637/47/1986.

WILLCOX, H. AND COFFEY, T. 1978. Environmental impacts of diflubenzuron (Dimilin) insecticide. Forest Insect and Disease Management, Northeastern Area, State and Private Forestry, U.S. Department of Agriculture, Forest Service, Broomall, Pennsylvania.

List of authors

Ph.D. Student Jens Abildtrup
The Royal Veterinary and Agricultural University
Department of Economics and Natural Resources,
Unit of Forestry, 23 Rolighedsvej.
DK-1958 Frederiksberg C, Denmark

Associate Professor Per Agrell
The Royal Veterinary and Agricultural University,
Department of Economics and Natural Resources,
Unit of Economics, 23 Rolighedsvej.
DK-1958 Frederiksberg C, Denmark

Ph.D. student Mette Asmild
The Royal Veterinary and Agricultural University
Department of Economics and Natural Resources,
Unit of Economics, 23 Rolighedsvej.
DK-1958 Frederiksberg C, Denmark

Professor Peter Berck
University of California, Berkeley
Department of Agricultural and Resource Economics
Berkeley, California 94720, USA

Professor Peter Bogetoft
The Royal Veterinary and Agricultural University
Department of Economics and Natural Resources,
Unit of Economics, 23 Rolighedsvej.
DK-1958 Frederiksberg C, Denmark

Ph.D. student Tove Enggrob Boon
Danish Forest and Landscape Research Institute
Hørsholm Kongevej 11
DK-2970 Hørsholm, Denmark

Ph.D. student Vilis Brukas
The Royal Veterinary and Agricultural University
Department of Economics and Natural Resources,
Unit of Forestry, 23 Rolighedsvej.
DK-1958 Frederiksberg C, Denmark

Professor Joseph Buongiorno
University of Wisconsin,
Department of Forest Ecology and Management,
1630 Linden Drive, Madison,
WI 53706, USA

Professor J. Keith Gilless
University of California, Berkeley
College of Natural Resources,
145 Mulford Hall, 3114 Berkeley,
CA 94720, USA

Professor Finn Helles
The Royal Veterinary and Agricultural University
Department of Economics and Natural Resources,
Unit of Forestry, 23 Rolighedsvej.
DK-1958 Frederiksberg C, Denmark

Ph.D. student Nico Hjortsø
The Royal Veterinary and Agricultural University
Department of Economics and Natural Resources,
Unit of Forestry, 23 Rolighedsvej.
DK-1958 Frederiksberg C, Denmark

Associate Professor Per Holten-Andersen
The Royal Veterinary and Agricultural University
Department of Economics and Natural Resources,
Unit of Forestry, 23 Rolighedsvej.
DK-1958 Frederiksberg C, Denmark

Associate Professor Jens Leth Hougaard
Copenhagen Business School,
Department of Operations Management, 60 Howitzvej, 1.
DK-2000 Frederiksberg C, Denmark

Ph.D. student Jan Juul Jensen
National Environment Research Institute,
399 Frederiksborgvej, P.O.Box 358,
DK-4000 Roskilde, Denmark

Dr. Jyrki Kangas
University of Joensuu
Finnish Forest Research Institute,
Kannus Research Station, P.O.Box 44,
FIN-69101 Kannus, Finland

Professor Pekka Korhonen
Helsinki School of Economics and Business Administration,
Basic Research Institute, P.O.Box 1210,
00101 Helsinki, Finland

Associate Professor Dorte Kronborg
Copenhagen Business School
10 Julius Thomsens Plads,
DK-1925 Frederiksberg C, Denmark

Ph.D. student Mikko Kurttila
University of Joensuu
Finnish Forest Research Institute, P.O.Box 18,
FIN-01301 Vantaa, Finland

Assistant Professor Ching-Rong Lin
General Education Course Center
Ming Chuan University
250 Sec. 5 Chung Shan N. Rd.
Taipei, Taiwan, ROC.

Associate Professor Henrik Meilby
The Royal Veterinary and Agricultural University
Department of Economics and Natural Resources,
Unit of Forestry, 23 Rolighedsvej.
DK-1958 Frederiksberg C, Denmark

Professor Colin Price
University of Wales, Bangor,
School of Agricultural & Forest Sciences,
Deiniol Road, Bangor,
Gwynedd, LL57 2UW, United Kingdom

Professor Ye Qi
University of California, Berkeley
Department of Environmental Science, Policy and Management
CA 94720-3310, USA

Ph.D. student Niels Strange
The Royal Veterinary and Agricultural University
Department of Economics and Natural Resources,
Unit of Forestry, 23 Rolighedsvej.
DK-1958 Frederiksberg C, Denmark

Associate Professor Peter Tarp
The Royal Veterinary and Agricultural University
Department of Economics and Natural Resources,
Unit of Forestry, 23 Rolighedsvej.
DK-1958 Frederiksberg C, Denmark

Research Assistant Lars Wichmann
The Royal Veterinary and Agricultural University
Department of Economics and Natural Resources,
Unit of Forestry, 23 Rolighedsvej.
DK-1958 Frederiksberg C, Denmark

Associate Professor, Dr. Stanislaw Zajac
The Forest Research Institute
Department of Forest Economics and Policy
ul. Bitwy Warszawskiej 1920 roku Nr 3,
00-973 Warszawa, Poland

FORESTRY SCIENCES

1. P. Baas (ed.): *New Perspectives in Wood Anatomy.* Published on the Occasion of the 50th Anniversary of the International Association of Wood Anatomists. 1982 ISBN 90-247-2526-7
2. C.F.L. Prins (ed.): *Production, Marketing and Use of Finger-Jointed Sawnwood.* Proceedings of an International Seminar Organized by the Timber Committee of the UNECE (Halmar, Norway, 1980). 1982 ISBN 90-247-2569-0
3. R.A.A. Oldeman (ed.): *Tropical Hardwood Utilization.* Practice and Prospects. 1982 ISBN 90-247-2581-X
4. P. den Ouden (in collaboration with B.K. Boom): *Manual of Cultivated Conifers.* Hardy in the Cold- and Warm-Temperate Zone. 3rd ed., 1982 ISBN Hb 90-247-2148-2; Pb 90-247-2644-1
5. J.M. Bonga and D.J. Durzan (eds.): *Tissue Culture in Forestry.* 1982 ISBN 90-247-2660-3
6. T. Satoo: *Forest Biomass.* Rev. ed. by H.A.I. Madgwick. 1982 ISBN 90-247-2710-3
7. Tran Van Nao (ed.): *Forest Fire Prevention and Control.* Proceedings of an International Seminar Organized by the Timber Committee of the UNECE (Warsaw, Poland, 1981). 1982 ISBN 90-247-3050-3
8. J.J. Douglas: *A Re-Appraisal of Forestry Development in Developing Countries.* 1983 ISBN 90-247-2830-4
9. J.C. Gordon and C.T. Wheeler (eds.): *Biological Nitrogen Fixation in Forest Ecosystems.* Foundations and Applications. 1983 ISBN 90-247-2849-5
10. M. NÇmeth: *Virus, Mycoplasma and Rickettsia Diseases of Fruit Trees.* Rev. (English) ed., 1986 ISBN 90-247-2868-1
11. M.L. Duryea and T.D. Landis (eds.): *Forest Nursery Manual.* Production of Bareroot Seedlings. 1984; 2nd printing 1987 ISBN Hb 90-247-2913-0; Pb 90-247-2914-9
12. F.C. Hummel: *Forest Policy.* A Contribution to Resource Development. 1984 ISBN 90-247-2883-5
13. P.D. Manion (ed.): *Scleroderris Canker of Conifers.* Proceedings of an International Symposium on Scleroderris Canker of Conifers (Syracuse, USA, 1983). 1984 ISBN 90-247-2912-2
14. M.L. Duryea and G.N. Brown (eds.): *Seedling Physiology and Reforestation Success.* Proceedings of the Physiology Working Group, Technical Session, Society of American Foresters National Convention (Portland, Oregon, USA, 1983). 1984 ISBN 90-247-2949-1
15. K.A.G. Staaf and N.A. Wiksten (eds.): *Tree Harvesting Techniques.* 1984 ISBN 90-247-2994-7
16. J.D. Boyd: *Biophysical Control of Microfibril Orientation in Plant Cell Walls.* Aquatic and Terrestrial Plants Including Trees. 1985 ISBN 90-247-3101-1
17. W.P.K. Findlay (ed.): *Preservation of Timber in the Tropics.* 1985 ISBN 90-247-3112-7
18. I. Samset: *Winch and Cable Systems.* 1985 ISBN 90-247-3205-0
19. R.A. Leary: *Interaction Theory in Forest Ecology and Management.* 1985 ISBN 90-247-3220-4
20. S.P. Gessel (ed.): *Forest Site and Productivity.* 1986 ISBN 90-247-3284-0
21. T.C. Hennessey, P.M. Dougherty, S.V. Kossuth and J.D. Johnson (eds.): *Stress Physiology and Forest Productivity.* Proceedings of the Physiology Working Group, Technical Session, Society of American Foresters National Convention (Fort Collins, Colorado, USA, 1985). 1986 ISBN 90-247-3359-6

FORESTRY SCIENCES

22. K.R. Shepherd: *Plantation Silviculture.* 1986 ISBN 90-247-3379-0
23. S. Sohlberg and V.E. Sokolov (eds.): *Practical Application of Remote Sensing in Forestry.* Proceedings of a Seminar on the Practical Application of Remote Sensing in Forestry (Jnkping, Sweden, 1985). 1986 ISBN 90-247-3392-8
24. J.M. Bonga and D.J. Durzan (eds.): *Cell and Tissue Culure in Forestry.* Volume 1: General Principles and Biotechnology. 1987 ISBN 90-247-3430-4
25. J.M. Bonga and D.J. Durzan (eds.): *Cell and Tissue Culure in Forestry.* Volume 2: Specific Principles and Methods: Growth and Development. 1987 ISBN 90-247-3431-2
26. J.M. Bonga and D.J. Durzan (eds.): *Cell and Tissue Culure in Forestry.* Volume 3: Case Histories: Gymnosperms, Angiosperms and Palms. 1987 ISBN 90-247-3432-0
Set ISBN (Volumes 24–26) 90-247-3433-9
27. E.G. Richards (ed.): *Forestry and the Forest Industries: Past and Future.* Major Developments in the Forest and Forest Industries Sector Since 1947 in Europe, the USSR and North America. In Commemoration of the 40th Anniversary of the Timber Committee of the UNECE. 1987 ISBN 90-247-3592-0
28. S.V. Kossuth and S.D. Ross (eds.): *Hormonal Control of Tree Growth.* Proceedings of the Physiology Working Group, Technical Session, Society of American Foresters National Convention (Birmingham, Alabama, USA, 1986). 1987 ISBN 90-247-3621-8
29. U. Sundberg and C.R. Silversides: *Operational Efficiency in Forestry.* Volume 1: Analysis. 1988 ISBN 90-247-3683-8
30. M.R. Ahuja (ed.): *Somatic Cell Genetics of Woody Plants.* Proceedings of the IUFRO Working Party S2.04-07 Somatic Cell Genetics (Grosshansdorf, Germany, 1987). 1988 ISBN 90-247-3728-1
31. P.K.R. Nair (ed.): *Agroforestry Systems in the Tropics.* 1989 ISBN 90-247-3790-7
32. C.R. Silversides and U. Sundberg: *Operational Efficiency in Forestry.* Volume 2: Practice. 1989 ISBN 0-7923-0063-7
Set ISBN (Volumes 29 and 32) 90-247-3684-6
33. T.L. White and G.R. Hodge (eds.): *Predicting Breeding Values with Applications in Forest Tree Improvement.* 1989 ISBN 0-7923-0460-8
34. H.J. Welch: *The Conifer Manual.* Volume 1. 1991 ISBN 0-7923-0616-3
35. P.K.R. Nair, H.L. Gholz, M.L. Duryea (eds.): *Agroforestry Education and Training.* Present and Future. 1990 ISBN 0-7923-0864-6
36. M.L. Duryea and P.M. Dougherty (eds.): *Forest Regeneration Manual.* 1991 ISBN 0-7923-0960-X
37. J.J.A. Janssen: *Mechanical Properties of Bamboo.* 1991 ISBN 0-7923-1260-0
38. J.M. Bonga and P. Von Aderkas: *In Vitro Culture of Trees.* 1992 ISBN 0-7923-1540-5
39. L. Fins, S.T. Friedman and J.V. Brotschol (eds.): *Handbook of Quantitative Forest Genetics.* 1992 ISBN 0-7923-1568-5
40. M.J. Kelty, B.C. Larson and C.D. Oliver (eds.): *The Ecology and Silviculture of Mixed-Species Forests.* A Festschrift for David M. Smith. 1992 ISBN 0-7923-1643-6
41. M.R. Ahuja (ed.): *Micropropagation of Woody Plants.* 1992 ISBN 0-7923-1807-2
42. W.T. Adams, S.H. Strauss, D.L. Copes and A.R. Griffin (eds.): *Population Genetics of Forest Trees.* Proceedings of an International Symposium (Corvallis, Oregon, USA, 1990). 1992 ISBN 0-7923-1857-9

FORESTRY SCIENCES

43. R.T. Prinsley (ed.): *The Role of Trees in Sustainable Agriculture.* 1993 ISBN 0-7923-2030-1
44. S.M. Jain, P.K. Gupta and R.J. Newton (eds.): *Somatic Embryogenesis in Woody Plants,* Volume 3: Gymnosperms. 1995 ISBN 0-7923-2938-4
45. S.M. Jain, P.K. Gupta and R.J. Newton (eds.): *Somatic Embryogenesis in Woody Plants,* Volume 1: History, Molecular and Biochemical Aspects, and Applications. 1995 ISBN 0-7923-3035-8
46. S.M. Jain, P.K Gupta and R.J. Newton (eds.): *Somatic Embryogenesis in Woody Plants,* Volume 2: Angiosperms. 1995 ISBN 0-7923-3070-6
    Set ISBN (Volumes 44–46) 0-7923-2939-2
47. F.L. Sinclair (ed.): *Agroforestry: Science, Policy and Practice.* Selected Papers from the Agroforestry Sessions of the IUFRO 20th World Congress (Tampere, Finland, 6–12 August 1995). 1995 ISBN 0-7923-3696-8
48. J.H. Goldammer and V.V. Furyaev (eds.): *Fire in Ecosystems of Boreal Eurasia.* 1996 ISBN 0-7923-4137-6
49. M.R. Ahuja, W. Boerjan and D.B. Neale (eds.): *Somatic Cell Genetics and Molecular Genetics of Trees.* 1996 ISBN 0-7923-4179-1
50. H.L. Gholz, K. Nakane and H. Shimoda (eds.): *The Use of Remote Sensing in the Modeling of Forest Productivity.* 1996 ISBN 0-7923-4278-X
51. P. Bachmann, M. Köhl and R. Päivinen (eds.): *Assessment of Biodiversity for Improved Forest Planning.* Proceedings of the Conference on Assessment of Biodiversity for Improved Planning (Monte Verità, Switzerland, 1996). 1998 ISBN 0-7923-4872-9
52. G.M.J. Mohren, K. Kramer and S. Sabaté (eds.): *Impacts of Global Change on Tree Physiology and Forest Ecosystems.* Proceedings of the International Conference on Impacts of Global Change on Tree Physiology and Forest Ecosystems (Wageningen, The Netherlands, 26–29 November 1996). 1997 ISBN 0-7923-4921-0
53. P.K.R. Nair and C.R. Latt (eds.): *Directions in Tropical Agroforestry Research.* Selected Papers from the Symposium on Tropical Agroforestry organized in connection with the annual meeting of USA (Indianapolis, U.S.A., 5 November 1996). 1998 ISBN 0-7923-5035-9
54. K. Sassa (ed.): *Environmental Forest Science.* Proceedings of the IUFRO Division 8 Conference Environmental Forest Science. (Kyoto, Japan, 19–23 October 1998). 1998 ISBN 0-7923-5280-7
55. S.M. Jain, P.K Gupta and R.J. Newton (eds.): *Somatic Embryogenesis in Woody Plants,* Volume 4. 1999 ISBN 0-7923-5340-4
56. J.R. Boyle, J.K. Vinjum, K. Kavanagh, E.J. Jensen (eds.): *Planted Forests: Contribution to the Quest for Sustainable Societies.* 1999 ISBN 0-7923-5468-0
57. K. von Gadow and G. Hui: *Modelling Forest Development.* 1999 ISBN 0-7923-5488-5
58. J. Abildtrup, F. Helles, P. Holten-Andersen, J.F. Larsen and B.J. Thorsen (eds.): *Modern Time Series Analysis in Forest Products Markets.* 1999 ISBN 0-7923-5524-5
59. S.M. Jain, P.K. Gupta and R.J. Newton (eds.): *Somatic Embrygenesis in Woody Plants*, Volume 5. 1999 ISBN 0-7923-5553-9
60. D. Auclair and C. Dupraz (eds.): *Agroforestry for Sustainable Land-Use.* Fundamental Research and Modelling with Emphasis on Temperate and Mediterranean Applications. 1999 ISBN 0-7923-5799-X

FORESTRY SCIENCES

61. F. Helles, P. Holten-Andersen and L. Wichmann (eds.): *Multiple Use of Forests and Other Natural Resources.* Aspects of Theory and Application. 1999 ISBN 0-7923-5859-7

KLUWER ACADEMIC PUBLISHERS – DORDRECHT / BOSTON / LONDON